BIOMETEOROLOGY

THE IMPACT OF THE WEATHER AND CLIMATE ON HUMANS AND THEIR ENVIRONMENT (ANIMALS AND PLANTS)

HEYDEN INTERNATIONAL TOPICS IN

Editor: L. C. Thomas

RTL/2 Design and Philosophy	J. G. P. BARNES
Introduction to PASCAL	C. A. G. WEBSTER
Computer Applications in Shipbuilding Technology	C. KUO
An Introduction to Computer-based Library Systems	L. A. TEDD
Analysis with Ion-Selective Electrodes	P. L. BAILEY
Analysis of Drugs of Abuse	E. BERMAN
Toxic Metals and their Analysis	E. BERMAN
Environmental Pollution Analysis	P. D. GOULDEN
Paper and Thin Layer Chromatograp in Environmental Analysis	M. E. GETZ
Computers and Instrumentation	A. CARRICK
Colour: Its Measurement, Computation and Application	G. J. CHAMBERLIN and D. G. CHAMBERLIN
Biometeorology	S. W. TROMP

CE

BIOMETEOROLOGY

THE IMPACT OF THE WEATHER AND CLIMATE ON HUMANS AND THEIR ENVIRONMENT (ANIMALS AND PLANTS)

S. W. TROMP

Biometeorological Research Centre,
Leiden, The Netherlands

LONDON · PHILADELPHIA · RHEINE

Heyden & Son Ltd., Spectrum House, Hillview Gardens, London NW4 2JQ
Heyden & Son Inc., 247 South 41st Street, Philadelphia, PA 19104, USA
Heyden & Son GmbH, Münsterstrasse 22, 4440 Rheine, West Germany

Tromp, S W
Biometeorology. – (Heyden international topics in science).
1. Bioclimatology
I. Title
574.5'42 QH543 79-41516
ISBN 0-85501-453-9

Printed in Great Britain by W & J Mackay Ltd, Chatham

CONTENTS

This book is dedicated to all my biometeorological friends who assisted me with their scientific knowledge and their constructive comments; and above all to my co-worker Mrs. Janneke Bouma, without whose unflagging assistance, critical eye and helpful suggestions this practical guide for environmentalists would never have been completed.

I wish to commemorate particularly my friend Professor J. H. Plokker, formerly Professor of Psychiatry at Utrecht University, for his many stimulating conversations and his great assistance in our studies on the influence of weather and climate on mental processes in man.

I am grateful to Mrs. Willy van Vianen who typed most cheerfully the whole manuscript which was often written in almost non-decipherable hieroglyphics.

I dedicate the book also to my wife who enabled me to work continuously on this fascinating subject during the evenings and weekends.

FOREWORD

One of the objectives of this series of monographs is to publish general reviews of selected areas of scientific advancement, for which a wider readership is envisaged than that for the more specialized monographs. This is the first of these more general reviews, and it is especially appropriate that it should cover biometeorology, a subject little known to most scientists and yet of direct interest to all, scientists and laymen alike.

The question 'What is biometeorology?' has been posed recently both in the technical literature and by the media. Even a glance at the contents list of this monograph would answer that query, and by the end of the book the reader will be well aware of the all-embracing ramifications of the subject and of its international impact, as indicated by over a thousand references to scientific papers from all parts of the world.

This volume describes the effects of the weather and climate on all living things, and many of the documented effects will surprise those unfamiliar with the subject. The discussion covers not only the obvious effects, such as the influence of climate on the crops that can be grown, the animals that can be reared, and the economic yields that can be obtained at a given location, but also unexpected topics like the geographical distribution of disease; the effect of month of conception on proneness to disease, on mental development, on temperament, and on the sex-ratios of the offspring; the effects of altitude on physical performance and on disease; and the therapeutic applications of biometeorological effects.

Dr. Tromp, an enthusiastic worker in the field as becomes his position as Director of the Biometeorological Research Centre at Leiden, The Netherlands, is not content with just reviewing the broad field of biometeorological investigations, but also indicates practical steps which can be taken to minimize deleterious meteorotropic effects and also how the data which have been gathered can be used to aid developing countries to increase agricultural productivity, control pests and the spread of diseases both animal and human, and to create a better environment for humans. Some of these suggestions, particularly those on the design of buildings and their environments, are also relevant in the more developed countries.

Biometeorology is a science which is concerned with the effects of weather and climate on all living things, and is shown to affect all of us from the moment of conception to the time when it influences death. This book will therefore be of interest to all, but especially to those, like doctors, horticulturalists, farmers and architects, who are especially concerned with the welfare of people, plants and animals.

L. C. Thomas
Salisbury
July 1979

PREFACE

Since 1930 several books have been published on biometeorology but the first comprehensive textbook *Medical Biometeorology* (Elsevier Publ. Co., Amsterdam) was prepared by the present author in 1963 in cooperation with 26 contributors. The 4400 references given in that textbook indicated already the wide scope of this relatively new environmental science. However, after 1963 such a vast development took place, particularly in the Anglo-Saxon countries and the USSR, that a new textbook was soon required covering the period 1963–1973. In this textbook *Progress in Human biometeorology* (Swets & Zeitlinger Publ. Co., Amsterdam) consisting of Parts IA, IB, II and III and totalling 1300 pages, the number of references increased by more than 3500, bringing the total number of important studies in the field of human biometeorology to almost 8000 in 15 years. Due to this enormous increase in scientific material about 100 new contributors were required to review the entire field of human biometeorology. Most of the topics were of special interest to biologists, chemists, engineers, meteorologists, physicians, veterinarians, psychologists, sociologists and many other professions dealing with the influence of our physical environment on the health, performance capability and behaviour of man. It is for this group of scientists that the present book was written, in a not too-technical style; if technical words had to be used then short explanations have been given.

The most up to date review of research in these areas is to be found in *Biometeorology Review 1973–1978*, a two-volume work devoted to the most significant recent results in both human and animal biometeorology and published like this monograph, by Heyden & Son. Therefore these two books should be considered the first stage in seeking to extend the knowledge given in this present basic overview, and the previous books referred to above constitute a comprehensive treatise on the earlier findings in biometeorological research.

Apart from the people mentioned in my dedication I wish to express my particular gratitude to Elsevier Publishing Co., and Swets & Zeitlinger Publishing Co. with whom I was able to prepare eight handbooks on various aspects of biometeorology. I am most grateful to the authors who contributed to *Progress in Animal Biometeorology* and *Progress in Plant Biometeorology*, in particular to F. T. Abushama, C. W. Alliston, P. D. Altland, W. Bianca, D. P. Cardinali, J. L. Cloudsley Thompson, J. D. Deaton, D. F. Dowling, W. Ferguson, F. E. J. Fry, R. F. Grover, W. O. Haufe, J. C. Hoffman, J. C. D. Hutchinson, K. L. Knox, D. L. Ingram, H. D. and K. G. Johnson, W. V. Macfarlane, G. M. O. Maloiy, J. L. Monteith, P. R. Morrison, L. E. Mount, R. O. Rawson, F. N. Reece, J. D. Rising, D. Robertshaw, H. S. Siegel, D. Smidt, L. P. Smith, J. Steinbach, C. J. Thwaites, M. Van Kampen, H. A. W. Venter, A. J. F. Webster, W. O. Wilson, N. T. M. Yeates and M. K. Yousef.

I also wish to thank Professor G. Melvyn Howe and William Heinemann Medical Books Ltd. who permitted me to reproduce parts of my paper in *Environmental Medicine*, Chapter V. I am most grateful for the permission given by Ciba-Geigy Ltd., Basle, to reproduce four illustrations prepared by Dr. Netter for the Ciba collection of *Medical Illustrations*, Volume 1, *Nervous System*. Without the assistance of all these experts this monograph would have been most difficult to complete.

S. W. Tromp
Wassenaar
July 1979

CHAPTER 1

INTRODUCTION

1.1. DEFINITION AND CLASSIFICATION OF BIOMETEOROLOGY

In 1956, during the Foundation of the International Society of Biometeorology (I.S.B.), at UNESCO Headquarters in Paris, biometeorology (in the German and Latin countries better known as bioclimatology) was briefly defined as 'the study of the direct and indirect interrelations between the geophysical and geochemical environment of the atmosphere and living organisms, plants, animals and man'. In 1970 it was recognized by the members of I.S.B., representing 56 countries, that the given definition does not cover all aspects of modern biometeorology. The definition was changed as follows: 'Biometeorology is the study of the direct and indirect effects (of an irregular, fluctuating or rhythmic nature) of the physical, chemical and physicochemical micro- and macro-environments, of both the earth's atmosphere and of similar extra-terrestrial environments, on physico-chemical systems in general and on living organisms (plants, animals and man) in particular'.

Both I.S.B., and the World Meteorological Organization (W.M.O.) decided to change the name 'bioclimatology' to 'biometeorology', because 'climatology' is part of the modern concept of 'meteorology'. Despite this international decision, the old name of bioclimatology is still being used particularly in Latin countries.

Biometeorology can be classified into six main groups (see Table 1.1): phytological, zoological, human, cosmic, space and paleobiometeorology. Each of these branches can be briefly described as follows:

1.1.1. Phytological or plant biometeorology

This is the study of the influence of the weather and climate on the development and distribution of both healthy and diseased plants, for general phytological, agricultural and forestry purposes, and also of the effects of both weather and

TABLE 1.1
Classification of General Biometeorology

Phytological or Plant Biometeorology	Zoological Biometeorology	Human Biometeorology	Cosmic Biometeorology	Space Biometeorology	Paleo-biometeorology
Physiological Agricultural Forest Pathological	Physiological Entomological Avian Veterinary Pathological	Physiological Sociological Pathological Architectural Urban Nautical			

climate on small plant organisms responsible for the development of plant, animal and human diseases. The branch of phytological biometeorology dealing with agricultural plants only is known as agricultural biometeorology. It can be defined as the study of the development, growth and yield of crops, in terms of the annual course of the climatic elements in an area. It is also concerned with the geographical location of plants consequent upon established crop–weather relationships.

1.1.2. Zoological biometeorology

This is the study of the influence of the weather and climate on animals in general, on milk production, on the reproduction and coating of cattle, etc. (veterinary biometeorology), on the egg-laying rhythm, etc. of birds (avian biometeorology) but in particular of the effect of the weather and climate on insects and terrestrial arthropoda, on their physiological activities and development and on the outbreak of plant, animal and human diseases as a result of these developments, such as the abundant development of locusts, malaria mosquitoes, etc. (entomological biometeorology).

1.1.3. Human biometeorology

This is the study of the influence of the weather and climate both on healthy man and on diseased subjects. A detailed classification of human biometeorology is given in Table 1.2.

(a) *Physiological biometeorology* describes the influence of the weather and climate (both macro and micro) on the physiological processes in healthy subjects. These influences of our physical environment are observed both in the free atmosphere and in houses, buildings, underground tunnels, space capsules etc.

(b) *Sociological biometeorology* is the study of the social significance of both favourable and unfavourable meteorological conditions on the health, behaviour and cultural activities of population groups.

(c) *Pathological biometeorology* is the study of the influence of the weather and climate on the various physiological and pathological phenomena associated with the diseases of man, such as the period of an outbreak of disease, and its intensity and geographical distribution.

(d) *Architectural and urban biometeorology* describes the influence of the microclimates in houses and cities on the health of man and the effect of architectural constructions and town planning on these microclimates; it also covers the best location and construction of sanatoria.

(e) *Nautical biometeorology* is the study of the influence of the weather and climate on animals and man living on a moving ship and the effects on its cargo. Of special importance for the general practitioner are the results obtained in acclimatization biometeorology (a branch of physiological biometeorology which studies the various physiological processes involved during adaptation to extreme conditions of temperature, humidity, etc. e.g. in the tropics or polar regions), in air pollution biometeorology (the influence of pollen, spores, dust and chemical air pollution on man), and the results of the various methods applied in Climatotherapy.

1.1.4. Cosmic biometeorology

This describes the possible influences of extra-terrestrial factors such as variations in solar activity, cosmic radiation, etc., on living organisms on earth.

1.1.5. Space biometeorology

This is closely related to space research. It is the study of the microclimatic conditions in space vehicles and of the biological effects (both observed and expected) of extra-terrestrial physical environments on man and animals on the moon and the planets and during space flights.

1.1.6. Paleo-biometeorology

This is the study of the influence of climatic conditions in the past (millions of years ago) on the evolutional development and geographic distribution of plants, animals and man.

TABLE 1.2
Classification of Human Biometeorology

Physiological Biometeorology	Social Biometeorology	Pathological Biometeorology	Architectural and Urban Biometeorology	Nautical Biometeorology
General Geographical Ethnological Acclimatization	General Psychological Archaeological	General Air pollution Geographical Climatotherapy Pharmacological	General Architectural Urban Sanatorium	General Cargo

1.2. HISTORIC DEVELOPMENT OF HUMAN BIOMETEOROLOGY

Since Hippocrates wrote his book *Airs, Waters and Places*, around 400 BC, it took almost 2300 years before the first really scientific studies were published on the effects of the weather and climate on the health of man and his diseases. However, the main development of this new border science of biometeorology started only in the second part of this century. Therefore biometeorology is a young and at the same time a very old science, young by modern scientific standards, old if we consider the strong belief of man, from the earliest historic times to the present day, in the great influence of the weather and climate on man's behaviour and happiness and on the origin of his diseases. When the Greek and Roman civilizations were at their zenith, important political decisions were not carried into effect and war campaigns were often postponed

Fig. 1.1 The Tower of the Winds in Athens, Greece

until the weather gods had been consulted. In the Temple of the Winds (Fig. 1.1) at Athens in Greece a little marble octagonal tower was built, the eight sides of which represent the eight principal winds in Greece. On each side a human figure is sculptured symbolizing the character and influence on man of the particular wind it faces. Around 400 BC the Greeks and in particular Hippocrates, the nestor of Medical Sciences who lived on the island of Kos in Eastern Greece, realized the great significance of weather and climate for healthy man and diseased subjects. They developed a special weather calendar for the different seasons. The oldest known calendar seems to have been made by the Greek astronomer Meton in 432 BC.

In his book *Airs, Waters and Places*, Hippocrates gave biometeorological advice to future physicians. One of his statements is 'Whoever wishes to pursue properly the science of medicine must proceed thus. First he ought to consider what effects each season of the year can produce: for the seasons are not alike, but differ widely both in themselves and at their changes. One should be especially on one's guard against the most violent changes of the seasons and, unless compelled, one should neither purge nor apply cautery or knife to the bowels, until at least ten days have passed'.

At the beginning of the twentieth century the first modern development of biometeorology was stimulated by many reports, particularly from German physicians, claiming a striking relationship between changes in the weather and the outbreak and development of diseases. However, most of these reports were based on empirical studies only and were not based on deeper physiological studies.

The more scientific developments started about 1930 in Germany under the leadership of Professor B. de Rudder, Director of the Department of Pediatrics at the University of Frankfurt,[1,2] and in the USA under Professor W. F. Petersen, Professor of Internal Medicine and Head of the Department of Pathology and Bacteriology at the University of Illinois in Chicago.[3] These studies had a rather empirical character but De Rudder in particular introduced modern statistical methods for the evaluation of his data. A highly scientific and in particular an experimental approach of biometeorology started after 1930, particularly in the Anglo-Saxon countries, and mainly for military purposes. In large climatic chambers groups of people were studied while they walked on so-called tread-mills, and at the same time a large number of physiological parameters were studied in relation to temperature, windspeed, humidity, type of food consumed, clothing used etc. Recently also recordings have been made in the open air using telemetric devices to record the body functions of a subject, or group of subjects, under natural conditions in the open air. These climatic chamber studies are particularly advanced in Great Britain, Canada, the USA, Australia, Japan and the USSR. On a minor scale certain aspects of human biometeorology are also studied in France and Germany.

The tremendously fast development of biometeorology started after World

War II.[4-11] It was realized that large scale military operations, like the invasion of Normandy or the war in the Western Desert in North Africa, required a thorough knowledge of meteorology and of the effects of extreme weather and climatic conditions on the soldiers involved. The rapid development of space research also had a great impact on the international recognition of biometeorology as a leading environmental science. Finally the founding of the I.S.B. in 1956 was an important event which demonstrated the significance of the study of the effects of the physical environment on plants, animals and man. The following section, describing some of the major indirect effects of the weather and climate on all human activities, will demonstrate this significance even better.

1.3. SIGNIFICANCE OF THE WEATHER AND CLIMATE FOR HUMAN ACTIVITIES

Whereas in the following sections only the direct effects of the weather and climate on the living world and the practical applications of these effects will be described, a short review of the indirect effects and their impact on human activities is essential because all human activities are affected, and often even controlled, by these indirect effects.

Some of the sociological aspects of human life are also effected directly by meteorological conditions. These mechanisms are briefly discussed in Chapter 8.

The indirect effects of weather and climate can be grouped into four larger categories:

(a) Influence on the way of living (our clothing, housing and urban planning, sports, behaviour etc.);
(b) Influence on food habits (kind, quality, chemical composition and quantity of food);
(c) Influence on agents causing or transmitting diseases (development of bacteria and viruses, arthropods, insects, etc.);
(d) Influence on socio-economic relationships. It is this last group of meteorological effects which seriously influences human and animal life. It has usually also a very great economic impact.

The indirect socio-economic effects can be sub-divided into at least five different categories:

(i) Influence on traffic and travel (air, road and sea);
(ii) Influence on agriculture and forestry (particularly storm-damage);
(iii) Influence on dairy farming (quality of meat, milk production, egg production, wool quality and quantity, construction of stables etc.: part of these influences are direct physiological effects and will also be discussed in Section 11.1);
(iv) Influence on energy expenditure (heating, light and industries);
(v) Influence on riots, invasion of armies etc.

It is beyond the scope of the present book to devote too much space to these indirect effects but a short review of some of the major disasters of recent years and of their economic consequences seems to be warranted.

We will first recall two major historic biometeorological events in the past. After the introduction in Europe by Sir Frances Drake of the potato from Chile and Peru in the 16th century it became, during the 18th century, an essential and cheap food product in Western Europe. However, a serious potato disease (potato blight) developed between 1845 and 1850, destroying most of the potato harvests in northern Europe. This caused a terrible famine in western Europe, particularly in Ireland with its very poor population. It triggered the beginning of a large scale migration of the Irish people to the USA.

We now know which particular meteorological conditions are responsible for this disease and warning systems have been developed, particularly by Bourke from the Irish Meteorological Service, to enable farmers to fight the disease before it fully develops.

A second important historic event was the military landing in Normandy in 1944 which enabled the allied forces to liberate Western Europe. Very careful planning by a large group of meteorologists made the landing a success despite the weather conditions which were rather unfavourable in that period of the year.

One of the most threatening effects of meteorological events on human life and property are the effects of tropical cyclones, atmospheric depressions in the tropics, also known as typhoons or hurricanes. They are characterized by winds and turbulence of extreme violence, heavy and prolonged rainfall, river flooding and storm surges. In fact they are the most dangerous of all natural destructive phenomena. In many cases the extreme winds are operating several hours before the onset of flooding and storm surge. In addition to killing people and destroying buildings, floods and storm surge create national disasters. The storm surges cause a piling up of sea water and a rise in water level of several metres. This is for example very common during tropical cyclones in the Bay of Bengal. It may cause financial losses equal to a year's social and economic progress in a developing country. With this consideration in mind in 1964 the United Nations Economic and Social Commission for Asia and the Pacific (ESCAP) instructed the W.M.O. to organize symposia to discuss ways to reduce the enormous damage caused by tropical cyclones.

One of the most recent devastating cyclones occurred on the east coast of Southern India in the state Andhra Pradesh around 20th November 1977 with wind velocities of 145 km h^{-1} accompanied by tidal waves several metres high. During this cyclone at least 10 000 people were killed, thousands of cattle drowned and the entire rice crop was destroyed. More than three million people became homeless. A similar drama occurred in 1864 when a heavy cyclone and tidal waves killed 35 000 people.

The great impact of the weather and climate on mankind may also become evident from the following examples. In recent years unfavourable meteorological conditions affected a large part of the world. Particularly in

1974, the abnormal weather conditions affected almost the entire world. The abnormally heavy rainfall in western Europe during the last three months of that year destroyed practically the entire potato and sugarbeet harvests in that area. The harvest of wheat was also far below normal.

The very hot and dry weather during the following summer in 1975 caused even more damage. In 20 provinces of Spain, particularly in Andalusia and Castile, more than half of the autumn grass seeds did not develop. The serious shortage of grass in 1975 forced the peasants to kill thousands of their cattle. In the Spanish rivers thousands of trout and salmon died because of the low water level of the rivers after the long dry summer. In the USA, a record breaking number of tornadoes occurred around April 1974, causing 315 deaths, 45 000 injuries and 500 million dollars property damage. It was the worst tornado spell in the history of the USA. In Bangladesh millions of people lost all their property (in 1974) because of the heavy cyclones accompanied by serious tidal waves and floods. In Northern Australia the entire city of Darwin was flattened by cyclone Tracy on 25th December 1974. Forty-five people were killed and 25 000 people lost their houses and property.

In Switzerland, Austria and Italy, during the very warm winter of 1974/75, an extraordinary large number of snow avalanches occurred which killed several hundred skiers.

Tromp,[12-15] found in The Netherlands and Norway considerable changes in the physico-chemical properties of the blood of healthy population groups during 1974. In Brazil, in the Parana district, one of the greatest coffee plantation disasters occurred during the night of 17th–18th July 1975 when 900 million coffee bushes were destroyed by one heavy night frost. This event caused a worldwide rise of more than 100% in the price of coffee, which lasted for many years. The disaster struck particularly the very poor farmers working in these plantations.

In the USSR the wheat harvest was a complete failure in 1975, due to the abnormally dry summer. This forced the Russians to buy 10 million tons of wheat from the USA, which considerably influenced the political relationship between the two countries.

In the so-called Sahel area in North Africa (a zone between 14 and 18° N. including Mauretania, Senegal, Guinea, Sudan, Kenya and Yemen) and in India and Ceylon, five years of terrible drought killed hundreds of thousands of men and about 20 million goats, sheep, cattle and dromedaries.

Due to the drought the Sahara desert moved gradually southwards creating new deserts in former reasonably fertile agricultural areas and grasslands.* In

*According to the American Environmental Institute (Worldwatch) the Sahara desert moved 90 km southwards in 20 years; in north Africa each year a few thousand square kilometres of fertile land are transformed into barren desert country. In order to discuss ways to stop this unfavourable development of great parts of the surface of the earth, the United Nations organized a special conference in August 1977 in Nairobi (Kenya), the UNCOD (United Nations Conference on Desertification).

1976 again an extremely hot summer occurred, the hottest summer of the last 250 years. In England the water supply for factories and drinking water, was terribly limited, some houses being without drinking water for seventeen hours each. In southern England large devastating forest fires occurred as a result of the drought. The restoration of the lost trees will cost millions of pounds, and take tens of years.

In The Netherlands due to the long lasting heat an epidemic of botulism occurred, caused by the bacterium *Clostridium botulinum* which develops particularly in water which has a temperature over 20°C and which is deficient in oxygen. The bacteria produce a poisonous substance which damages the nerve connections with the muscles and which paralyzes animals swimming in this water. Thousands of ducks and other aquatic animals died in the canals and lakes of The Netherlands.

In the human population the very hot summer caused a drastic increase in blood pressure and changes in the physico-chemical properties of the blood.[14] At the end of December 1976 and in January 1977 very cold weather (accompanied by very heavy snowfall) prevailed in northern Europe and the USA. Sweden in particular was seriously hit, with temperatures of −50°C; heavy blizzards caused snowdrifts up to 4m high which blocked almost all traffic. In western Europe very heavy snowfall occurred in England.

The USA (in particular the mid-west and the eastern parts of the USA) was most severely hit in January and February 1977. Apart from the general discomfort, the closing of factories (about 90 000 in the Mid-West only) caused a considerable amount of temporary unemployment (in the state of Ohio alone 750 000 people were affected) and energy production was completely disrupted for a very long time. It forced the Government to change many of its long term plans for energy improvement.

According to preliminary estimates this heavy frost, accompanied by blizzards of 120 km h^{-1}, has cost the USA about six billion dollars if one reckons the higher imports of oil, gas and food, the higher prices for domestically produced fuel, fruits and vegetables and lost wages. The great shortage of natural gas was due to the 50% increase required for heating in this very cold spell. The Mississippi and Ohio rivers were frozen and hundreds of oil tankers could not move, increasing the energy disaster.

The strangest paradox of this cold winter occurred in the western part of the USA where very little snow had fallen. Supplies of water for drinking and irrigation became very scarce. It is estimated that as a result farm losses in California in the summer of 1977 will amount to one million dollars or even more.

Contrary to conditions in northern Europe, in Switzerland the snow was melting even above 2000 m. The World Skating Championships, due to be held in February 1977, had to be moved to The Netherlands because it was impossible to produce sufficient ice in Davos at 1500 m.

After the very warm months of January and February 1977 in Western

Europe in contrast to Northern Europe, which was comparable with the usual climate during spring, many agricultural plants and fruit trees developed so rapidly that the sudden very cold weather in April, accompanied by severe night-frosts and snow, damaged the vegetable and fruit harvests in very large areas, particularly in southern France. The potato crop, strawberries, grapes, peaches and prune trees in southern France were badly hit, causing losses of millions of francs for the farmers.

The lack of rain or snow in China during the last six months of 1977, particularly near the Yellow and Hoeai rivers, forced the Chinese to order more than six million tons of wheat from the West which was a severe blow for their political aspirations.

An important, initially indirect, but later direct effect of long lasting heat waves, for instance in the USA, is the triggering of violent riots particularly among the poorer black population in the USA. The long-lasting oppressive heat in the poorly built negro slums in Washington, Detroit, Los Angeles and other large cities during summer, often with temperatures above 35 °C and very high humidity, causes a malfunction of the human thermoregulation mechanism (see Section 3.3), and will seriously affect human behaviour, particularly in uneducated, non-disciplinary subjects. Usually only a very minor cause of dissatisfaction may blow up the temper of such subjects, particularly if they work, live or play together. Examples of such events are the terrible riots in Los Angeles in the USA on 16th August 1965 causing 32 deaths, 800 wounded, destruction of more than 1000 shops and houses and a total damage of more than three million dollars. A similar riot occurred in Chicago on 12th July 1966 after a very long heat wave, with temperatures over 38 °C, which also killed more than 1000 people in New York due to hyperthermia. The actual reason for dissatisfaction in Chicago was the closure by the police of fire brigade hydrants which were opened by negro boys to cool their bodies. More than 3000 policemen were required to suppress the revolt and to stop more than 20 large fires in the city. Similar dramatic riots occurred in seven large cities in the North East part of the USA around 25th July 1967. In Newark (south of New York), with 55% of black inhabitants, the result of the riots were 26 dead, 1300 wounded and more than 15 million dollars of property damage. The initial cause was a negro driver who was annoyed about a slowly driving police-car and started to hinder the police-car. The extremely hot summer made everybody highly irritated and initiated the riots which had to be suppressed by 13 000 federal soldiers.

Although poor, extremely hot, living quarters and lack of education of children may stimulate these riots, similar riots may occur in so-called ‘highly cultured populations’ among well paid labourers. In the evening of June 13th 1970 after some extremely hot days a violent riot occurred in Amsterdam, in The Netherlands, amongst the non-organized building workers who were collecting their holiday salary. They objected that 2% administrative costs were deducted. The riots which started with 30 men increased rapidly to 1500 persons who plundered and destroyed 60 shops in the centre of Amsterdam causing half a million guilders damage. Only a very large police force from other cities could suppress the riot.

The present brief survey of the most serious consequences of abnormal weather conditions in the world may show clearly the great impact of weather and climate on human activities and the economic developments in the world. In fact these indirect effects, combined with the direct effects on health and behaviour of man, explain why biometeorology is one of the most important environmental sciences for investigating the effect of weather and climate upon most activities of mankind.

1.4. CHEMICAL AND PHYSICAL PROPERTIES OF THE ATMOSPHERE

1.4.1. Normal chemical components of the atmosphere

The atmosphere is made up of a mixture of different elementary gases: nitrogen (78.08%), oxygen (20.95%), argon (0.93%), neon (0.0018%), helium (0.0005%), krypton (0.0001%), hydrogen (0.00005%), xenon (8×10^{-6}%), ozone (1×10^{-6}%) and radon (i.e. radium emanation, 6×10^{-18}%). Except for ozone (O_3) and radon, these percentages are practically the same throughout the atmosphere, at least up to 150 km altitude. Only the partial pressure is different at different altitudes.

Most people are acquainted with the fresh and penetrating odour of ozone (mainly caused by the oxidation of nitrogen oxides and organic substances), which is often noticed after electrical discharges in the laboratory or in nature after lightning storms; still very little is known about its real biometeorological significance. It is employed for purifying and deodorizing air in buildings or underground railways, for sterilizing drinking-water, etc. In other words, there is no doubt that, in larger concentrations which we can smell, ozone has a biological effect. In nature, near the surface of the earth, the concentrations are quite small. It is chiefly due to the studies of the German scientists, Cauer, Daubert, Effenberger and Warmbt that we know considerably more about the distribution of ozone in the lower atmosphere of central Europe and the meteorological factors affecting this distribution. According to Cauer and Effenberger,[17] the ozone content varies between 20 and 50 $\mu g\ m^{-3}$.

The ozone content increases with height and reaches a maximum in the warm layer of the stratosphere (see Section 1.4.3) at an elevation of about 25 km.

According to recent satellite observations in the stratosphere by Ghazi[16] in Cologne (1970–71) the ozone concentration increases from the equator towards high latitude (65 °N) followed by small decreases towards the north pole. In the northern hemisphere at high latitudes a maximum of ozone concentration occurs during spring, a minimum during the fall. In the southern hemisphere the total ozone and its seasonal variations are smaller than in the northern hemisphere. Changes in the tropics are usually small.

Even in the ozone layer or ozonosphere of the higher atmosphere the maximum ozone concentration is 10–20 ppm. The ozone in the higher atmosphere is produced by the irradiation of oxygen molecules by ultraviolet light (with wavelengths less than 2420 Å = 242 nm) in the region between 10 and 50 km above the earth's surface. The irradiation produces oxygen atoms (O) which unite with oxygen molecules (O_2) to form ozone (O_3). Above 50 km the oxygen becomes too diffuse for molecules to interact. Below 10 km, turbulence of the air will bring small amounts of ozone to the earth's surface.

Organic evolution of land species of animals could not proceed until plant photosynthesis had brought enough oxygen into the earth's atmosphere. This did not come about until the early palaeozoic age. The ozone layer absorbs all ultraviolet (UV) radiation with wavelengths less than 290 nm (UV-C) which is common above 2000 m altitude in mountains, and thus prevents the most actinic solar radiation from penetrating deeper into the earth's atmosphere and to the ground. This absorption also accounts for the temperature maximum at 50 km. The highest ozone concentration occurs towards the polar regions, the lowest near the equator.

Destruction of the earth's ozone layer would create a most dangerous situation both for the animal world and for man. This explains the extensive studies which are made at present on the effects of certain industrial products and high altitude aeroplane flights on the ozone layer. The following considerations may explain this even better. Several chemical processes in the stratosphere combined with UV irradiation are responsible for modifications of the ozone layer.

(a) Upward diffusion of nitrous oxide (N_2O) released during bacterial denitrification. This nitrous oxide is converted in the stratosphere, by reaction with ozone, under the influence of radiation of wavelengths of less than 310 nm, into nitric oxide (NO) and nitrogen dioxide (NO_2) causing a reduction in the ozone layer.

(b) The gases nitrogen and oxygen of air in contact with very hot surfaces, e.g. motor vehicles or aircraft engines, are converted into nitric oxide. Also nuclear hydrogen bombs create much nitric oxide which uses ozone for its disintegration which decreases the ozone concentration. It has been estimated that a fleet of 500 supersonic transport aircraft would cause a 16% ozone depletion in the northern hemisphere. These problems are studied at present by the Climatic Impact Assessment Program (CIAP) of the US Department of Transportation.

(c) About 50 years ago refrigerant fluids were developed containing chlorofluoromethanes (CCl_2F_2 and CCl_3F, known as fluorocarbon (or Freon)-12 and -11). In 1973 the production of fluorocarbon-12 and -11 was 500 000 and 300 000 tons respectively and all this will eventually be released into the atmosphere, causing a reduction of ozone in the ozone layer as a result of photochemical reactions in the stratosphere.[18,19]

A reduction of 10% in the ozone concentration would result in an increase of about 20% in ultraviolet B (erythemal radiation) at the surface of the earth. It has been estimated that a 5% decrease in ozone would cause a 10% increase in skin cancer, which is mainly caused by excessive ultraviolet radiation (see Section 5.2). The solar energy absorbed by ozone is released to other molecules and acts as the most important heat source in the stratosphere. Therefore, changes of ozone distribution with altitude will have a direct influence on the temperature structure of the stratosphere. It will influence both the transmission of solar radiation downward and the upward escape of, mainly infrared, radiation from the earth.

In 1976 the W.M.O. estimated an average temperature decrease of up to 10°C in the upper stratosphere due to the reduction of ozone as a result of man's activities. These processes may create worldwide changes in climate. It is in view of these considerations that a permanent international study group under the leadership of W.M.O. is studying at present all aspects of stratospheric environment relevant to ozone.

Various meteorological factors affect the distribution of ozone near the surface of the earth. The ozone content of the lower atmosphere has a daily rhythm usually with a maximum around 2 p.m. and a minimum at 8 a.m. In May and following summer months daily fluctuations up to 14 $\mu g\ m^{-3}$ (21 against 35 $\mu g\ m^{-3}$) have been observed, the differences being small in winter (only few $\mu g\ m^{-3}$).

In winter the ozone content is usually lower than in summer, but airmass influxes may change this rule completely.

The ozone content is very low (particularly in winter) on land under stagnant air conditions, with fog and during inversions (see Section 1.5); it increases with increased turbulence and thermal convection in the air (particularly in summer).

During days with high atmospheric pressure (in western Europe often accompanied by fog and inversions) we usually find low ozone values, whereas days with thunderstorms, lightning and snowstorms have high values. The ozone content also varies with the dominating direction of the wind. In north west Germany (according to Effenberger[17]) the average ozone content with east wind is 29.9 $\mu g\ m^{-3}$, with a west wind 42.6 $\mu g\ m^{-3}$, with a south wind 51.4 $\mu g\ m^{-3}$, with a north wind 41.3 $\mu g\ m^{-3}$. The low value with an east wind, due to the low ozone content of continental air masses in western Europe, is partly the result of the oxidizing effect of plants and forests in central and eastern Europe.

High ozone values are usually observed in north west Europe during the passage of cold fronts and the influx of polar air masses, whereas warm fronts with tropical air are usually characterized by low values.

Contrary to the case of ozone the amount of radon decreases in general with height. The vertical distribution of natural radioactive products in the atmosphere is controlled by the degree of turbulence and by temperature inversions (see Section 1.5). Therefore the concentration fluctuates considerably at different levels of the troposphere (see Section 1.4.3). Further details can be found in the paper by Reiter.[20]

Apart from the gases already mentioned the atmosphere contains carbon dioxide (0.03%) and particularly large masses of water vapour. Water vapour is important for the transportation of water through the atmosphere and because of its capacity to absorb infrared solar radiation with a wavelength longer than 800 nm. On a clear humid day it absorbs less than 15% of the solar infrared radiation. However, it absorbs the infrared re-radiation of the earth (wavelength 4–50 μm) almost completely. If the partial water vapour pressure at sea level is considered to be one, it is only 0.68 at 1 km altitude, 0.41 at 2 km, 0.11 at 5 km, and 0.013 at 8 km.

Carbon dioxide is also an important heat regulator of the atmosphere. It has the capacity to permit the sun's rays to reach the earth's surface but absorbs the reflected heat rays (the so-called *Greenhouse effect*). Although at present the amount of carbon dioxide in the atmosphere is estimated to be 0.03% (according to Kreutz[21] near the earth's surface 0.0438%) it may have been considerably higher in many parts of the world during geological periods of volcanic activity. According to a theory of Arrhenius[22] an increase of the amount of carbon dioxide in the atmosphere to 0.08% would increase the temperatures, e.g. in the Arctic, by 8–9 °C, a decrease in the concentration to 0.015% would lower the temperatures, e.g. at 45° latitude by 4–5 °C. However, more recent studies by Ångstrom,[23] Rubens[24] and Ladenburg[24] seem to contradict the assumptions of Arrhenius.

Carpenter,[24] who studied the carbon dioxide content of the atmosphere in about 1937 before the period of intensive industrialization, believed in a great constancy of the concentration. He analysed eleven hundred samples of outdoors air in New Hampshire (Boston) and Baltimore (USA). All these analyses gave approximately the same carbon dioxide concentration of 0.03%. This value did not change significantly during the different seasons or from year to year. However, in more recent studies Carpenter's observations were not confirmed. Considerable changes may occur depending on the degree of industrialization in different parts of the world.

For example, Callendar[24] believed that the amount of carbon dioxide in the atmosphere has increased since 1900 by about 10% due to the burning of fossil fuel by man.* However, according to Junge[26] the collected data are insufficient and not consistent enough to prove a worldwide increase. The great scattering

* A recent report of an Advisory Committee of the American Academy of Sciences assumes that if the present day consumption of gas and oil is replaced by coal the amount of carbon dioxide in the atmosphere would increase 4–8 times from the year 2000–2150. This increased concentration would allow the solar radiation to reach the earth's surface but it would absorb the reflected radiation, which would cause a rise in temperature of 6–9 °C in 175–200 years. This 'greenhouse effect' might be responsible for a drastic change in our climate.

of the data values is probably due to differences in the location of the sampling point, the time of the day and day of the year that the sample was collected, the sampling height, windspeed etc.[27] This makes it very difficult to compare the data of different research workers. Studies by Bray[28] indicated also that the observed increases were not always statistically significant.

Still the general concensus of opinion amongst chemists today seems to be that, at least in industrialized areas, the increase is real. More recent studies by Callendar,[29] using only reliable sources, indicated that prior to 1900 the average carbon dioxide content of the atmosphere in the northern hemisphere was 290 ppm, in 1920 about 305 ppm, in 1960 about 325 ppm (about a 13% increase). Observations near the south pole indicated a constant annual increase in the carbon dioxide content of 1.3 ppm. According to Munn and Bolin[30] the carbon dioxide content of the atmosphere has steadily increased since 1880 (about 0.7 ppm per year). In the first 3 km of the atmosphere the concentration is about 15 ppm and decreases to 7 ppm below the tropopause (see Section 1.4.3). According to Hagemann *et al.*[31] up to 80 km altitude the mixing of atmospheric gases is sufficiently intensive to inhibit gravitational separation of the gases. He found a constant mixing ratio of 311 ± 2 ppm up to 30 km altitude.

After the International Geophysical Year the existence of season fluctuations of the carbon dioxide concentration was discovered.[31] In the southern hemisphere there was almost no seasonal variation. In the north Pacific and in the northern Arctic the concentration showed a marked seasonal variation with an amplitude of 5 ppm. The maximum occurred in the spring at the onset of the growing season, a minimum occurred in the fall. The variation is probably due to biological cycles of the terrestrial biosphere.

1.4.2. Trace elements and other polluting substances of the atmosphere

Apart from the above mentioned small concentrations of the normal gases in the atmosphere, like argon, neon, helium etc. many other trace substances also occur in the atmosphere. These have been described in the handbook by Junge.[26] Table 1.3 shows some of these important trace gases. Only some of the known trace substances will be discussed.

The distribution of one of the most important trace gases, ozone and its significance for human life has already been discussed. Apart from natural occurrence ozone plays an important role in atmospheric pollution.

Since World War II the population of Los Angeles (USA) has been suffering from the effects of heavy smog (see Section 1.5). During such smog periods the population suffers from inflammation of the eyes and of the mucous membranes of nose and throat; plants and vegetables suffer heavily; small cracks appear in vulcanized rubber (e.g. of car tyres). Haagen-Smit[32] was able to demonstrate that this smoggy atmosphere is rich in oxidizing substances and organic material which derived from decomposed hydrocarbons. One of the

TABLE 1.3
Concentrations of different trace gases at ground level[26]

Gas		Concentration µg m^{-3} STP
Ozone	O_3	0–100
Hydrogen	H_2	36–90
Carbon dioxide	CO_2	$4–8 \times 10^5$
Carbon monoxide	CO	1–20
Methane	CH_4	$10–10^2$
Formaldehyde	CH_2O	0–16
Nitrous oxide	N_2O	$5–12 \times 10^2$
Nitrogen dioxide	NO_2	0–6
Ammonia	NH_3	0–15
Sulphur dioxide	SO_2	0–50
Hydrogen sulfide	H_2S	3–30
Chlorine	Cl_2	1–5
Iodine	I_2	0.05–0.5

most important oxidizing substances of smog proves to be ozone. Owing to the oxidation of hydrocarbons, a great variety of substances are formed such as peroxides, aldehydes, butenes, nitric acid etc. which are responsible for the irritation of the membranes of nose and throat. Studies in the Los Angeles area [32-34] have shown that ozone is practically absent during the night but is formed by sunlight as a result of the following chemical reactions: nitric oxide (NO) is formed from the exhaust gases of more than 2 million cars in Los Angeles, which also result in the evaporative loss of 1000–1350 tons of hydrocarbons each year. With radiation of wavelength shorter than 360 nm, nitrogen dioxide (NO_2) is formed, and this plus active hydrocarbons plus radiation produces ozone. According to Haagen-Smith,[32-34] the presence of alkyl or aryl groups in the hydrocarbons increases the ozone formation considerably. On the other hand, sulfur dioxide and nitrogen dioxide in a smog area may reduce the ozone concentration, because these substances are oxidized by ozone into sulfuric acid and nitrates.

Recent studies point to the effect of ozone on tobacco leaf. In the USA, in Connecticut valley, at Beltsville in North Carolina, in Ontario (Canada) and in other places, tobacco leaf damage, known as 'weather fleck', often creates a serious problem in the production of cigar-wrapper tobacco. Rubber strip tests and an ozone recorder indicated that highly significant correlations existed between the ozone level and the appearance of weather fleck during periods of abnormal concentrations of ozone. Artificial ozone in climatic chambers produces the same effect. Similar damage is observed on the upper-leaf surfaces of beans and grapes. Whereas toxic products of the oxidation of hydrocarbons during smog inflict damage on the lower leaf surface, ozone damage appears on the upper-leaf surface only. Although average values did not exceed six parts per hundred million (60 ppb), maximum values of 500 ppb were recorded during certain sunny warm periods of the day. Certain species have a threshold value of 200 ppb for 3 h of exposure before fresh damage develops.

Extensive studies have been made on the biological effects of ozone. At the beginning of this century Schwarzenbach, Barlow and others[35] demonstrated that ozone caused irritation and oedema of the lungs, and ultimately death if inhaled in strong concentrations. With concentrations of 0.05% death follows in 2 h, with 1% after 1 h. Other reported effects of ozone are reduction of general metabolism, changes in potassium and calcium level in the blood serum, increased susceptibility to respiratory infections etc.[15,36]

In the lower part of the atmosphere many toxic trace substances occur due to motor traffic. They are in part hydrocarbons related to gasoline and in part are gases resulting when the fuel is broken down into a series of cracking products. In the neat exhaust gases carbon dioxide, water, hydrogen (from 1 to 3%), carbon monoxide (up to 4%), aromatic substances, aldehydes (e.g. formaldehyde), considerable quantities of nitrogen oxides (such as NO), alcoholic substances (up to 0.01%), sulfur dioxide (0.07%) and hydrogen sulfide (0.01%) have been found.

Also lead products used in present-day motor spirit (about 0.4 ml l^{-1}) reach the atmosphere in the form of small crystals of lead chloride and bromide (of 2–100 nm diameter) and as lead tetra-alkyl through direct evaporation of petrol. According to Preis[37] the average lead content of air in Zürich in 1955 amounted to 2–5 $\mu g\ m^{-3}$, in traffic tunnels 20 $\mu g\ m^{-3}$, in garages as much as 40 $\mu g\ m^{-3}$. In Los Angeles, on clear days the lead content of the atmosphere is 2 $\mu g\ m^{-3}$ (with carbon monoxide 3–5 ppm), on 'smog' days 42 $\mu g\ m^{-3}$ Pb with 23 ppm of carbon monoxide. The data show that carbon monoxide is the most abundant exhaust product. The toxic effects of this are generally known. Wilkins[38] recorded around 1960 in the streets of Los Angeles carbon monoxide concentrations of 58–160 ppm; in London 1–50 ppm (average 15). These concentrations are of course much higher at the present time. The concentration decreases with height. For example, in the centre of London 10 ppm were found at 4 ft against 3.2 ppm, at 100 ft. This supports the view that the main source of carbon monoxide is road traffic and not chimney smoke. It is interesting to note that, during the heavy smog period in London of 4th–7th January 1956, the average carbon monoxide concentration amounted to 8 ppm only, probably due to the greatly reduced road traffic. With ordinary traffic in western Europe the carbon monoxide content is 5–20 ppm; with heavy traffic 50–80 ppm. In tunnels for road traffic it may increase to 600 ppm.

In contrast to gasoline traffic, the carbon monoxide production in exhaust gases from diesel engine traffic is very small or non-existent. The hydrocarbon content is also low. On the other hand diesel engines produce large amounts of soot and other, odorous, substances which are often irritating to the nasal and laryngeal mucous membranes and increase smog conditions at low levels in the atmosphere. Owing to the incomplete combustion of gasoline, acetylene and other related compounds are formed which, on account of high pressure in the diesel engine, are transformed into polycyclic compounds (arenes) such as pyrene, fluoranthrene, 3,4-benzpyrene, 1,12-benzperylene, anthanthrene and coronene.

It is known that 3,4-benzpyrene in particular has carcinogenic properties if applied to the skin. Waller[39,40] found in various cities in the UK an average 3,4-benzpyrene content of 10–50 ng m^{-3}. Kotin found in the same period 30 ng m^{-3} of benzpyrene 5 m above the ground in Los Angeles on smog days. The amount fluctuates with the season. In London 25 m above the street the content in summer was 20 ng m^{-3}, in January 150 ng m^{-3}. In Sheffield during heavy fog in February 330 ng m^{-3} was measured. A similar seasonal fluctuation is found in the atmosphere in floating soot, which contains 100 ppm of benzpyrene in summer against 300 ppm in January. This would suggest that the heating of buildings in winter is partly responsible for the high winter values. According to studies by Kotin,[41] the chief producer of benzpyrene is the lightly-loaded petrol engine running in low gear. Diesel engines produce little or no benzpyrene, provided there is complete combustion. Otherwise these engines also produce benzpyrene. Another important meteorological factor in the benzpyrene problem is the influence of light. Benzpyrene is very stable in dark surroundings, but quickly decomposes in sunlight. The presence of oxidizing substances like ozone also accelerates decomposition.

Particularly near the coast the atmosphere contains traces of iodine, magnesium chloride and salt (NaCl), all three important compounds which may affect the thyroid gland and nervous system of the human body. The presence of salt is particularly noticeable on foggy days near the coast, when electric instruments in the open air show very considerable losses of current due to decreased insulation. The salt dissolved in the water droplets changes the water into a good conductor. Cauer[42] in particular has considerably increased our knowledge of the daily fluctuations of various trace substances in the atmosphere. He could demonstrate in western Europe that most of the iodine in the atmosphere does not originate from evaporating sea-water but is due to the burning of large masses of seaweed rich in iodine, as is customary in various coastal areas of western Europe: Brittany (France), Shetland Isles, Stavanger area (Norway), area near Coruña (Spain), northern and western coasts of Eire, but also along the Black Sea coast and in the Kola Peninsular (USSR). The seaweeds contain 0.0008–0.0012% of their dry weight in iodine. The iodine of the air is carried with various air masses as far as central Europe. On the French coast (in Brittany) the average iodine content is 0.4 $\mu g\ m^{-3}$. Shortly after the large-scale burning of seaweed in this area it increases to 6 $\mu g\ m^{-3}$. In central Europe the average content is about 0.05 $\mu g\ m^{-3}$, but up to 1934 (after which date Chile saltpetre production caused a heavy drop in the world market price of iodine), it was much higher (0.6 $\mu g\ m^{-3}$). During periods of weed burning along the coast of western Europe and strong air mass influxes over the European continent it rapidly increases from 0.05 to 0.6 $\mu g\ m^{-3}$. Calculations suggest that iodine which actually originated from evaporated sea-water is only 2.5 ng m^{-3} (in central Europe).

Polar continental and polar maritime air masses are usually poor in iodine, tropical maritime air usually has the highest values; in foggy areas volatile iodine in the air rapidly decreases, but is concentrated in the water droplets; solar radiation together with ozone usually oxidize the iodine ions in the water droplets into volatile iodine pentoxide. These various factors explain the great variation in the concentration of iodine in the air.

Few accurate data are available on the magnesium chloride content of the air. According to Cauer[42] the average content is about 3.1 μg m^{-3} with a maximum of 65 μg m^{-3}. The main source seems to be evaporating sea-water and strong west winds (at least in western Europe) which carry those water droplets far inland.

1.4.3. Physical structure of the higher atmosphere

From the surface of the earth to space the earth's atmosphere does not decrease gradually in temperature and density; it has a shell structure. In Fig. 1.2 the names of these different shells are given in accordance with the classification of the W.M.O. The Figure shows the physical structure of the lowest 100 km.

At 100 km the density of the air is only one thousandth that of the air at the surface of the earth. The temperature is represented as a function of air pressure. To about 10 km altitude (higher in the tropics and lower near the poles), the so-called troposphere, the air temperature decreases with height. This is the shell in which our main weather conditions take place. The top of this layer is called the tropopause.

Above the troposphere lies a region known as the stratosphere, marked by an almost complete lack of clouds and slight turbulence. It is divided into three layers: a lower cold isothermic layer 12–35 km above sea level with fairly constant temperatures (around −50 °C), followed by a warm layer (up to 60 km above sea level) and then a very cold upper layer (60 to 80 km above sea level) with temperatures down to −80 °C. The upper layer is also called the mesosphere, the top being called the mesopause. The warm layer lies near the top of a region in the atmosphere with maximum ozone concentration (the ozone layer or ozonosphere already mentioned above) and with maximum absorption of ultraviolet radiation. Some meteorologists believe that ozone plays an important role in stratospheric circulation and therefore may considerably influence the general weather conditions in the world.

In and above the mesosphere lies the ionosphere characterized by strongly ionized air causing the reflection of radio waves (Heaviside layer) and the various auroras (northern lights) in the atmosphere.

The lowest reflecting layer, of the ionosphere, known as the D-layer, ranges from 60 to 90 km. It has a charged particle density of 5×10^2 cm^{-3}. It reflects low frequency radio waves but absorbs medium and high-frequency waves. The

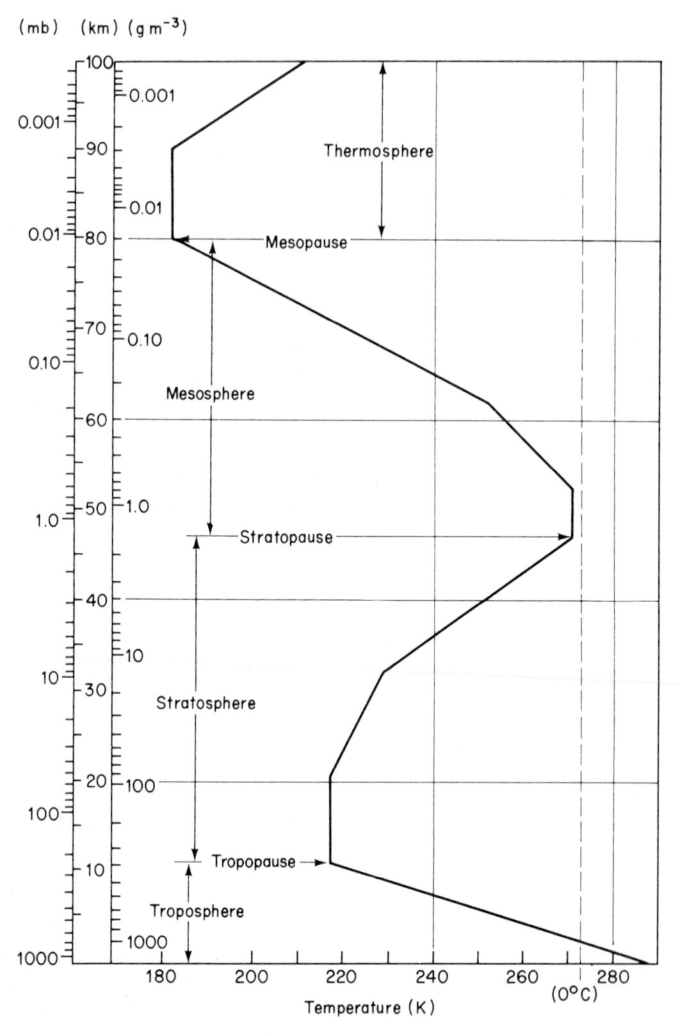

Fig. 1.2 Vertical temperature distribution of the standard atmosphere and nomenclature for layers (from: H. E. Landsberg in *Progress in Human Biometeorology*, Vol. 1, Part IA, p 47)

layer disappears as soon as the sun sets and during periods of pronounced sun flares. The ionization of the layer may be caused by dissociation of oxygen or ozone molecules due to intense ultraviolet radiation as a result of eruptions in the sun's chromosphere. The part in the atmosphere above 80 km is also known as the thermosphere. The temperature rises above 90 km.

Between 90 and 160 km a so-called E-layer occurs, which strongly reflects medium- and high-frequency radio waves. This ionizing layer, which contains 2×10^5 to 2×10^3 cm^{-3} ions, is usually explained as being produced by ultraviolet photons from the sun acting on oxygen and nitrogen molecules. The layer becomes thinner during the night.

Above the E-layer various other layers are distinguished: the E_2-region (at about 150 km height), the F_1- and F_2-regions, the G-region (around 400 km) with an electron density of 5×10^5 cm^{-3}. It is at a height of between 80 and 300 km that the northern lights are usually formed as a result of the bombardment of oxygen and nitrogen molecules by electron and proton showers from the sun.

The various layers are seriously disturbed after solar eruptions, which may cause widespread disruption and interference with radio-communications and strong geo-magnetic disturbances. Above the ionosphere the density of the atmosphere becomes extremely low. This outermost sphere is often named the exosphere after L. Spitzer. Satellite studies have demonstrated the existence of a zone with very intensive radiation known as Van Allen radiation, after the first investigator J. A. Van Allen.[44] The radiation seems to be due to protons and electrons concentrated in several belts around the earth. The lowest belt reaches a maximum intensity at about 2400 km altitude; the upper belt has a zone of maximum radiation at 32 000 km. There is evidence for a third and lower belt with an extremely intense radiation fatal to human beings if exposed to it for more than one hour. It seems to be caused by direct emission from the sun and by the decay of cosmic ray neutrons, created within the earth's atmosphere, that are directed outward.

1.4.4. Physico-chemical properties of the planetary atmospheres (Table 1.4)

Since the beginning of space research scientific interest in the physico-chemical properties of the atmospheres of the planets of our solar system has increased enormously, in view of possible landings of spacecraft on these planets in the near future and the growing interest in the possibility of life on other planets. In Table 1.3 a summary is given of the different planets and their main properties.

Particularly interesting observations were made on the planet Mars, after the landing of the American space craft Viking 1 and Viking 2 on the surface on 20th July and 4th September 1976, about 20 years after the launching of the first Russian sputnik. Our knowledge about the atmosphere of this planet has increased enormously as a result. The most important findings can be summarized as follows: the surface temperature fluctuates at the landing-site between −93 and −13 °C, at the equator the temperature was −33 °C, on the south pole temperatures of −133 °C were measured; water vapour is extremely rare particularly on the southern hemisphere; near the poles there is a layer of water-ice contrary to previous speculations. The average atmospheric pressure on Mars was about 7.65 mbar (an atmospheric pressure observed on earth at an altitude of 30–35 km) against 1013 on the surface of the earth (equal to 760 mm Hg); the usual atmospheric pressure changes on Mars are very small, about 0.1 mbar per Mars-day; the windspeeds fluctuated only between 0 and 10 m s^{-1}. The gravity on Mars is only 0.4 g; in the lowest layers of the atmosphere of Mars

TABLE 1.4
Planets of the solar system and some of their physical properties

Planet	Average distance to sun (Earth=1)	Equatorial diameter (Earth=1)	Mass (Earth=1)	Density (g cm^{-3})	Rotation period (in days=d, hours=h and minutes=min)	Moons	Most important components of the planetary atmosphere
Mercury	0.40	0.38	0.056	5.50	58.6 d	1	helium, argon, neon
Venus	0.73	0.95	0.815	5.27	243.0 d	?	97% CO_2
Earth	1.00 (149.6 × 10^6 km)	1.00 (6378.16 km)	1.000 (5.97×10^{24} kg)	5.52	23 h 56 min	1	78% N_2 21% O_2
Mars	1.52	0.53	0.107	3.94	24 h 37 min	2	90% CO_2
Jupiter	5.50	11.19	317.9	1.33	9 h 55 min	12	hydrogen, helium
Saturn	9.52	9.47	95.1	0.69	10 h 14 min	10	hydrogen, helium
Uranus	19.16	3.69	14.5	1.68	10 h 48 min	5	hydrogen, helium
Neptune	29.99	3.92	17.3	1.59	15 h 48 min	2	hydrogen, helium
Pluto	39.37	0.5	0.18	7.70	6 d 0 h 5 min	?	?

the chemical composition was as follows, nitrogen 2–3% (78% on earth), oxygen 0.3–0.4% (~21% on earth), carbon dioxide 95% (0.03% on earth), argon 1–2% (0.93% on earth); at great height in the atmosphere of Mars, the ionosphere consists mainly of ionized oxygen; carbon dioxide occurs at one ninth the concentration of oxygen; on earth nitrogen is the dominating gas, on Mars carbon dioxide. Thus the atmosphere of Mars is entirely different from that of the earth and so even if some kind of life would exist on Mars it will be entirely different from that on the earth.

Much less is known about Mercury although the Mariner 10 passed this planet on 29th March 1974 at a distance of only 400 miles. Apparently Mercury has an extremely thin atmosphere of helium, argon, neon and a few other gases in very small concentrations. The presence of such an atmosphere is difficult to understand because of the low gravity of the planet (0.38 g), which normally would be too weak to prevent a gaseous envelope from escaping into space. Also the magnetic field is very weak, about 1% of the earth's. As Mercury is only 60 million km from the sun (the earth is 150 million) the temperatures are much higher and differ considerably due to the great eccentricity of its orbit. At the point where Mercury is nearest to the sun the temperature is twice that of the farthest point. At the first point temperatures of ~420°C were measured, which dropped during the night to −180°C.

The planet Venus was studied by the American spacecraft Mariner 2 (27th August 1972) and Mariner 5 and by Venus 1 to Venus 8 from the USSR. The first landing on Venus, after a trip of 293 000 000 km during 109 days, was on 15th December 1970 by Venus 7, followed by the Soviet automatic interplanetary station Venus 8 which made a soft landing on the day-side of Venus on 22nd July 1972. However, the first atmospheric measurements had already been made on 18th October 1967 by Venus 4 after the failure of Venusnik in February 1961.

According to these various studies the composition of the atmosphere is entirely different from the earth: 97% carbon dioxide, maximum 2% nitrogen, 0.01% oxygen, 0.05% water and 0.01–0.1% ammonia. The atmospheric pressure on the surface of Venus was 95–100 atmospheres. The surface temperature was about 465 °C.

Very little is known about the great planets Jupiter and Saturn. The preliminary studies by two American unmanned spacecrafts had already started in 1968, but it was only on 4th December 1972 that Pioneer 10 was launched. Pioneer 11 on 6th April 1973. These studies revealed that Jupiter and Saturn are huge, low density, rapidly rotating planets.

The major constituent of the atmosphere of Jupiter and Saturn is hydrogen. Also helium (He), methane (CH_4) and ammonia (NH_3) have been reported. The quantities observed are He 0.132%, CH_4 0.0006%, and NH_3 0.00015%. The pressure near the surface of these planets is about 3.0 atm, at 75 km 0.1, at 125 km 0.01, at 175 km about 0.001 atm. According to recent reports[45] by Russian astronomers Jupiter is not a strongly cooling planet although it loses 24 times more heat radiation than it receives from the sun, probably due to internal nuclear reactions. The Russian scientists believe that Jupiter is warming up and that after 3 milliard years it will be a star like the sun. Mars, Earth, Venus and Mercury have been compared with huge thermodynamic engines with solar radiation as their only source of energy. The total energy received from the sun is balanced by their emitted energy. For the giant planets Jupiter and Saturn it is different. The infrared emissions alone from these planets exceed by factors of 2.3–2.7 the solar energy absorbed, indicating the existence of internal heat sources.

It is evident that future spaceflights will reveal new facts which may alter entirely our present concepts about the planets, their origin, geology and meteorology.

1.5. METEOROLOGICAL CONCEPTS IN BIOMETEOROLOGY

Biometeorology makes use of a number of concepts. Weather comprises the day-to-day changes in meteorological conditions including temperature, rainfall, snow etc. whereas climate relates to average weather conditions based on 30 or more years of observations. Macroclimate relates to the climatic conditions of the atmosphere above specified areas, whereas microclimate is a term used to designate the climatic conditions directly surrounding the living organism both near the surface of the earth and in a limited niche, e.g. in a room, factory or mine. The differences between the two types of climate are usually considerable as it is the microclimate which is the main determining factor for the health of man. A correlation between biological events and the macroclimate may lead to erroneous results. In 1919 V. and J. Bjerkness, two Norwegian

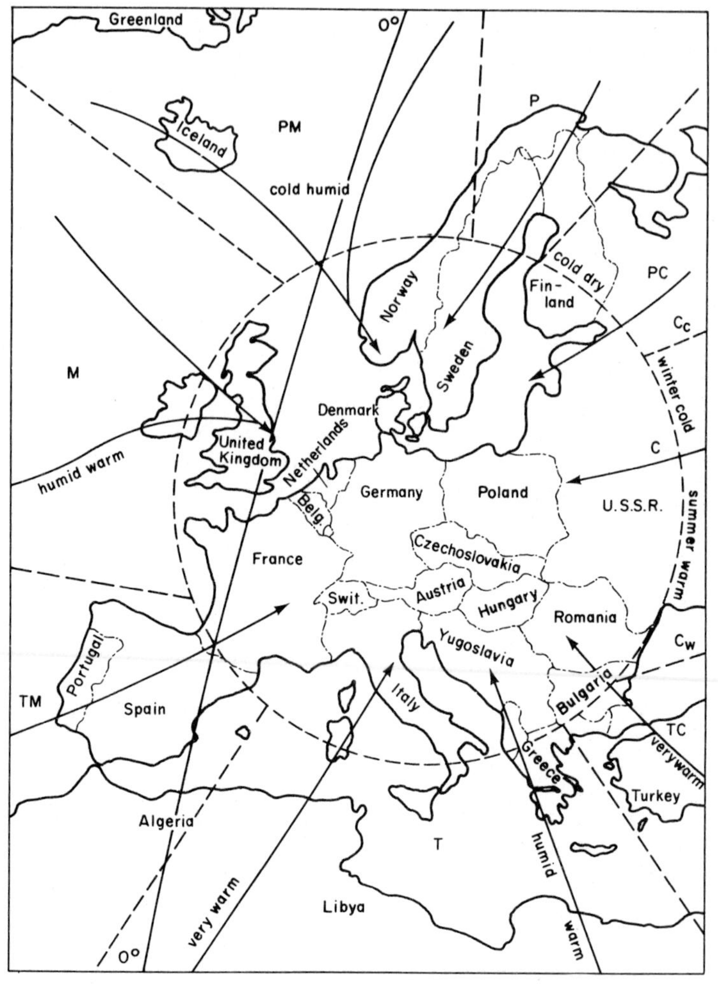

Fig. 1.3 Air masses in western Europe, their average temperature and humidity (based on a map of Lincke)

meteorologists, introduced the concept of air masses. The term air mass is used to denote large masses of air, hundreds to thousands of kilometres long and several kilometres high, with roughly the same physical and chemical properties. Depending on their source area, meteorologists distinguish between polar, tropical and other types of air masses (Fig. 1.3). If two air masses meet, e.g. polar and tropical, they usually do not mix readily but remain separated by a boundary plane known as a frontal surface. The line of intersection of this surface with the surface of the earth is called a front. If warm air moves towards and over an area with cold air there is a warm front; if cold air moves toward and usually beneath an area of warm air there is a cold front. If a cold front overtakes a warm front the warm front is pushed aloft. When the warm air is lifted off the earth's surface the resulting phenomenon is known as an occlusion. Usually temperatures in the atmosphere decrease with increasing height. Sometimes, however, at a certain level temperatures may increase with increasing height; this phenomenon is known as an inversion. Inversions are especially common during periods, particularly in winter, when atmospheric pressure is high and there is little air turbulence. They usually cause dense fog. If an inversion is sufficiently strong to prevent the smoke from industries and/or domestic chimneys from dispersing, smog (smoke-fog) is likely to form. It is a dense fog, polluted with soot particles and various volatile chemical substances; it is particularly harmful to bronchitic patients. Over 4000 deaths were attributed to the London smog of 1952. One other important concept in biometeorology is meteorotropism, i.e. the sensitivity of the living organism to changing meteorological environments.

From a biometeorological point of view it is necessary to distinguish between three fundamentally different climates (mountain climate, forest climate and maritime climate) each with its own specific physico-chemical characteristics and with different biological effects on the living organism.

1.5.1. Mountain climates

Particularly above 1500 m altitude climates are characterized by the following physico-chemical properties.

(a) *Reduced partial oxygen pressure.**
At sea level the partial pressure of oxygen is 159 mm mercury against a total barometric pressure of 760 mm, whereas at 2000 m altitude the oxygen pressure is reduced to about 125 mm. As a result the oxygen binding substance in the human blood, haemoglobin, will be less saturated with oxygen at high altitude.

(b) *Differences in quality and intensity of solar radiation.*
This is particularly true for the invisible ultraviolet part of the solar spectrum with wavelengths of 290–380 nm. Three kinds of ultraviolet are distinguished: UV-A (λ† = 315–380 nm); UV-B (λ = 290–315 nm) or Dorno radiation (named after the biometeorologist Dorno); UV-C (λ less than 290 nm). In UV-B and in particular in the interval 297–302 nm, there are strong biological (e.g. anti-rachitic) effects; UV-C is observed only at altitudes in excess of 2000 m.

(c) *Average daily temperatures are generally lower.*

Apart from these three principal differences with low land climates the following differences have also been observed: there is usually less atmospheric

*Air is composed of different gases, each with its own pressure. The total pressure of all gases together is about 760 mm Hg. The pressures of the separate gases are known as 'partial pressures'.
†λ = wavelength.

turbulence and less water vapour; the heat loss from the earth (terrestrial radiation) is in consequence greater than at lower altitudes. The ozone content is often higher; the number of large ions is reduced (from about 10^4 to 10^2), but the number of small ions is increased; there is also a reduction in the dust content of the atmosphere and in air pollution in general (this applies to both allergens and chemical pollutants).

1.5.2. Forest climates

Forest climates are characterized by the following physico-chemical factors.

(a) *Photochemical action of sunlight.* The action of solar radiation on the chlorophyll bodies in the cytoplasm of plant cells. A process known as assimilation or photosynthesis is described in Section 2.1. During this process carbon dioxide and water vapour are absorbed by the plant and converted into carbohydrates and oxygen. As a result the air around trees becomes progressively poorer in carbon dioxide and richer in oxygen during daylight hours.

(b) *Humidity.* This is usually increased.

(c) *Wind speed, general air movement* and the amount of direct or reflected solar radiation reaching the ground are considerably reduced; the general cooling during clear nights, which is very large in open fields, is small in a forest, and this prevents the formation of ground mist.

(d) *Electrical conductivity* of the air and the number of small ions is reduced, but the number of large ions is increased; the electrostatic potential gradient (i.e. the electric potential difference between a point in the atmosphere and the surface of the earth per metre altitude difference) and the daily fluctuations are very small.

(e) *Pollution.* Chemical air pollution is considerably reduced in forests as a result of the absorption of man-made chemical pollutants; on the other hand the amount of natural pollutants* is increased.

1.5.3. Maritime climates

In temperate lands they are typically ones in which the difference in average temperature between summer and winter (i.e. the range of temperature) is less than 15°C. A typical continental climate has differences in excess of 20°C. Some of the most striking physico-chemical characteristics of the maritime climate are:

(a) *Considerable turbulence*: due to the usually high to strong winds along the coast. This causes appreciable cooling of the human body. Atmospheric cooling or wind chill is a very important physiological stimulus. It is a

*The term 'natural' pollutants refers to pollutants which derive from living organisms, e.g. pollen, spores, micro-organisms, etc.

function of temperature and windspeed and may be calculated using the formula of Bedford and others:[48-51]

$$h = \Delta t(9.9 + 10.9\sqrt{V} - V) \qquad (1.1)$$

where h = the amount of heat loss in kcal m^{-2} h^{-1}
Δt = difference in temperature between subject and environment in °C
V = wind speed in m s^{-1}

In cold polar areas the formula of Wilson[52,53] is probably more applicable:

$$H = (\sqrt{100V} + 10.45 - V)(33 - t) \qquad (1.2)$$

where H = heat loss in kcal m^{-2} min^{-1}
t = ambient temperature in °C
V = windspeed in m s^{-1}

(b) *High intensity of reflected solar energy.* The amount of solar energy reflected by water may be twice that reflected by grass, particularly with respect to the shorter ultraviolet rays. Sand reflects about four times more strongly than grass. Strong reflection of the infrared radiation by the environment gives rise to excessive heat of the skin on days when there is little air movement.
(c) *Differences in temperature.* Both diurnal and seasonal variations are smaller near the coast than inland. Winter temperatures are relatively high, summer temperatures relatively low and there are fewer frosty days.
(d) *Humidity* is greater near the coast but, owing to the increase of air movement, foggy conditions tend to clear up more rapidly there than inland.
(e) *Trace elements.* There are higher concentrations of ozone, iodine, magnesium chloride and sodium chloride above the sea.
(f) *Air pollutants.* Dust, pollen and chemicals, are generally rare when there are sea breezes.
(g) *A different electrical field in the atmosphere above the sea.* The number of large ions is usually less above the sea than on land, and the total number of ions is usually considerably smaller than inland. Considering the physiological effects of each of the above mentioned parameters, which differ for mountain, forest and maritime climates it is evident that the meteorotropic effects caused by each of these climatic conditions will be entirely different.

1.5.4. Important meteorological factors

The following factors have to be recorded in biometeorological research.
(a) *Thermal factors:* temperature, combined with wind and humidity, determines the cooling index (see Section 1.5.3.). Particularly with high temperatures, humidity hampers regular sweating and evaporative cooling of the body.

(b) *Radiation*: infrared (heat) radiation, ultraviolet radiation (see Section 1.5.1.) and visible light. Both wavelength and intensity must be considered if meteorotropic phenomena are to be studied (see Section 5.2).
(c) *Wind speed and degree of atmospheric turbulence* are very important in connection with the cooling index.
(d) *Humidity* is a very important physiological factor as it determines with temperature not only the degree of human comfort (see Section 4.6), but also the local resistance against infections of the respiratory tract and the development of viruses and bacteria (see Sections 6.3 and 6.7).
(e) *Partial oxygen pressure* decreases with altitude and is particularly important for high altitude phenomena above 1500 m (see Section 5.3).
(f) *Electrostatic and electromagnetic fields and degree of ionization of the air.* Despite the many reports in recent years on important biological effects, in reality the reported effects are very small and often non-existent if rigorously controlled experiments are carried out. This is particularly true for the many claims concerning the respiratory effects of a surplus of positive or negative ions. The well established phenomena can only be demonstrated under very strictly controlled laboratory conditions (constant temperature, humidity and light; no metal surfaces, and constant smoke conditions) and even then the effects are very small as compared with those mentioned above.
(g) *Floating particles in the air.* They are either plant pollen, fungi spores, microbes, inorganic mineral fragments, trace elements or chemical pollutants (natural or industrial).

In space research the effects of strong magnetic fields, changes in gravity, cosmic and corpuscular solar radiation etc. also have to be considered. Of these complex, often simultaneously acting forces, the thermal and radiation ones are the most important for general biological and medical research.

1.6. METHODS APPLIED IN BIOMETEOROLOGY

If a preliminary study of the possible relationships between certain medical or biological phenomena and meteorological factors suggests causal relationships, then three methods, differing in principle, can be applied in biometeorology in order to establish whether the observed correlations are statistically significant. These are the empirical method, the experimental method and the method of prediction.

1.6.1. The empirical method

Two fundamentally different methods have been used in empirical biometeorology, the indirect or medical geographical method and the direct biometeorological method.

(a) *The indirect or medical geographical method.* In countries where accurate mortality or morbidity figures are available and where the age structure of the population is also known, it is possible to prepare *geographical distribution maps* (Fig. 1.4). The basic map is usually an ordinary administrative map giving the boundaries of the smallest administrative units in a country, either municipalities, counties, boroughs, or the like. The standardized mortality or morbidity figures are given in each of these units, i.e. the

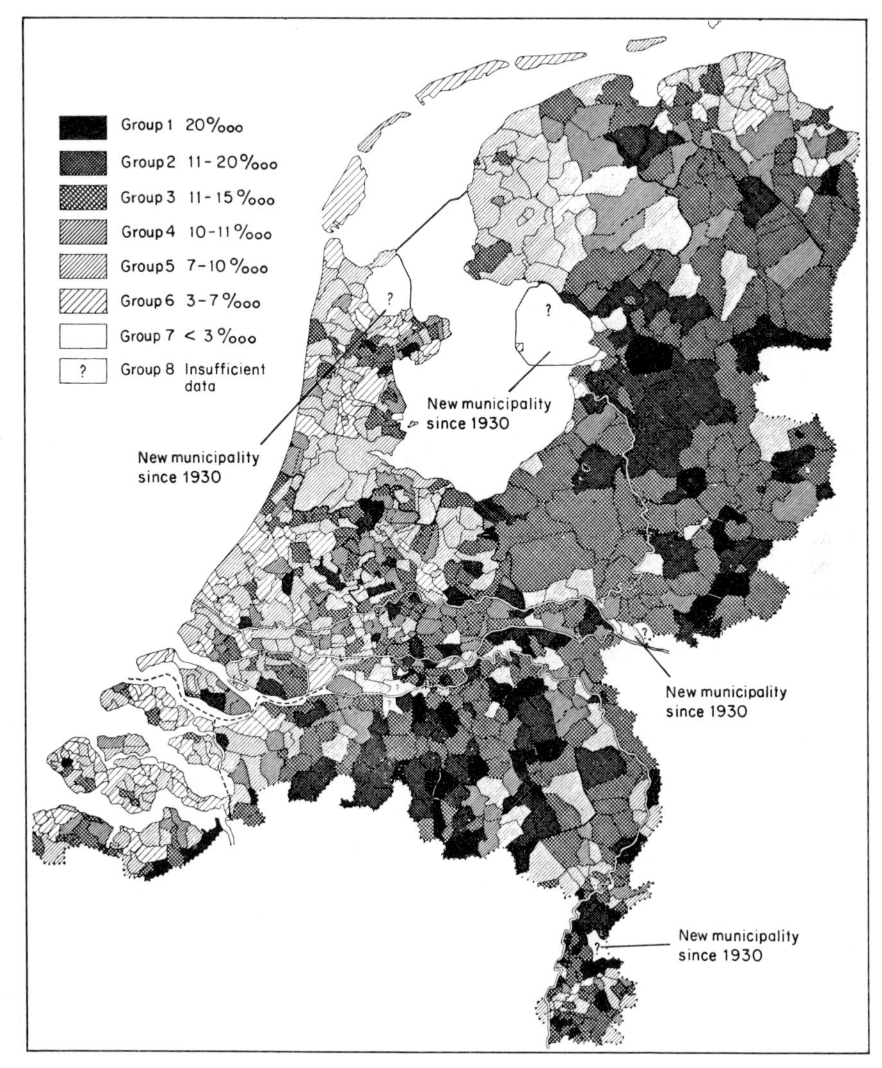

Fig. 1.4 Average yearly death rate from all types of bronchitis and pneumonia as primary cause of death (period 1900–1930) (based on data of The Netherlands Central Bureau of Statistics, compiled by J. C. Diehl and S. W. Tromp)

mortality or morbidity per 1000 (10 000 or 100 000) inhabitants of a special age group. For certain diseases the total of all age groups can be used; in other instances (for example cancer), when the incidence of the disease is predominantly developed in older or younger age groups, it is necessary to standardize on the population of that particular age group. Once the figures are plotted in each administrative unit of a geographical map of an area, certain mortality groups are selected which are indicated by the same colour or black and white hatching, e.g. the mortality figures 0–25% above the country average are given the same colour, 26–50% another colour, etc. As a rule, a colour system from light to dark is selected with increasing mortality or morbidity. Once such a map is prepared and it can be shown that differences in diagnostic accuracy, medical facilities, social or racial differences, food habits etc. cannot account for the observed geographical differences, then these medical geographical maps can be compared with climatic maps and one may find, that certain boundaries roughly coincide with climatic boundaries in the country.

(b) *Direct biometeorological method*. This is the more common method and offers a readier means of discovering possible relationships between weather, climate and diseases. Each biometeorological analysis of clinical data on an empirical basis requires two groups of observations, accurate meteorological data and clinical data.

The various meteorological parameters described in Section 1.5.4. should be recorded if possible in the centre of the area where the biometeorological studies are carried out. For some components, such as passages of weather fronts, air mass etc. the data from more distant areas can also be used.

Before a statistical analysis is carried out it is advisable to plot these various observations with special symbols in so-called biometeorological logs, i.e. charts giving in a compact form the daily changes of a large number of recorded meteorological parameters.

For the clinical or physiological data either one healthy subject or patient or a group of subjects or patients have to be studied daily for a very long period. The advantage of using only one person is the elimination of the problem of biological variations which arises if more persons are studied. On the other hand, the phenomena observed in one or only a few healthy persons or patients, particularly if they differ in age groups, sex and profession, may not be at all representative for the population as a whole. Therefore, a large group of healthy people or of patients suffering from one particular disease, if selected at random will give a better impression of the weather relationships in a larger population group than will a single patient. We may also discover a number of weather relationships which do not always show up in a single person. Once certain relationships have been established, it is advisable to study the individual cases separately in order to be able to carry out confirmatory experiments, as will be described (Section 1.6.2).

Apart from the severity of the complaint, other details must be indicated in special clinical enquiry sheets, such as the medical treatment applied, drugs used, possible psychosomatic influences, possible causes of exacerbation and the like.

Such data sheets are either filled in by the patient himself (which has proved to be very successful for example in the case of adult asthmatic patients), or by the nurse or doctor in charge of a ward in a hospital or clinic.

The daily data of all the patients are transferred to larger tables. From these tables the daily percentage of patients (of a constant group of persons) suffering from a certain disease or a particular physiological phenomenon can be calculated, and the daily severity of their conditions, can also be calculated. For this purpose the severity can be indicated with an imaginary figure, 1, 2, etc. The percentage and severity figures are represented graphically in large charts (clinical logs), the scale of one day being the same as in the biometeorological log.

A comparison of the biometeorological and clinical logs may give a first indication of existing correlations. It is not advisable to put all the data automatically into a computer for statistical analysis (which is an all too common procedure in modern science) because one may then overlook certain relationships. It is easier to distinguish visually between obvious and vague correlations and to indicate the plausability of a causal relationship.

Once certain apparent relationships are observed, the best mathematical method for a statistical analysis can be selected in consultation with a statistician. Despite all precautionary measures which are taken the medical records should be accepted with due reserve for the following reasons:

(a) errors in diagnosis because inhomogeneous diseases (for example schizophrenia or asthma) are grouped together under one name, whereas each variety of the same general disease may react differently to meteorological stimuli;
(b) errors in recording;
(c) errors due to the effects of medication with certain drugs;
(d) poor weather conditions which affect the nursing staff psychologically and therefore their rating scales;
(e) presence of one restless or difficult patient in a ward (particularly important in mental hospitals);
(f) errors due to differences in sex, age, profession, social status and food habits of the observed subjects;
(g) errors due to differences in the hour, month or year of observation, which may give different results because of the biological rhythm phenomena.

These and other factors affecting the outcome of a meteorological stimulus (Section 5.13) could create non-valid weather correlations. For example males and females may react oppositely and then by adding the data of both groups any effects would cancel and no relationship could be established.

1.6.2. The experimental method

Once a certain statistically significant correlation is found it is necessary to demonstrate whether the observed correlation is a causal one. Therefore an experiment should be devised, based on this correlation, which could demonstrate that the prediction is correct. Such experiments are often carried out in low pressure-climatic chambers which enable the observer to imitate all kinds of weather conditions and to confirm the observed empirical observations.

1.6.3. The method of prediction

If the previous methods confirm the existence of a relationship which seems to be a causal one, further confirmation can be obtained by predicting certain clinical or physiological events in relation to certain weather conditions or to certain climatic conditions during certain days or months of the year. If these predictions give results similar to the actual observed clinical data then it may be assumed that the meteorotropic relationships are really causal ones.

REFERENCES

1. B. De Rudder, *Med. Meteor.* **4**, 2 (1950).
2. *idem, Grundriss einer Meteorobiologie des Menschen,* Springer Verlag, Berlin, West Germany, 1952.
3. W. F. Petersen, *The Patient and Weather* (4 vols), Edwards, Ann Arbor, Michigan, USA, 1938.
4. H. Berg, *Wetter und Krankheiten,* Bouvier Verlag, Bonn, West Germany, 1948.
5. H. Muecher, *Arz. Forsch.* **7**, 123 (1957).
6. R. Reiter, in *Probleme der Bioklimatologie 6*, Akad. Verlagsgesellsch., Geest & Portig, Leipzig, East Germany, 1960.
7. D. Assmann, *Die Wetterfüligkeit des Menschen*, Gustav Fisher Verlag, Jena, East Germany, 1963.
8. F. Sargent, II, in *Medical Climatology*, (S. Licht, ed.), Elizabeth Licht, New Haven, USA, 1964.
9. *idem* and S. W. Tromp, *Technical Note 65, W.M.O. No. 160 TP 78*, W.M.O., Geneva, Switzerland, 1964.
10. S. W. Tromp (ed.) *Progress in Biometeorology, Divs. A–C*, Swets & Zeitlinger, Amsterdam, The Netherlands, 1972–1977.
11. V. Faust, *Biometeorologie*, Hippokrates Verlag, Stuttgart, West Germany, 1976.
12. S. W. Tromp, *J. Interdiscipl. Cycle Res.* **5**, 309 (1974).
13. *idem ibid.* **6**, 71 (1975).
14. *idem, Experientia* **32**, 126 (1976).
15. *idem, Medical Biometeorology*, Elsevier, Amsterdam, The Netherlands, 1963.
16. A. Ghazi, *J. Interdiscipl. Cycle Res.* **8**, 183 (1977).
17. Effenberger, in H. Cauer, *Arch. Phys. Therap.* **3**, 200 (1951).
18. *Halocarbons, Effects on Stratospheric Ozone*, Nat. Acad. Sci., Washington, USA, 1976.
19. *Halocarbons, Environmental Effects of Chlorofluoromethane Release*, Nat. Acad. Sci., Washington, USA, 1976.
20. R. Reiter, in ref. 10, Part 1A, 1976.
21. W. Kreutz, *Angew. Botan.* **23**, 89 (1941).

22. A. Arrhenius, *Phil. Mag.* **41**, 237 (1896).
23. A. J. Ångström, *Ann. Physik.* 3 (1900).
24. in H. Cauer, *Arch. Meteor. Geophys. Bioklimatol., Ser. B* **3–4**, 221–256 (1949).
25. *Advisory Cttee. Rep. Amer. Acad. Sci.*, summarized in U. Hampicke, *Umschau* **77**, 599 (1977).
26. C. E. Junge, *Air Chemistry and Radioactivity*, Academic Press, New York, 1963.
27. P. A. Hollingdale-Smith and L. C. Thomas, *Proc. Soc. Anal. Chem.* **10**, 205 (1973).
28. in ref. 26.
29. G. S. Callendar, *Tellus* **10**, 243 (1958).
30. R. E. Munn and B. Bolin, *Atm. Envir.* **5**, 363 (1971).
31. in G. J. Rasool, *Chemistry of the Lower Atmosphere*, Plenum, New York, 1973.
32. A. J. Haagen-Smit, *Ind. Eng. Chem.* **44**, 1342 (1952).
33. *idem ibid.* **48**, 65A (1956).
34. *idem*, E. F. Durley, M. Zaitlin, H. Hull and W. Noble, *Plant Physiol.* **27**, 18 (1952).
35. in L. Hill and M. Flack, *Proc. Roy. Soc. B.* **84**, 404 (1912).
36. in ref. 10, Div. B, pp. 78 and 204.
37. H. Preis, *Auto-Techn.*, **6**, 97 (1956).
38. E. T. Wilkins, *Roy. Soc. Arom. Hlth. J.* 677 (1956).
39. R. E. Waller, *Brit. J. Cancer* **13**, 1723 (1952).
40. *idem* and P. J. Lawther, *Brit. Med. J.* **II**, 1473 (1957).
41. P. Kotin, *A. M. A. Arch. Ind. Hlth.* **11**, 113 (1955).
42. H. Cauer, *Beihft. Z. Ver. Dtsch. Chem.* **34**, 1 (1939).
43. H. G. Landsberg, in ref. 10, Part 1A, p. 47, 1972.
44. J. A. Van Allen and L. A. Frank, *Nature (London)* **183**, 430 (1959).
45. W. Engelhardt, *Umschau* **77**, 505 (1977).
46. V. Bjerkness, J. Bjerkness, H. Solberg and T. Bergeron, *Phys. Hydrodyn.* 565 (1933).
47. F. Linke, *Z. Phys. Therap.* **37**, 217 (1929).
48. T. Bedford, *Basic Principles of Ventilation and Heating*, Lewis, London, 1948.
49. P. A. Siple, *Proc. Amer. Phil. Soc.* **89**, 177 (1945).
50. A. Court, *Bull. Amer. Meteor. Soc.* **29**, 487 (1949).
51. M. K. Thomas and D. W. Boyd, *Canad. Geograph.* **10**, 29 (1957).
52. O. Wilson, *Metabolism* **5**, 543 (1962).
53. *idem*, in *Biometeorology I*, Pergamon, Oxford, 1962, p. 411.

CHAPTER 2

INFLUENCE OF THE WEATHER AND CLIMATE ON PLANTS AND ITS PRACTICAL IMPLICATIONS

In this and following chapters various practical aspects of the study of meteorotropic phenomena in plants, man and animals will be reviewed. However, for a better understanding of these practical applications the more theoretical, physiological aspects of meteorotropic phenomena have first to be discussed.

In the present chapter we are dealing only with plants, the first group of living organisms in which the influence of weather and climate has been clearly observed and recognized. Every layman knows the great effect of water, temperature, radiation (artificial or solar), wind and altitude on plants, their development, distribution and diseases.

In many publications on Agricultural Biometeorology, and particular in *Progress in Biometeorology,*[1] the various effects of the weather and climate on plants have been fully described. So in the present chapter only some particularly important physiological and practical aspects will be briefly reviewed. As it seems to be more difficult to demonstrate meteorotropic effects in animals, and particularly in man, a more lengthy review of the observed meteorotropic phenomena will be given in later chapters.

2.1 THE MICRO-METEOROLOGICAL ENVIRONMENT OF PLANTS

An extensive survey of the problems related to the physical micro-environment of plants (not the soil conditions) was given recently by Schwerdtfeger *et al.*[1] Their studies were mainly carried out in the Institute for Atmospheric Science in southern Australia. One of the major parameters of plant life is its water balance. More than 80% of the weight of plant cells consists of water. Therefore practically all the processes of growth and development of plants are seriously affected by even minor changes in the water balance of a plant. The substances dissolved in the water of plants determine the osmotic pressure* which is important for the absorption of water into the plant cells and for their growth.

A second important factor in plant life is the ambient temperature. It affects almost every aspect of plant growth. This is largely due to its effects on the activity of enzyme systems. Different temperatures lead to quantitative differences in plant growth as a result of the different rates at which physiological processes proceed in response to temperature. This is different from the effects of light which are the result of the addition of a measurable amount of radiation energy.

Temperature effects vary with the age of the plant, its water content, its nutritional status, the preceding temperature history and the solar radiant energy levels.

A third factor is radiation either solar or artificial. The response of higher plants to day-length or to the number of hours of artificial light (photoperiod) is determined by the phytochrome system. Phytochromes are light sensitive pigments in plants. They absorb visible light of different wavelengths in different quantities. Some research workers believe that the phytochrome system is not the major factor mediating the response to day-length but that this response is mainly due to an endogenous circadian rhythm. Phytochrome is a bright blue protein pigment which occurs in most green plants in concentrations far lower than that of chlorophyll. It has two inter-convertible forms with absorption maxima in the red part of the spectrum at 660 nm and near the limit of vision at 730 nm. It regulates plant growth and development by differential gene activation and repression. The phytochrome system is responsible for many plant responses such as lengthening, unfolding of lamina, hair formation, changes in cell respiration, protein synthesis etc.

An important photoprocess in plants is known as 'assimilation' or photosynthesis in which green plants use the electromagnetic energy of light to produce carbohydrates and oxygen from carbon dioxide and water. This is possible due to the photochemical action of sunlight on the *chlorophyll bodies* in the cytoplasm of the plant cells. As a result, during the day the atmosphere in the immediate surroundings of plants or trees becomes poorer in carbon dioxide and richer in oxygen. This assimilation process depends on a number of factors: the structure, shape and number of chlorophyll bodies in the plants, temperature, humidity of the air, wind, and the intensity and prevailing wavelength distribution of sunlight. For most green plants photosynthetically active radiation falls within the wavelength range of 490–760 nm (green to red). However, plants must also be able to breath. In darkness a plant absorbs oxygen from the air and breaks down carbohydrates to carbon dioxide and water, a process known as respiration.

Net-photosynthesis represents the balance between photosynthetic and

*Osmosis is the diffusion of water or solvent molecules through a semi-permeable membrane, the molecules passing from the side of lower concentration to that of higher. The pressure thus developed is called osmotic pressure.

respiratory processes. These various processes have certain optimum temperatures which are different for different plant species. Owing to the respiration process most plants are able to produce 5–10 times their volume of carbon dioxide in 24 h. A rise in temperature of 10°C doubles or triples the carbon dioxide production. Therefore a rich vegetation surrounding the living quarters of man will affect the oxygen/carbon dioxide ratio of the air, which will be influenced by the temperature, the number of hours of sunshine and the windspeed in the immediate environment.

A fourth important parameter in plant life is wind. Not only animals and man experience great discomfort due to wind, because of its chilling effect (p. 27), or because of the desiccation effects of hot winds; plants too are subject to damage caused by excessive chilling and desiccation. It is for this reason that windbreaks (any structure reducing the windspeed) or shelter belts (rows of trees planted for wind protection) are applied to reduce these stresses, the reduction of which can be very beneficial for crop production.

Finally altitude affects plant growth considerably but not due to reduced partial oxygen pressure as in the case of animals and man (Section 5.3). Altitude changes the average temperature, the wavelength distribution of solar radiation, the humidity and the average windspeed, in other words it is not a direct environmental parameter of plants.

It is evident that apart from these natural physical factors air quality, i.e. air pollution, also considerably affects plant life. Waggoner *et al.*[2] showed the effects of air pollution on plants. Through the stomata a downward flux of polluted substances penetrates into the plants. Anti-transpirants, e.g. phenylmercuric acetate, have been used for protecting the plants against pollution as they cause the stomata to close. In Section 1.4.2 the influence of ozone on tobacco leaf causing weather fleck was described. Anti-transpirants, provided they are sprayed on the lower surface of the tobacco leaf, protect against ozone damage; spraying on the upper surface closes the stomata but does not protect against ozone damage.

This brief summary of the major natural physical factors affecting plant life shows clearly that micro-meteorological factors must have a profound influence on the growth and development of plants.

2.2. INFLUENCE OF THE METEOROLOGICAL ENVIRONMENT ON THE WATER BALANCE OF PLANTS

As 80% of plants consists of water, it plays an important role in metabolic processes such as photosynthesis and respiration in the plant cells. It increases the *turgor* (firmness of the cells) of plants and gives them the strength to support certain parts of the plant.

The availability of water for evaporation determines the energy budget of a leaf or crop. As the degree of evaporation is determined by the environmental temperature and by air movements around the plant, it is evident that the meteorological environment of plants is of great importance for their water balance.

2.3. SIGNIFICANCE OF THERMAL STIMULI FOR PLANT RESPONSES

The temperature of plants depends on various environmental factors such as air temperature, radiant energy, windspeed and humidity. For example the temperature of the leaf of a plant is determined by the leaf energy balance. The dispersion of thermal energy by transpiration or convection depends upon air temperature, humidity and windspeed and the general physiological condition of the leaf expressed through the resistance to vapour pressure.

Gates, Ratschke and others expressed these physiological conditions in complicated mathematical equations.[1] One of the factors in this formula is the aerodynamic resistance which depends on leafsize, its shape, the windspeed over the leaf and the leaf resistance, controlling the effect of water vapour from inside the leaf as well the influx of carbon dioxide. Both depend on the stomatal apertures of the leaf, which are larger at higher temperatures.

If leaves suffer from water stress, the stomata are not fully open. The leaf temperatures may rise above 35 °C due to lack of transpiration cooling. In tea bushes, for example, temperatures of 40–42 °C were measured when the ambient temperature was only 30–32 °C. The temperatures of buds and stems are also determined by an energy balance similar to that of leaves; the *cambium* temperature of exposed peach trees may be as much as 16 °C above the air temperature, and the recorded temperatures of watermelons and tomatoes in a desert valley in California were 15 °C higher than the ambient temperature.

In plant communities the energy balance depends on the canopy structure and density (which determines the penetration of radiation energy), on the transpiration in the various layers of plants and on turbulent transfer processes. The latter determine whether the heat exchange between canopy and environment is small. There is a tendency for a day time maximum heat exchange in the canopy. At night canopy temperatures tend to be similar to the air temperature. It is evident that seed germination and seedling growth are strongly affected by soil temperatures. Root temperatures are usually the same as those of the surrounding soil.

The effect of temperature on photosynthesis has already been discussed (Section 2.1). The previous thermal condition of a plant, the so-called preconditioning effect, is extremely important for it. Generally plants previously subjected to high temperatures for a certain period, subsequently require higher temperatures for optimal net-photosynthesis than plants subjected to lower temperatures. For the same reason in winter-hardened spruce, photosynthesis of forest shoots also continues at the same rate after exposure to sub-zero temperatures. For further details the reader is referred to the paper by Landsberg in *Progress in Plant Biometeorology*.[1]

2.4. EFFECTS OF SOLAR RADIATION ON PLANT GROWTH, CROP PRODUCTION AND FLOWERING

In Section 2.1 the significance of phytochromes, the light sensitive pigments in plants which are able to register fluctuations in solar radiation and to distinguish between differences in spectral energy distribution, has been discussed. They regulate plant growth and many other plant responses to light stimulation. According to Stanhill[3] there is a close relationship between solar radiation, water requirements and the amount of dry matter production in a crop. In humid climates, where precipitation exceeds or equals the evapotranspiration of crops, the method of Penman[4] and the equations of Monteith[5] enable us to estimate, within 10% of the measured values, the water loss from crops and natural vegetation. The measured values should be averaged over periods of from 10 to 30 days. Such estimates, in areas with trustworthy climatological data, are used for irrigation or drainage schemes. In arid zones also, where water loss consistently exceeds water supply, good results can be obtained by using the equations of Penman.[4] These estimates can be used for the long-term planning and managing of crop irrigation.[6] De Wit[7] developed equations to estimate the potential photosynthesis of a crop canopy on the basis of global solar radiation measurements. These and other mathematical approaches have been used for glasshouse crops in western Europe in order to introduce automatic irrigation systems, the operation of which is triggered by a signal from an integrating solar radiation sensor. Similar empirical studies were made on potential dry matter production of crops for routine management. However, here the estimates are less certain.

Crop yields have also been studied mathematically in relation to solar radiation. Very high correlations have been reported. It is evident that practical applications of the above mentioned solar/crop relationships require continuous measurements of the solar spectrum between 290 and 725 nm. Garner and Allard[8] first demonstrated the effect of day-length on flowering plants. To initiate flowering the plant must receive a specific minimum number of photoinductive cycles after its juvenile phase. This minimum requirement varies with the plant species. According to several authors a light intensity of as little as 10 lux has a photoperiodic effect on plants. One distinguishes between short- and long-day flowering plants. For the short-day plants a day-length of less than 12 h is required, for long-day plants more than 14 h. For further details reference should be made to the various botanical textbooks (e.g. Butterfass[9]).

2.5. EFFECTS OF WIND

The application of both windbreaks and shelterbelts to reduce windstress effects was mentioned in Section 2.1. The effectiveness of a windbreak depends on its height, porosity and length. The higher the windbreak the greater will be the distance of its influence.

A disadvantage of windbreaks is that they significantly reduce solar radiation effects, unless they have a north-south orientation. During daytime the air temperatures are usually higher in sheltered than in open fields because of the reduction of turbulent air currents. Also humidity and vapour pressure are increased in sheltered areas because the transpired and evaporated water is not easily transported away from the source. According to Lemon[10] and others low windspeed in sheltered areas reduces the carbon dioxide supply and causes decreased photosynthesis. Nocturnal carbon dioxide concentrations will be high behind shelters particularly with light winds; the net assimilation of carbon dioxide is increased. The liberated carbon dioxide will accumulate over, and concentrate within, a crop canopy. These various shelter effects will affect the microclimate behind shelters and the growth and production of crops; i.e. plant growth and yield are usually greater behind shelters. According to Rosenberg[11] a major influence on plant behaviour seems to be a greater turgidity and wider stomatal apertures in the leaves of sheltered plants, resulting in greater water use and reduced evaporation. Another important aspect of windbreaks is also the reduction of soil water consumption. Brown and Rosenberg[12] succeeded in developing a complex equation to predict these various meteorological effects of a windbreak on vegetation.

2.6. PLANT DISEASES AND PLANT PESTS

2.6.1. Plant diseases

In most plant diseases and plant pests (damage caused by insects) meteorological factors play an important role. It is therefore not surprising that long before the study of meteorological effects on animals and man, pathological phytobiometeorology developed into an important practical branch of biometeorology. However, the major developments, and particularly the methods developed for predicting plant diseases, have a more recent character.

According to Schrödter,[13] in spite of all modern techniques the plant production losses due to plant diseases are still 25–40% and more. It has been estimated that the production losses in Europe are about 25%, in the USA 29%, in the USSR and China 30%, and in large areas of South America, Asia and Africa 40% and more. Part of these losses is due to the fact that for certain diseases no adequate preventive methods have yet been developed. This is particularly true for plant diseases due to viruses which are often carried by insects. Also for certain mycotic diseases (diseases caused by fungi) no efficient chemical control has yet been developed.

The aim of crop protection, using meteorotropic forecasting methods, is not only to prevent crop losses, but also to do so at minimum cost. Also no agricultural ecosystem should be interfered with unless the damage caused surpasses a certain economically determined threshold.

Some of the major plant diseases which have been studied in connection with their meteorotropic behaviour are potato blight (caused by *Phytophtora infestans*), leafspot disease of sugar beet (caused by *Cercospora beticola*), eyespot disease (caused by *Cercosporella herpotrichoides*) and downey mildew in tobacco (caused by *Peronospora tabacina*).

Review papers on the methods to be applied to forecast these diseases were prepared by Bourke[14-16] in Ireland, De Weille[17,18] and Zadoks[19] in The Netherlands, Schrödter and Ullrich[20] in Germany, Journet *et al.*[21] in France, Rotem[22,23] in the Mediterranean area, Van der Plant,[24] Waggoner and Horsfall[25] and Wallin[26] in the USA and by many other experts in the field of plant diseases.

In the past rather simple empirical methods have been used; for example Gullach and Wallin[27] found that leaf-spot disease of sugar beet was related to certain types of air masses in the USA and they could distinguish between dangerous and harmless weather by studying the 500 mb contours on a synoptic weather map.

However, in view of the many complex weather factors which can affect the development and spreading of plant diseases it is only in recent years, after the introduction of mathematical formulae and mathematical models, that plant disease forecasting has become a really scientific discipline and has reached a high degree of accuracy.

The development of *Crop Indices* (describing mathematically the relationship between plants and meteorological parameters) by Baier,[28,29] Chapman and Brown,[30] Nicholson and Bryant,[31] Oullet and Sherk,[32] Penman,[33] Robertson[34] and many others, facilitated the design of plant disease models.

In these various models most of the following meteorological parameters should be considered.[35]

(a) *High ambient temperatures* may set limits to the development of plant diseases. For example temperatures above 35 °C may kill the fungicide *Exobasidium vexians* (the organism causing blister blight in tea) in the hot low country estates in Java whereas it is common in up-country tea estates. Temperatures over 25 °C prevent spore formation in *Phytophtora infestans*, the potato blight fungus. In the southern USA high January temperatures stimulate germination of over-wintering spores of the blue mould *Peronospora tabacina* causing downey mildew of tobacco.

(b) *Low temperatures* are favourable to the longevity of several kinds of fungal spores. Temperatures of 13 °C may favour the germination of conidia of *Phytophtora infestans*. Very low temperatures either stop or kill the mycelium of fungi.

(c) *Moisture*. Leaf wetness is important in several stages in the life cycle of fungi. *Phytophtora infestans*, for example, requires at least 13 h of leaf wetness. It is evident that dew, fog and the duration of precipitation, are important factors for the development of these diseases. Not the macroclimatic but the microclimatic-humidity conditions have to be studied continuously.

(d) *Wind* is important for the dispersal of spores. For example yellow rust spore clouds cover distances exceeding 1000 km. Wind also influences the humidity of the air and the air temperature.

(e) *Radiation*, both visible (380–780 nm) and UV (290–380 nm), may have epidemiological significance. For example germination of blister blight spores is favoured by faint light; *Phytophtora* sporulation takes place at night under favourable humidity conditions. Small doses of ultraviolet light stimulate the germinative power of *Phytophtora*, large doses minimize the infectivity.

(f) *Cloudiness* is important because of the significance of leaf wetness. It is experienced by certain plants as an important meteorological parameter. On cloudy days also little ultraviolet light is received.

A further discussion of the highly complex mathematical models, which have been developed to incorporate all these factors and which determine the date of action against certain plant diseases, is beyond the scope of the present book. Readers interested in more details are directed to Reference 1, pp. 444–445.

Apart from the diseases caused by fungi, bacteria, viruses and air pollution (see Section 2.1) plants are often seriously damaged, both directly and indirectly, by insects, e.g. the direct destructive action of locust swarms (p. 294). Insects are also carriers of pathogens which may cause plant diseases.

2.6.2. Pest control methods

Four different methods have been developed for the control of plant pests:

(a) *Chemical control*: Various chemical pesticides have been developed, their efficacy depends on temperature and humidity;

(b) *Biological control* which depends on the exploitation of another organism as a parasite or a predator of the pest in the crop; the efficiency also depends on temperature and humidity;

(c) *Integrated control*, a combination of pesticides and biological control, in which the natural enemies of the pests are spared;

(d) *Cultural control*, usually genetic control in which genetic material of a pest controls its own species, e.g. by liberating sterile males to mate with normal females.

Further details on pest control in relation to meteorological factors have been summarized by Haufe.[36]

This brief review of the role of timely forecasting in the fight against plant diseases and plant pests, sufficiently indicates the great economic importance of plant biometeorology for the food production of the world population.

2.7. SPECIFIC METEOROLOGICAL CONDITIONS FOR MAJOR FOOD CROPS

The previous sections clearly indicate that the major food crops are closely bound to the specific physico-chemical environments in different parts of the world. In some instances the crops can develop in a great variety of environments, in most instances, however, their development is determined by specific geological factors (such as the type of soil in which they grow, its chemical composition, porosity, permeability, pH etc.) and a number of specific meteorological factors, such as a certain altitude, temperature (average, minimum and maximum), moisture (dew, fog, precipitation), solar radiation (duration and solar spectrum), cloudiness and windspeed (average, minimum and maximum). In Table 2.1 optimum geological and meteorological conditions for 30 major food and industrial crops in the world have been summarized in alphabetical order.

2.8. INFLUENCE OF THE WEATHER AND CLIMATE ON THE DEVELOPMENT OF FORESTS

It is generally accepted that the great changes in forest types in western Europe and the Americas during and after the glacial periods occurred in response to the drastic climatic changes accompanying these periods. The exact physiological mechanisms involved are still being studied by a great number of experts, plant physiologists, geographers, geologists etc.

It is known that in any block of forest the growth of trees varies from place to place even between neighbouring trees. Much of this variation is due to genetic factors and to differences in the soil. However, in many instances the differences are also due to different microclimatic factors. For example, the height of a tree in relation to its neighbours has a major influence on the amount and quality of light that it receives. Once the growth of a tree stays behind that of its neighbours this effect will increase.

Also the air temperature and the air movement between trees will change from the centre of a forest to the edges. The same applies to differences in the amount of rain which reaches the smaller trees depending on their locations in a forest. Some of the meteorological characteristics of a forest climate and its biological effects have already been given in Section 1.5.2. Further information on the effects of weather and climate on forests can be found in the books by Geiger,[37] Taylor[38] and others.

2.9. INFLUENCE OF THE PLANT ENVIRONMENT ON HUMANS

Plants can affect the physical environment of man in at least five different ways:

(a) they affect the general composition of the air;
(b) they influence the degree of air pollution in man's immediate environment;
(c) they affect the noise level;
(d) they affect the amount of solar radiation reaching man and animals and the living quarters of man;
(e) they affect the degree of air turbulence in the immediate environment of man.

TABLE 2.1
Relationships between weather and climate and 30 major food and industrial crops (compiled by S. W. Tromp)[a]

Name of crop	Biometeorological factors					Geological factors (type of soil, composition, porosity, pH)
	Altitude and general remarks	Temperature	Moisture	Solar radiation and cloudiness	Windspeed	
Bamboo	Low and high altitudes (over 2400 m)	Usually warm tropical climates	Humid	Much solar radiation	—	Wide range of soil types
Banana	Musa: two types *Eumusa* series *Australimusa* series (Pacific) Low altitudes.	Warm tropical or sub-tropical climates. Optimal temp. 26 °C. Practical min. 15 °C, growth ceases at 10 °C. Frost causes death. Low temp. delays emergence of leaves and reduces leaf size.	High humidity and even distribution of rainfall. At least 5 cm rainfall/month and not more than 125 cm.	Much solar radiation.	Low windspeed. Strong winds damage crop seriously.	Growth on great variety of porous, well drained soils, preferably rich in humus, nitrogen and potassium. Poorly drained heavy clay soil is unfavourable. Acid soils favour Panama disease.
Cacao	*Theobroma cacao L.* Low altitudes. Many, meteorotropically determined, plant diseases and pests occur.	Warm humid tropical climates. Average temp. 10–21 °C, max. 30 °C. At high temp. leaf development reduced. Production of flowers and fruit very sensitive to low temp.	High humidity. Preferably no dry season. In countries with distinct dry and wet season main crop produced in wet season.	For young seedlings shade trees required against too strong solar radiation (probably due to high leaf temp.). Mature cacao trees can resist full sunlight.	Windbreaks (trees) required.	Wide range of porous, nutrient-rich soils. Requires usually no fertilizer.

[a]In the table above only the optimum meteorological and geological conditions for each crop have been summarized. The data are based on various botanical books, but in particular the different volumes of the *Encyclopaedia of The Netherlands East Indies* (in Dutch), edited by D. G. Stibbe during the period 1917–1939, with the assistance of G. Spat, J. Paulus and others; and verbal information from various agricultural experts. However, since 1945 many new developments have taken place, partly summarized in the book by C. N. Williams: *The Agronomy of the Major Tropical Plants*, O.U.P., London, 1975. Many new species have been developed which are able to grow beyond the meteorological and geological conditions mentioned above. Also new methods of manuring in the field increased the growth range in recent years. The statements made above represent only the optimum conditions for the growth of the original species in a natural environment without special artificial manuring. By porous soil is meant a well-drained soil.

TABLE 2.1—contd

Name of crop	Biometeorological factors					Geological factors (type of soil, composition, porosity, pH)
	Altitude and general remarks	Temperature	Moisture	Solar radiation and cloudiness	Windspeed	
Caout-chouc or Rubber	*Ficus elastica* and *Hevea braziliensis* Preferably low altitudes.	Warm humid tropical climates. Temperature affects rate of growth of tree.	High humidity. Dry season should not exceed 3 months. Preferably 2700–5000 mm rain/year. Even short dry season reduces latex yield.	Much solar radiation	Very susceptible to strong winds.	Well drained acid (pH 4–6.5) soil. Young seedlings sensitive to low pH. If pH 8.0 growth retardation. Mature trees grow on wide range of soils.
Cassava or tapioca	*Manihot utilissima* Tubers grow up to 1000 m, but preferably in low land. Tubers very rich (25%) in starch.	Warm tropical and subtropical climates. Average yearly temperature 25–27 °C.	High humidity. Rainfall 175–250 mm. In east Africa only 75 mm.	Much solar radiation. Tuber initiation requires short-day photoperiods.	—	Loose texture of soil is favourable. Strong impoverishing action on the soil.
Cin-chona	Preferably 1600–2400 m. Producing quinine products.	Cool climate. Average temperature 12–13 °C.	Humid climate.	Medium solar radiation.	—	Porous soil.
Citrus fruits lemon, orange, grape-fruit, etc.	Usually low altitudes. In Columbia altitude of navel orange 1100–1200 m, in Spain 0–1750 m.	Subtropical climates. Temperature 12 °C.	Rather dry climate. 125 mm/ year rain. Sensitive to salt spray near the coast (Israel).	Much solar radiation (usually not available at high altitude in the tropics).	—	Light porous soil.
Cocos	*Cocos nucifera* Low altitudes (Max. 1000 m).	Warm tropical and subtropical climate between latitude 18 °N and 18 °S.	Rainy periods. Preferably 125–375 mm rain.	Much solar radiation.	Low wind speed.	Coral sands, fertile well-drained clay or volcanic soil, rich in K and Na, pH 5–8.
Coffee	*Coffea liberica* *Coffea robusta* *Coffea arabica* *Coffea excelsa* Altitude 800–1000 m. *liberica* and	Growing everywhere between latitude 25 °N and 30 °S. *Arabica* requires	Sufficient rain required (at least 1500 mm/ year), but dry period required for	Insufficient sun gives poor crops.	Sensitive to strong winds. Windbreaks (trees) required.	Deep 1–2 m, porous, well-drained soils, rich in humus. Clay or sandy soils are not recommended.

TABLE 2.1—contd

Name of crop	Biometeorological factors					Geological factors (type of soil, composition, porosity, pH)
	Altitude and general remarks	Temperature	Moisture	Solar radiation and cloudiness	Windspeed	
	robusta grow in tropical lowland; *Arabica* at higher altitudes with cooler climates. Steep slopes give poor crops.	temperature 13–27 °C, optimum 24 °C. Mountains with rather low temperature unfavourable. Night frosts have devastating effects.	flowering. In Kenya moisture supply critical 10–17 weeks after flowering for berry formation.			
Cotton	Low altitudes. 25% of growth factors determined by meteorological factors.	Warm tropical or subtropical climate.	Dry climate. Monsoons with heavy rains are most unfavourable Drainage of soil to the roots is required. Rain during flowering or fruit formation causes serious crop damage.	Much solar radiation required. It determines the formation of pods (long seed reservoirs).	Low windspeeds.	Very porous, sandy soils, rich in humus, often near coasts or rivers. Clay soils give poor crops. 75% of growth factors determined by soil condition and fertilizers.
Fruit trees (apple, pear, cherries, etc.)	Usually low altitudes. Weather conditions immediately prior to harvest are very important.	Warm weather at end of ripening determines colour, taste and sugar content. Night frost before flowering is most damaging.	No excess. rainfall	Much solar radiation at end of ripening.	Strong wind or storm before harvest is damaging.	Very porous soil with great permeability.
Grapes (for food con-sumption and wine).	For wine in northern hemisphere weather conditions in July and early August most important.	Weather conditions (particularly temperature) during previous year in August determining the number of flowerbuds.	Sufficient rain in July and first half of August in current year.	Sufficient sunshine in July and first half of August in current year.	Storm before harvest may be damaging.	Porous soils.

TABLE 2.1—contd

Name of crop	Biometeorological factors: Altitude and general remarks	Temperature	Moisture	Solar radiation and cloudiness	Windspeed	Geological factors (type of soil, composition, porosity, pH)
Groundnuts	*Arachis hypogaea* Usually low altitude, but can grow up to 1500 m.	Warm tropical or subtropical climates.	Humid climate with normal or seasonal rainfall.	Much solar radiation.	—	Grows in many soils, but preferably in light porous soils.
Maize	Indian corn. At low and higher altitudes. Usually harvested at a late date.	In high latitude countries and at high altitude temperature is a dominant factor.	Needs plenty of water up to and after flowering.	Extensive cloud cover in autumn is a limiting factor.	Strong winds shortly before harvesting have damaging effect.	Deep, porous soils (no heavy clay soils).
Oil palm	*Elaeis guineensis* 0–400 m.	Optimal temperature 24–26 °C. Seedling growth arrested at 15 °C.	Rather steady humidity (not more than 50–70%) and equally distributed rainfall during the year (no long dry season). In west Africa rainfall exceeds evaporation with 400 mm/year.	Much solar radiation.	—	Required 1–1½ m deep well-drained porous soil, rich in humus and other minerals. In east Sumatra loamy soils near rivers and peat soils. Palm cannot grow with stagnant high groundwater.
Olives	*Olea europea* Low altitude.	Warm subtropical climates.	Dry climates.	Much solar radiation.	—	Porous, dry limey soils, rich in nitrogen.
Papaya	*Carica papaya* At low altitudes fruit-bearing after 6 months; at 600 m (South Africa) 9–11 months.	Tropical and subtropical (frost-free) climates. Temperature preferably >19 °C. Lower temperature gives small fruit size and low quality.	Humid climates.	Much solar radiation.	Strong winds have unfavourable effects.	Good drained soils, rich in humus.
Pepper	Low altitude.	Grows between 20 °N	Much rain without a	Light shadow	Low wind velocity.	Young, well-drained forest soil, rich in

TABLE 2.1—contd

Name of crop	Biometeorological factors					Geological factors (type of soil, composition, porosity, pH)
	Altitude and general remarks	Temperature	Moisture	Solar radiation and cloudiness	Windspeed	
		and 20 °S latitude; requires high temperatures.	long dry period. Particularly areas near equator are favourable.	trees required.		humus.
Pine-apple	*Ananas comosa.* Low altitude, at high altitude smaller plants, less green and low sugar content (probably due to lower temperature).	Warm tropical and subtropical climates. With lower temperature time of crop formation is longer. South Africa: maximum 40–47 °C minimum −1 to +5 °C Hawaii: maximum 32 °C minimum 10 °C Malaya: maximum 29 °C minimum 26 °C Australia: maximum 32 °C minimum 11 °C.	High humidity. Same rain conditions as bananas, but can grow also under drier conditions. Rainfall in major pineapple areas varies between 60–250 mm (optimal condition 100–150 mm). In drier areas often dew compensation for lack of rainfall.	Much solar radiation. Fruits often covered with straw when approaching maturity.	Little effect of wind.	Grows on wide range of well-drained soils but particularly on laterite, volcanic and peat soils. Calcareous soils and soils with high pH are unsuitable. Very acid soils rich in Mn require extra iron application. Plants require little P.
Potatoes	Low altitudes in northern countries or high altitude in the tropics.	Preferably cool climates.	Humid climate but not excessive rain.	Preferably much solar radiation.	—	Preferably well-drained soils.
Rice	*Oryza sativa* Two main varieties *indica* and *japonica.* Main source of starch production. Preferably growing on flooded terraces	Grows from equatorial tropics to high latitudes (up to 49 °N). In latter case only a summer crop. Rice requires at	Rainy areas, preferably irrigated fields or terraces. Planting in tropics in wet season.	Solar radiation very important for starch formation. Fast growing varieties must be	Strong winds during time of ear maturity cause serious damage. Therefore	Grows on a great variety of moist soils, with an impermeable layer just below the worked horizon preventing downward movement of water.

TABLE 2.1—contd

Name of crop	Biometeorological factors					Geological factors (type of soil, composition, porosity, pH)
	Altitude and general remarks	Temperature	Moisture	Solar radiation and cloudiness	Windspeed	
Rice (contd)	on gentle mountain slopes under a water cover of 5–10 cm. In flat areas irrigation required. Climate affects mainly length of time of development.	least 4 months an average temperature of 20 °C. Tropically adapted varieties require 25–35 °C. In warm areas more than one harvest/year. Low temperatures do not favour tillering.		planted in the dry season with much solar radiation.	varieties with short stem are preferable.	Depending on the variety pH is 3.5–8.5. Moderate acid soils produce low nutrient rice. Silica prevents Blast Disease and reduces Mn and Al toxicity in rice.
Rye	Usually low altitude.	Cold parts of temperate climates. Rather cold resistant. Humid tropic or warm areas with short cold summers are unfavourable.	Excessive rain and night frost during flowering has devastating effects.	Medium solar radiation.	—	Particularly rather dry, light, porous loess soils or loamy sandy soils are very favourable.
Sago	*Metroxylon* Very rich in starch. Low altitude.	Warm climate.	Humid climate with rainy season.	Much solar radiation.	—	Humid, often muddy (not too deep) soils.
Sisal	*Agave* Usually low altitude but can grow up to 1800 m. Important fibre crop.	Warm climates.	Long dry periods alternating with rain.	Much solar radiation.	—	Porous soil.
Soya bean	Grows in many climates, both in the tropics and at higher latitudes, provided the summer is warm.	Growth period (15 weeks) should be warm. Night frosts are devastating.	Rainfall during sowing and growth period must be at least 550 mm, alternating with sunny	Sunny periods alternating with rain. Number of required sunhours differs for	—	Deep, porous soils, rich in calcium, pH 6–6.5. Heavy clay soils are unfavourable.

TABLE 2.1—contd

Name of crop	Biometeorological factors					Geological factors (type of soil, composition, porosity, pH)
	Altitude and general remarks	Temperature	Moisture	Solar radiation and cloudiness	Windspeed	
			periods. Final period of crop development must be dry.	different species.		
Sugar beet	*Beta vulgaris* Low and high altitude.	Temperate climates.	No excessive rain.	Much solar radiation.	—	Preferably heavily manured, drained soil.
Sugar cane	*Saccharum officinarum* Low altitude.	Similar to bananas, i.e. warm tropical and subtropical climates. Optimal temperature 26–38 °C. Minimum 10 °C. Extension of stem most sensitive to temperature. At maturity slight cooling favours sugar accumulation. Frost usually damages crop.	Humid (but not excessive) climates. Sufficient (125–150 mm) rain required, particularly during germination, tillering and main period of growth, but short dry season prior to harvesting is required permitting development of high sucrose content in the stalks; during dry periods irrigation necessary. Excessively humid climate causes high tonnage of cane with low sucrose content.	Very much solar radiation. High and low sunlight may affect yield and sugar content in a ratio 3:1. Photoperiod is very important. Sugar cane is short-day plant, flowering reduces sugar yields and cane growth. Many varieties therefore require specific photoperiods.	Low wind velocity. Strong winds damage crops.	Porous, well drained fertile soil.
Tea main varieties *Sinensis* and *assamica*	*Camellia sinensis* Preferably 500–1500 m. At lower altitude too many poor quality leaves.	Humid tropical to sub-humid and temperate climates. Temperature 21–32 °C.	Much rain (1150–1400 mm/year). If monthly rainfall remains below 50 mm for	Much solar radiation; shady trees not essential.	With sufficient soil moisture windbreaks or shelter trees	Wide range of soil types. Preferably deep, porous, young forest soils (at the beginning of a plantation), rich in humus, Al

TABLE 2.1—contd

Name of crop	Biometeorological factors					Geological factors (type of soil, composition, porosity, pH)
	Altitude and general remarks	Temperature	Moisture	Solar radiation and cloudiness	Windspeed	
Tea (contd)		Optimal soil temperature 25 °C. Frost has unfavourable effect.	several months production declines severely. No stagnant water near roots (preferably hill slopes).		increase yields.	and low pH (4.0–4.5). Tea contains 17 000 parts of Al per million.
Tobacco	*Nicotiana tabacum* Low altitude or low hills.	Mainly in subtropical climates, but also grown in northern countries with warm summers (France, Germany, central Europe).	Humid climate with 135–210 mm rain. During rainy season no sharp distinction between dry and rainy periods.	Much solar radiation.	Strong winds are unfavourable.	Preferably porous, often sandy soil rich in humus or peat, or weathered, fine, volcanic material.
Wheat	One of the most important crops for carbohydrate and starch production. Mostly low altitudes, but can grow also in cool mountainous areas in the tropics. Same climatic rules as other cereals (rye, barley and oats).	Low temperatures at the time of tillering causes low yields. At high latitudes temperature is a limiting factor.	Short period of drought at the time of tillering (when plant shoots appear on stalk) produces low yields. Yields depend on rainfall preceding growth period. In dry countries moisture is the limiting factor. There is a direct relationship between yield and rainfall.	Optimal growth with sufficient solar radiation.	Very strong wind shortly before harvesting has damaging effect.	Deep porous soils, rich in lime and humus, retaining moisture of winter-rain or snow. Moist light clay, loess or loamy soils are particularly favourable.

2.9.1. Composition of the air

The chemical changes of the air near plants and trees, due to assimilation and respiration of plants, were discussed in Section 2.1.

2.9.2. Effect on air pollution

Foy,[39,40] Baier and Chan,[41] Went[42,43] and others pointed out that plants could act as air polluters, air pollutant removers and air filters, air conditioners and air pollution detectors.

(a) *Air polluting effects*: Foy[39,40] pointed out that phytotoxic gaseous emanations and root exudates from certain trees and shrubs may inhibit the growth of the surrounding vegetation. It is estimated that approximately one thousand million tons of volatile organic substances are released each year by the vegetation over the whole world. According to Went[42,43] a great part of these organic volatile substances are terpenes. They occur in concentrations of 2–20 parts per thousand million (ppb). They have considerable effects on other plants. After their release they are photochemically altered by sunlight, forming peroxides, ozonides etc. The volatilization of the terpenes and other plant products causes the blue haze and veil clouds which play a very important role in the heat balance of the earth and of the atmosphere. They absorb, according to Went,[42,43] about 20% of all solar radiation reaching the earth's surface.

(b) *Air pollutant removers and filters*: Baier and Chan[41] pointed out that the toxic effects of ozone to plants, and in turn the removal of ozone from the air by plants, indicates that plants can play an important role both as air pollution removers, and as air cleansing factors. They described as an example the cleansing function of a forest. If a mass of polluted air which contains 150 ppb of ozone passes over a forest of trees 4–5 m tall (concentration actually found on a polluted day in Connecticut) after one hour the air filtered to the forest floor would have only 60–90 ppb of ozone, the rest having been taken up by the canopy of leaves. After 8 h near the forest floor only 30 ppb were recorded. Tall trees and plants with large stomatal pores have the greatest removal capacity.

(c) *Air conditioners*: plants are able to modify the environmental temperature, the airflow, moisture and chemical composition of the air. They act in a way like air conditioners in a building. Their capacity to remove air pollution enables them to replace polluted air by good clean air. Moisture on the leaves traps the dust particles and holds them until they are washed away by the rain. Many plants are able to replace an obnoxious odour with a stronger, pleasant smell.

(d) *Air pollution detectors*: plants are very sensitive to smog, which makes them excellent early indicators of air pollution. As different plants are sensitive to different toxic air pollutants they can be used to identify toxic components even weeks after a pollution episode.

2.9.3. Effect on noise level

Shrubs and trees are able to reduce the noise of cities, factories, airports, highways etc. Plants break up sound waves, change their direction and reduce their intensity. The noise reflected by large concrete walls or buildings can be reduced threefold by planting trees between the concrete surfaces. Plants screen the noise of moving vehicles. According to Kämpfer[44] a multi-layer forest of 200–250 m reduces traffic noise by 35–45 decibel; but plants also have an additional beneficial effect because the traffic cannot be seen anymore.

2.9.4. Effect on solar radiation received

The effects of forests and other high vegetation on the physical environment of man (radiation, humidity, air movements etc.) have already been discussed (p. 26).

2.9.5. Effect on air turbulence

The various effects of shelterbelts have already been discussed (Sections 2.1 and 2.5). They reduced both the chilling and the desiccation effects of wind in cold and hot areas respectively.

Summarizing it can be stated that plants and forests have a great significance for man not only because of their importance as major food and industrial products but also in view of their beneficial effects on his immediate environment.

REFERENCES

1. L. P. Smith (Ed), *Progress in Plant Biometeorology*, Div. C, Vol. 1 of *Progress in Biometeorology*, S. W. Tromp (Ed), Swets & Zeitlinger, Amsterdam, 1975.
2. P. E. Waggoner, *Bioscience* **21**, 455 (1971).
3. G. Stanhill, *J. Hort. Sci.* **33**, 264 (1958).
4. H. L. Penman, *Proc. Roy. Soc. Ser. A*, 120 (1948).
5. J. L. Monteith, *Proc. Symp. Soc. Expl. Biol.* **19**, 205 (1965).
6. G. Stanhill and D. W. Puckridge, in ref. 1.
7. C. T. De Wit, *Agric. Res. Rep., Pudoc, Wageningen* **663**, 57 (1965).
8. W. W. Garner and H. A. Allard, *J. Agric. Res.* **18**, 553 (1920).
9. T. Butterfass, *Wachstums- und Entwicklungsphysiologie der Pflanze*, Quelle and Meyer, Heidelberg, West Germany, 1970.
10. E. Lemon, in *Physiological Aspects of Crop Yield*, Easton *et al.* (Eds), Amer. Soc. Agron., 1970 p. 117.
11. N. J. Rosenberg, *Agric. Meteor.* **9**, 241 (1972).
12. K. W. Brown and N. J. Rosenberg, in ref. 1, p. 128.
13. H. Schrödter, in ref. 1.
14. P. M. A. Bourke, in *Proc. Region. Training Sem. Agrometeor., May 1968* Agric. Univ. Wageningen, The Netherlands, 1968, p. 129.
15. *idem ibid.* p. 199.
16. *idem, Ann. Rev. Phytopathol.* **8**, 345 (1970).
17. G. A. De Weille, *Agric. Meteor.* **2**, 1 (1965).
18. *idem*, in ref. 14, p. 69.
19. J. G. Zadoks, *ibid.*, p. 179.
20. H. Schrödter and J. Ullrich, *Agric. Meteor.* **4**, 119 (1967).

21. P. Journet, P. Latard, G. Paitier and J. Bouchet, *Min. Agric. Bull. Techn. Inform.* **238**, 191 (1969).
22. J. Rotem, in ref. 14, p. 125.
23. *idem ibid.*, p. 195.
24. J. E. Van der Plant, *Plant Diseases, Epidemics and Control*, Academic Press, London, 1963.
25. P. P. E. Waggoner and J. G. Horsfall, *Bull. Conn. Agric. Expl. Stat.* **698**, 80 pp. (1969).
26. J. R. Wallin, *Amer. Agric. Advis. Serv. Pubn.* **86**, 149 (1967).
27. C. B. Gullach and J. R. Wallin, *Int. J. Biometeorol.* **14**, 349 (1970).
28. W. Baier and G. W. Robertson, *Canad. J. Plant Sci.* **47**, 617 (1967).
29. W. Baier, *ibid.* **51**, 255 (1971).
30. L. J. Chapman and D. M. Brown, *Rep. No. 3*, Canad. Land Inventory, Ottawa, 1966.
31. J. Nicholson and D. G. Bryant, *Publ. 1299*, Canad. Forest Service, 1972.
32. C. E. Oullet and L. C. Sherk, *Canad. J. Plant Sci.* **47**, 231 (1967).
33. H. L. Penman, *Tech. Commun. 53*, Commonwealth Bureau of Soils, CCAB, Farnham Royal, Bucks., UK, 1963.
34. G. W. Robertson, in *Plant Response to Climatic Factors*, R. O. Slatyer (Ed), *Proc. Uppsala Symp. 1970*, UNESCO, *Ecol. Conserv.* **5**, 327 (1973).
35. G. A. De Weille, *Agric. Meteor.* **2**, 1 (1965).
36. W. O. Haufe, in ref. 1, p. 287.
37. R. Geiger, *Das Klima der Bodennahen Luftschicht* Vieweg, Braunschweig, 1961.
38. J. A. Taylor, *Research Papers in Forest Meteorology*, Proc. 8th Aberystwyth, UK., Symp. March 1970, 1972.
39. C. L. Foy and S. W. Bingham, *Resid. Rev.* **29**, 105 (1969).
40. C. L. Foy, in *Agronomy and Health*, Amer. Soc. Agron. Spec. Pubn. No. 16, 1970, p. 37.
41. W. Baier and A. Chan, in *Progress in Biometeorology* Div. A, Vol. 1, Part III, S. W. Tromp (Ed), Swets & Zeitlinger, Amsterdam, The Netherlands, 1972.
42. F. W. Went, *Nature (London)* **187**, 642 (1960).
43. *idem, Desert Res. Prepr. Ser. No. 31*, Univ. Nevada, Reno, 1966.
44. M. Kämpfer, *Vegetationskunde*, Naturschutz, Landschaftspflämpfer, Bad Godesberg, 1967.

CHAPTER 3

CENTRES IN THE HUMAN BODY REGISTERING METEOROLOGICAL STIMULI

3.1. SIMPLE METHODS TO DEMONSTRATE THE EFFECTS OF THE WEATHER AND CLIMATE ON HUMANS

Whereas it is relatively easy to demonstrate the influence of the weather and climate on plants, it is much more difficult to collect convincing evidence for meteorotropic effects on animals and humans.

One of the simplest ways to demonstrate meteorotropic effects on humans is the study of the quantity of 24 h urinary samples of healthy subjects, provided that fluid consumption is kept constant. Even slight changes in the cooling index of the atmosphere (p. 27) are followed, within a few hours, by a number of easily measurable urinary parameters. For example a slight fall of temperature is accompanied by an increased urinary output, increased excretion of 17-ketosteroids, a decrease of chloride, sodium, urea, hexosamine, a rise in pH etc.

A rise in temperature (or decrease in cooling) causes the reverse effects. In subjects with a good thermoregulation (p. 66) these changes occur within 12 h. In poorly regulated subjects there is a considerable delay. These meteorotropic effects occur if the subject stays only one hour, sometimes even less, in the open air. Even if he stays in an air conditioned environment afterwards he will continue to have the cooling or warming-up pattern for several hours.

A more sophisticated method is, for example, the study of the thyroid of rats at cellular level after subjecting them to a thermal change in the environment and sacrificing them afterwards. In *Medical Biometeorology*[1] on p. 270 several examples are given. Similar drastic changes in cellular structure can be seen if rats are kept in complete darkness or if they are subjected to ultraviolet light.

During the last 50 years many, relatively simple, laboratory tests were carried out indicating physiological changes after short term weather or long term seasonal changes which are caused by changes in the thermal and solar radiation environment (both infrared and ultraviolet radiation).

3.2. PRINCIPAL REGISTERING CENTRES

The various stimuli of our physical environment are registered by the skin, the respiratory tract (lungs and throat), the nose, the eyes and directly by the nervous system of the body. In Table 3.1 a summary is given of the various mechanisms involved in these different registration centres.

TABLE 3.1
Principal centres of the human body registering meteorological stimuli[1]

I. Skin

1. Thermal effects: through conduction, convection or infrared radiation (780–>7000 nm wavelength) recorded by thermoreceptors in skin and hypothalamus, effects counterbalanced by vasodilatation or constriction or sweating. According to Hardy[1], the human skin registers radiation heat of 15×10^{-5} g cal cm^{-2} s^{-1} causing a rise of skin temperature of 0.003 °C.
   Effects on:
   (a) *Hormonal functions:* of pituitary (antidiuretic, thyrotrophic and gonadotrophic hormones), thyroid and adrenal gland (17-ketosteroids) and pancreas (insulin production and blood sugar level).
   (b) *Blood:* changes in albumen and globulin levels and blood cell composition.
   (c) *Electrolyte balance:* pH and other substances in urine. Cold stress decreases excretion in urine of chloride, sodium, urea, hexosamines; causes rise in pH.
   (d) *Liver function:* cold stress increases the liver enzymes (SGOT, SGPT and SLDH), respiration of liver cells etc.
   (e) Other physiological processes affected by (a)–(d).
2. Ultraviolet radiation effects: particularly UV-B (290–315 nm), interval 297–302 nm strong anti-rachitic effect; UV-C (<290 nm) particularly occurring above 2000 m. Absorption of UV by skin proteins increases below 295 nm.
   Effects are:
   (a) Melanin oxidation.
   (b) Increased vitamin D and histamine in the skin.
   (c) Increased gastric acid secretion.
   (d) Blood: increased haemoglobin, Ca, Mg and phosphate level.
   (e) Increased protein metabolism.
   (f) Hormones: thyroid (hyperthyroidism) and adrenal gland, gonadotrophic functions.
   (g) Direct lethal effects on bacteria (indirect effects on man).
3. Changes in acidity of the skin
   (a) by aerosols;
   (b) by factors affecting sweat production and evaporation.

II. Lungs and throat (Respiratory tract)

1. Temperature and humidity affecting mucous membrane.
   Dry air:
   (a) drying and decreased elasticity of mucous membranes (leading to microfissures) and decreased ciliary activity (inefficient removal of dust);

TABLE 3.1—contd

II. Lungs and throat (Respiratory tract)—contd

(b) decreased mucous (and antibody) production;
(c) decreased blood flow and warming-up of inhaled air.

Cooling:
(a) decreased permeability of membrane.
(b) constriction of blood capillaries. Both temperature and humidity affect survival and penetration of bacteria and viruses. Gram-negative bacteria increase with high humidity, gram-positive and influenza virus with low humidity and low windspeed.

2. Ionization: surplus of positive ions in air, particularly CO^{+}_2 mol (conc. of 10^4–10^5 ions $cm^{-2}\ s^{-1}$), causes decreased ciliary activities in trachea (from 1400 to 1100 min^{-1}), decreased mucous transport, dry throat, headaches; surplus of negative ions (particularly O^{-}_2 mol) causes increased ciliary activity (from 1100–1700 min^{-1}). Possible cause: the effect of CO^{+}_2 on cell enzymes, releasing serotonin, causing contraction of smooth muscles in trachea.

3. Acidity of the air.
   (a) pH >8.0: increased permeability, inflation of mucous membrane cells, decreased ciliary activity.
   (b) pH <7.0: shrinkage of cells.
   (c) pH <5.0: increased incidence of bronchitis.

4. Decreased partial oxygen pressure. In mountains or low-pressure climatic chambers from 159 mm Hg (sea-level) to 125 mm (2000 m), causing reduced oxygen saturation of blood from 96% (sea-level) to 85% (3000 m). 30% increase of UV radiation.
   High altitude affects above 1500 m:
   (a) increased lung ventilation;
   (b) increased adrenal activity;
   (c) increased heart and pulse rate;
   (d) changes in composition of blood cells (e.g. increased Hb);
   (e) increased peripheral blood circulation;
   (f) increased thermoregulatory efficiency;
   (g) greater balance between the autonomic nervous systems;
   (h) increase of the fibrinogen-content of blood;
   (i) reduced stomach acid production;
   (j) lowered resistance to toxic drugs.
   Clinical applications: asthma, bronchitis, whooping-cough, rhinitis, rheumatoid arthritis, peripheral blood circulation problems, certain forms of migraine and eczema, phantom pains.

5. Increased partial oxygen pressure. Treatment of gas gangrene, CO poisoning, emphysema, myocardial infarction.

6. Trace components.
   (a) Ozone: in low concentration killing bacteria, in high concentration irritating mucous membranes and increasing incidence of infections; changes in red blood cells (spherocyte formation) after inhalation of 0.25 ppm;[2] three weeks inhalation during 7 h/day causes increased neonatal death in mice.
   (b) Salt (NaCl): fine aerosols with $\frac{1}{4}$–$\frac{1}{2}$% hypotonic salt solution (<0.9% NaCl) cause swelling of mucous membranes. Hypertonic solutions cause shrinkage.
   (c) Radon effects: studies by Engelmann[3] suggest biol. effects by inhaling radon aerosols (<80 Mache U.) or drinking radon water (800 MU l^{-1}); stimul. of cell functions, oxidizing processes, lowering of Syst.B.P., dilatation of periph. arteries, acceleration of endocrines.

TABLE 3.1—contd

II. Lungs and throat (Respiratory tract)—contd

7. Air pollution.
   (a) Gases: (SO_2, CO_2 benzpyrene, etc.).
   (b) Particles:
      (i) Organic (pollen, spores), causing allergic reactions.
      (ii) Inorganic: mineral dust (silicosis, etc.).
   (c) Aerosols: strong increase of physico-chemical action due to increased action surface.

III. Nose

1. Direct stimulation of the olfactory nerve by volatile substances and corresponding stimulation of part of rhinencephalon, affecting emotions, behaviour, hormonal and cardio-vasomotor processes, intestinal function, etc.
2. Acidity of nasal mucosa. Normally pH 5.5–6.5; it affects membrane permeability. pH decreases with rest, inhalation of warm air, diluted acid, aerosol sprays etc. pH rises during rhinitis and common cold; it furthers development of gram-negative bacteria in nasal mucosa.
3. See effects described under 'Respiratory tract'.

IV. Eyes

1. Direct overstimulation of eyes by direct solar radiation (particularly in spring and early autumn) causing acute conjunctivitis.
2. Cold stress, particularly combined with strong atmospheric turbulence triggers acute glaucoma.
3. Changing acidity of the eye membranes by aerosols, thermal stress, wind etc. causing blepharitis, conjunctivitis, hordeolum.
4. Retinal detachments. In The Netherlands, particularly in summer (maximum June) minimum in winter.[16] In Switzerland maximum March–May, minimum November–January.[17]
5. Light-flickering during strong sunlight with certain frequencies causes epileptic attacks.
6. Benoit[20]-Miline[21-23] Effect. Influence of sunlight, through eye-nerves, on pituitary, hypothalamus, production of gonadotropic hormones, thyroid and adrenal function, biological rhythms, etc.
7. Hollwich[24,25] Effect. Long absence of light (caves, blindness) changes carbohydrate metabolism, urinary volume, blood sugar level, reduction of pituitary and adrenal function.

V. Direct effects on the nervous system

1. Electrostatic (pulsating) and electromagnetic fields, affecting plants,[4–6] bees,[7,8] golden hamsters, nerves,[9,10] reaction speed,[11] organic and inorganic colloidal fluids.[12–15]
2. Microseismic effects (amplitude 1–20 μm and frequency 6–8 s^{-1}) on eel.[18]
3. Direct olfactory stimulation.
4. Rohracher effect.[19] Natural mechanical vibrations of the living organism, with frequencies 6–12 s^{-1}, ampl. 1–5 nm. Frequency increases to 14 s^{-1} during hyperthyroidism, fever; fibrations missing in poikilothermic animals.
5. Cosmic ray effects. Recent studies suggest direct biological effects at a cellular level due to cosmic rays.[26–31]

TABLE 3.1—contd

VI. Important biometeorological stimuli during space flights
1. Gravity fields. Acceleration of 2 g and more seems to affect electric potentials in plant tissue;[32] changed growth in mice, hamsters and birds;[32,33] 2 g acceleration during 11 days increases heart weight, decreases weight of spleen, haemoglobin and blood haemotocrit in animals and affects behaviour of man. Effects of very long periods of weightlessness on man have been studied extensively in space flights.
2. Magnetic fields. Very strong fields (>1500 Gauss) affect cell rotation,[34] growth of tissue,[35] growth of mice, haematological changes in blood of mice etc.;[36,37,39] influence on orientation of gastropods, planarias and drosophila.[38]
3. Corpuscular solar radiation. Electrons, protons, heavy cores of atoms etc. but particularly proton streams during solar flares, could have dangerous biological effects due to deep penetrative capacity.

3.3. LOCATION AND MECHANISM OF THE THERMOREGULATORY CENTRES IN HUMANS

Every subject knows from his own experience that if the temperature in a room changes 2 or 3 °C it is often hardly noticeable, but if the internal temperature in his body rises 2 or 3°C, causing a body temperature of 39° or 40°C, he is terribly sick, and may die if the temperature increases another degree. This is true both for animals and man although the body temperature differs for different animals. Living organisms are classified into two large groups on the basis of their thermal behaviour: poikilothermic or cold-blooded organisms (from the Greek *poikilos* = different) and homoiothermic, or warm-blooded, organisms (from the Greek *homoios* = equal or uniform).

The poikilothermic organisms comprise all animals (except birds and mammals) having a variable body temperature. Their temperature fluctuates rapidly with changing environmental temperature.

The homoiotherms (birds and mammals) are able to keep their body temperature fairly constant within certain limits. It is usually higher than the ambient temperature and they are therefore known as warm-blooded animals, in contrast to the cold-blooded poikilotherms. Although the individual temperatures of homoiotherms are fairly constant, marked differences are observed between different birds or mammals. According to Eisentraut birds have a higher body temperature than mammals. The temperature of the former usually exceeds 40 °C. This is due to their very high metabolism. During sudden influxes of cold air, birds will hide or fall asleep, slowing down their metabolism. Similar processes occur during hibernation.

Considerable fluctuations in temperature are observed in the lower mammals as well as in the poikilotherms. The temperature of bats, for example, may fluctuate between 41 and 20°C (during sleep).

The thermal condition of the living body can be compared to a heated room with enough windows to let heat out if the room is too warm, or to prevent it from getting cold by keeping the windows closed. In a similar way the human body of a homoiotherm keeps a thermal balance by the heat production and heat loss. The human body is provided with various mechanisms for maintaining almost perfect thermal balance. These thermostatic processes are assisted by the structure of the skin and the water content of the body.

3.3.1. The hypothalamus

One of the principal structures in the brain through which changes in weather and climate affect the body is the hypothalamus (Figs. 3.1 and 3.2) the principal

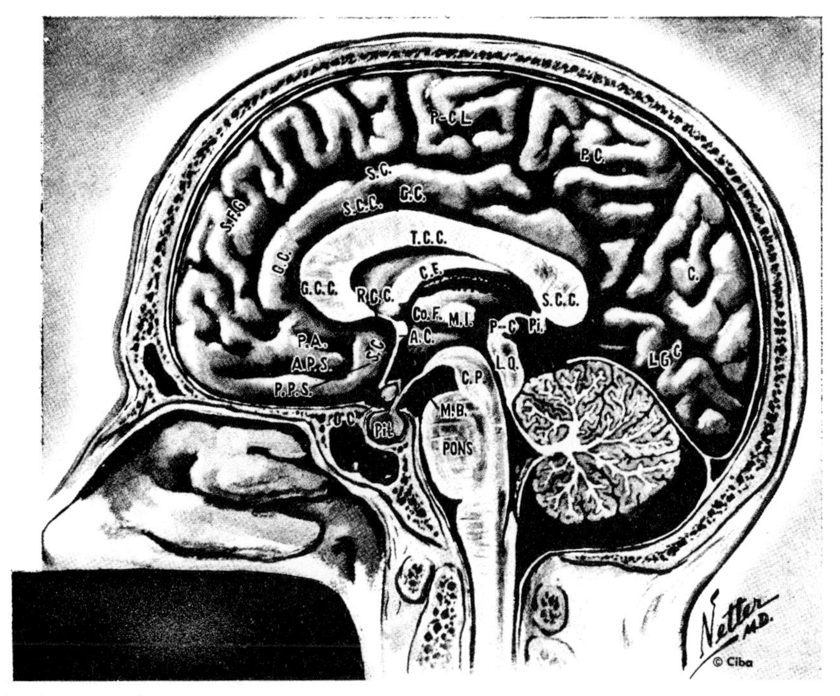

Fig. 3.1 Medial aspect of the human brain A.C. anterio commissure; A.P.S. anterior parolfactory sulcus; C. cuneus; Ca.F. calcarine fissure; C.F. body of fornix; C.P. cerebral peduncle; C.P.V. 3 choroid plexus of 3rd ventricle; Co.F. column of fornix; D.F.H. dentate fascia of hippocampus; F. G. fusiform gyrus; F.I. interpeduncular fossa; G.C. gyrus cinguli; G.C.C. genu of corpus callosum; H.G. hippocampal gyrus; I.T.G. inferior temporal gyrus; L.G. lingual gyres; L.O. lamina quadrigemina; M.I. massa intermedia; M.B. mammillary body; O.C. optic chiasm; O.R. optic recess; P.A. parol factory area; P.C. precuneus; P-C. posterior commissure; P.O.S. parieto-occipital fissue; P-C.L. paracentral lobe; Pi. pineal body; Pit. pituitary gland; P.P.S. posterior parolfactory sulcus; R.C.C. rostrum of corpus callosum; S.C. sulous cinguli; S.C. (P.F.) sulcus cinguli (pars frontalis); S.C. (P.M.) sulcus cinguli (pars marginalis); S.C.C. splenium of corpus callosum; S.C.G. subcallosal gyrus; S.F.G. superior frontal gyrus; T.C.C. trunk of corpus callosum; Th. thalamus; T.P. temporal pole; U. uncus. (from: Frank H. Netter, *The Ciba Collection of Medical Illustrations, Vol. 1, Nervous System,* 1958, p 40)

heat regulatory centre. It is located more or less between the thalamus and pituitary and has control of the autonomic nervous system and secretion of the anterior and posterior lobes of the pituitary. (See Figs. 3.4 and 3.5, pp. 67, 70). This takes place either through nerve fibres or by blood vessels carrying a chemical substance from the hypothalamus to the pituitary. The principal reasons for the assumption that the hypothalamus has a key position in all meteorotropic phenomena are:

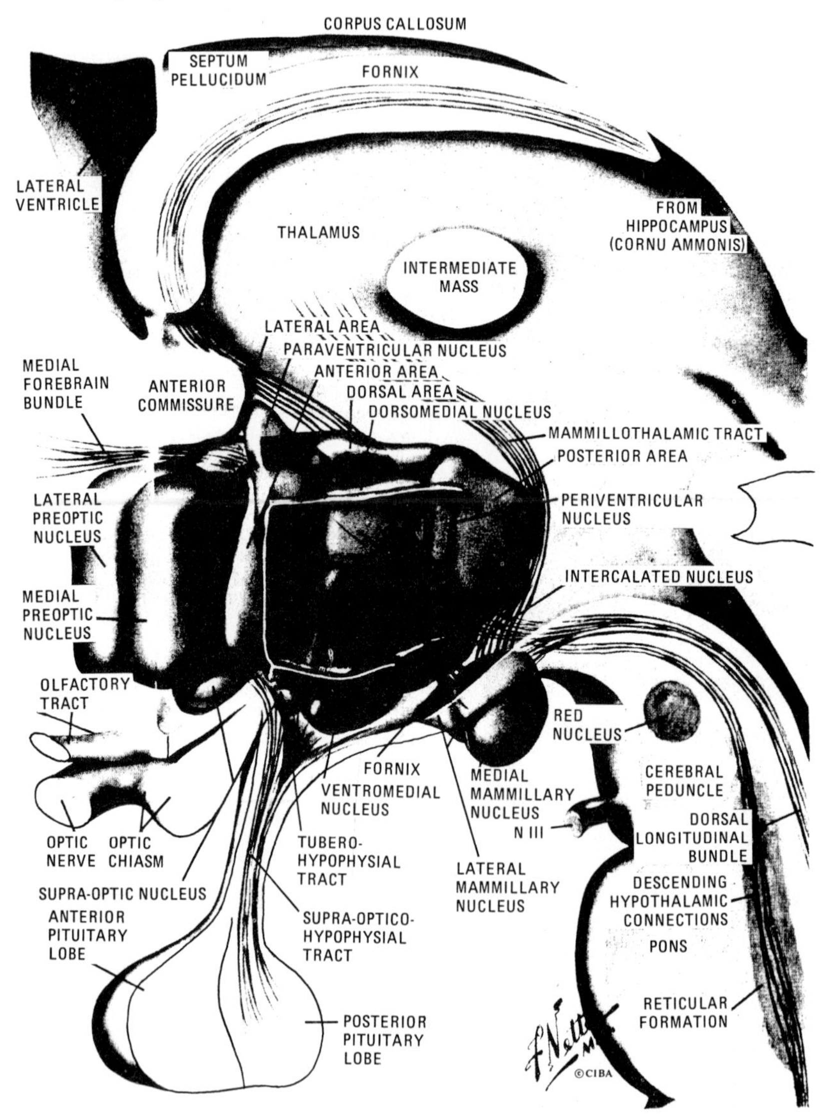

Fig. 3.2 General topography of the hypothalamic area (from: Frank H. Netter, *The Ciba Collection of Medical Illustrations, Vol. 1, Nervous System,* 1958, p 151)

(a) The same meteorological factor or group of factors affects a number of entirely different physiological and pathological processes in the human body, which suggests that one particular control centre in the body is affected by the meteorological stimuli and is able to influence several important physiological centres;

(b) The hypothalamus exerts control over the pituitary gland which in turn controls most of the hormonal processes in the body, which explains that the same meteorological trigger can affect many independent hormonal processes in the body.

The various mechanisms involved in heat regulation are described in many publications and have been summarized in *Medical Biometeorology*[1] (pp. 207–255) and *Progress in Human Biometeorology* (Part IB, pp. 413–461).[32] The most important aspects of these vital mechanisms for meteorotropic phenomena will be summarized in the following pages. The discussion of some of the mechanisms will be extended in Chapter 5. However the studies of the thermoregulatory mechanisms in man have greatly increased our knowledge of these mechanisms.

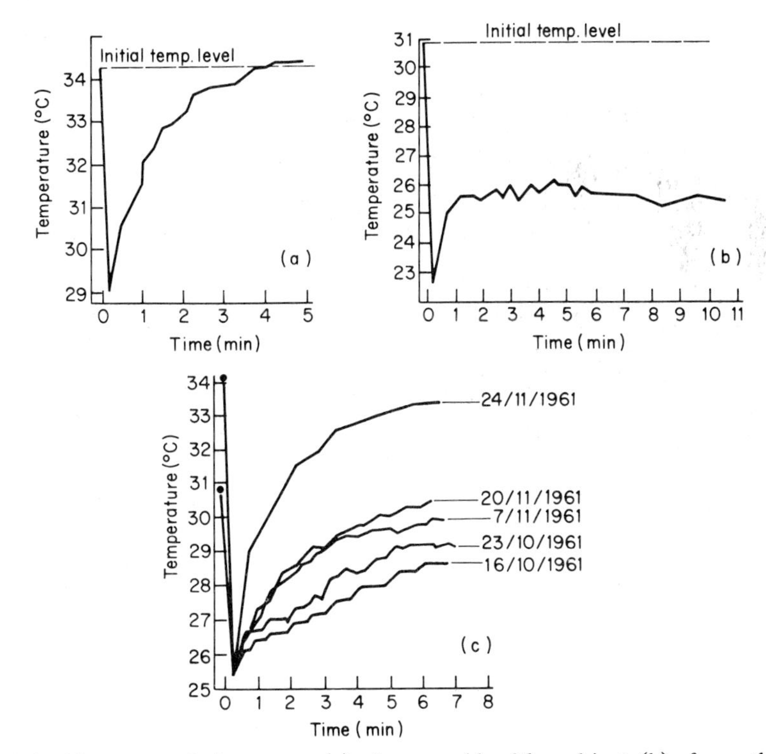

Fig. 3.3 Thermoregulation curves (a) of a normal healthy subject, (b) of an asthmatic patient and (c) improvement of the thermoregulation after a series of high altitude treatments

It is evident that the heat balance of the body is one of the most vital mechanisms to protect the body against thermal stresses of the meteorological environment. The body cells consist of about 70% water. Therefore a considerable amount of thermal energy is required to increase the warmth of this fluid through 1 °C. On the other hand, a considerable loss of heat is required to lower

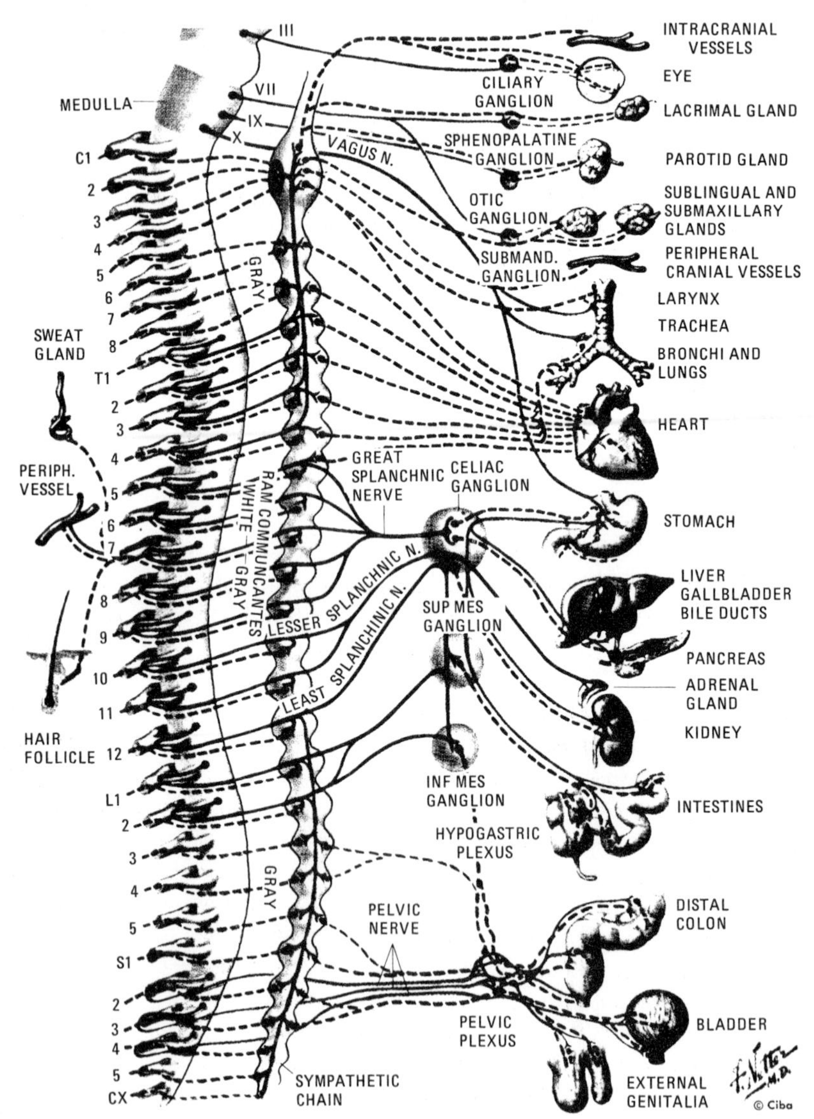

Fig. 3.4 The autonomic nervous systems and their control of various organs (from: Frank H. Netter, *The Ciba Collection of Medical Illustrations, Vol. 1, Nervous System,* 1958, p. 81)

the temperature by 1 °C. Therefore two opposing processes are required for the thermal balance of the body, i.e. heat production and heat loss.

3.3.2. Heat production

This occurs under resting conditions mainly in the skeletal muscles where exothermic oxidative and non-oxidative reactions are constantly going on. Muscular activity raises the body temperature 1–2 °C; lowering of temperature causes a 'tensing' of the muscles, mostly followed by synchronous contraction in the muscle fibres, known as shivering.

The normal heat production is known as basal metabolic rate (BMR). It depends on at least seven factors:

(a) surface area (depending on height and weight of the body);
(b) age and sex (greater in children than adults, greater in males than females);
(c) activity of ductless glands (e.g. thyroxine, regulated by the pituitary by means of the hormone thyrotrophin, stimulates the thyroid and general metabolism);
(d) adrenalin secretion (increases BMR);
(e) body temperature;
(f) food intake; and
(g) muscular activity.

3.3.3. Heat loss

This takes place in four different ways:

(a) *Conduction* through the skin. Depending on the structure and conductivity of the upper skin it amounts to about 11 cal m^{-2} h^{-1}°C^{-1}; a decrease in environmental temperature leads to vasoconstriction and lowering of the temperature of the skin surface.
(b) *Convection* represents 15% of the total heat loss. Owing to the low specific heat of air, 0.24 cal g^{-1}, the temperature of the air near the skin rises quickly. With normal convection currents in the air a thin layer of 1–2 mm of air in contact with the body becomes humid, warmed and lighter. As a result it rises and is replaced by cooler, drier air. At an air temperature of 34.5 °C, convection loss falls to zero; above 37 °C it becomes negative, i.e. the body warms up by convection; movement of the air increases convection; at medium velocity (up to 30 m s^{-1}) heat loss through convection increases roughly with the square root of the air speed.
(c) *Radiation* of heat amounts to about 60% of the total heat loss. The human skin, irrespective of its colour, behaves like a perfect 'black body'. The rate of heat exchange with the environment depends on the difference between the fourth powers of the absolute temperatures of the body and of the environment (Stephan's law), resulting in the emission of waves of 5–20 μm wavelength, maximum at 9μm. With rising temperature of the body the wavelength of the maximum radiation intensity becomes shorter and the

intensity of this maximum radiation is proportional to the 5th power of the absolute temperature.

(d) *Evaporation* of the water of the skin and lungs is another important mechanism of heat loss if the internal body becomes overheated. It amounts to about 25% of the total heat loss of a resting person. Water is lost through the skin either by insensible perspiration and/or by sweating.

(i) *Insensible perspiration*: this invisible and intangible loss amounts to about 600–800 ml each 24 h, which is about double the water loss at rest from the lungs. The amount lost is about the same over the whole body surface except for the following places, where it is considerably higher: face, neck, dorsum of hand, soles and palms (highest). The process of insensible perspiration consists of the passage of water through the epidermis (the outermost layer of the skin) by diffusion of tissue fluid. The quantity diffused increases slightly with rising environmental temperature and increased blood-flow and falls a little in humid surroundings.

(ii) *Sweating*: is caused by two kinds of sweat glands in man: the eccrine glands, distributed over the whole body, secrete a dilute water solution containing sodium chloride (0.1–0.37% according to Kuno,[42] 0.5–0.9% according to Hill[42]), urea (0.03%), sugar (0.004%) and during exercise, lactic acid (0.07%). These eccrine glands are densest on the palms and soles, next dense on the head and least dense on the trunk and extremities. A second group of sweat glands is known as apocrine glands, found in the armpit, around the nipples, etc. They secrete a fluid with a characteristic odour. The functioning of these sweat glands is controlled by cholinergic fibres of the sympathetic nervous system, i.e. they are stimulated by acetylcholine, not adrenaline. The actual secretion is produced by direct or reflex stimulation of the centres in the spinal cord, medulla, hypothalamus or cerebral cortex (see Fig. 3.1, p. 59).

Sweat secretion is increased by three factors:

(a) *rise of environmental temperature ('thermal sweating')*, e.g. by sudden influx of warm tropical air masses in summer, hot baths, high frequency electromagnetic induction (e.g. used in 'thermotherapy');

(b) *emotional states ('mental sweating')* when sweating is usually limited to the palms, soles and armpit and is caused by impulses from the higher brain centres; and

(c) *exercise* where both mental and thermal sweating are involved. The maximum rate of sweat secretion is usually 1–2 l h^{-1}. In extremely hot (shade temperature 34–38 °C) and dry regions (humidity less than 30%) the sweat output may be almost 80 l in 14 days and salt excretion is very high. However, during thermal acclimatization (Section 3.5), the salt content of the sweat decreases considerably.

As stated before, the various processes described above are controlled by the hypothalamus which consists of not clearly defined clusters of cells, the hypothalamic nuclei, which are abundantly supplied by blood capillaries (see Fig. 3.2). They are usually classified from the front toward the rear into the anterior nuclei (resp. supra-optic, paraventricular and infundibular nuclei), the middle nuclei occupying the middle part of the tuber cinereum (ventromedial, dorsomedial and lateral hypothalamic nuclei) and the posterior nuclei (mammillary-infundibular nuclei, the posterior part of the tuber cinereum and the medial and lateral nuclei of the mammillary body). The posterior part of the tuber cinereum functions as an orthosympathetic regulation centre. It causes secretion of adrenalin and contraction of the peripheral capillaries during cold stress preventing a too strong cooling of the body. The anterior nuclei, apart from regulating the secretion of the anti-diuretic hormone vasopressin, regulate the peripheral vasodilatation of the skin and sweating during excessive heat in the body.

The hypothalamic nuclei are stimulated by thermal receptors in the skin which enable us to experience differences in temperature between the body and its environment. Studies by Blix[43] and Donaldson[43] suggest that different centres exist for registration of cold and heat. They are known as Krause end-bulbs for cold and the Ruffini organs for warmth. It has been pointed out by Hensel, Stöhr, Kantner and others[43] that these thermoreceptors should not be considered as anatomically sharply defined bodies or end-plates. They are networks of nerves in various layers, one above the other.

Hensel, Zotterman and Dodt[44,45] were able to show that these nerve networks of both cold and heat receptors are characterized by continuous electric discharges (action potentials) even at a constant environmental temperature. The discharge of the cold fibres, however, suddenly increases during cold stress and decreases or disappears in the neighbourhood of heat radiation, whereas the heat receptors increase in discharge during heat stress. Hensel[44] summarized the differences between the two groups as follows:

Cold receptors	Heat receptors
Stationary discharge at constant temperature between 10 and 41 °C.	Between 20 and 47 °C.
Frequency maximum between 20 and 34 °C.	Maximum between 38 and 43 °C.
Stationary maximum frequency 10 s^{-1}.	3.7 s^{-1}.
Regular discharges at constant temperature.	Irregular discharges at constant temperature.
Rise in discharge frequency during rapid cooling.	Same during rapid heating.

Registration of changes in environmental temperature takes place as a result of changes in temperature of the peripheral blood capillaries, followed by minor alterations in the physico-chemical state of the blood circulating through the rich network of hypothalamic capillaries. At the same time, the hypothalamic nuclei receive information from the thermal receptors in the skin. Vogt and associates showed that noradrenaline, adrenaline and 5-hydroxytryptamine (5-HT or serotonin) occur in very small concentrations in the hypothalamus; Domer and colleagues found that excess of free 5-HT or catecholamines on one side of the anterior hypothalamus seemed to be a factor determining the level of body temperature (for further details see *Medical Biometeorology*,[1] and *Progress in Human Biometeorology*,[32,41]).

Several methods have been suggested to measure the total efficiency of the thermoregulatory mechanism in the human body. A simple method, the water-bath test developed by Tromp,[46] is the measurement of the temperature of the left-hand palm with a thermocouple. After cooling the hand in water to 10 °C the re-warming is measured every 15 s. In a well thermoregulated subject the re-warming curve reaches its initial temperature level in about 6 min, whereas in poorly thermoregulated subjects, e.g. those suffering from rheumatism or asthma, the initial level is only reached after 20 or more minutes (see Figs. 3.3a, b and c, p. 61).

3.4. CONTROL OF OTHER CENTRES IN THE BRAIN BY THE THERMOREGULATORY CENTRE

The hypothalamus, due to its location, is able to control or affect a number of important brain centres, in particular the autonomic nervous system, the pituitary or hypophysis (also known as the 'leader of the endocrine orchestra'), the rhinencephalon (also known as olfactory or emotional brain) and the epiphysis (or pineal gland).

For the non-physiologically trained reader it may be useful to say a few words about these very important physiological centres in our body as they play a vital role in meteorotropic phenomena (see Fig. 3.4, p. 62).

3.4.1. The nervous system

Both in animals and humans, is composed of nerve cells, known as neurons, 4–100 μm large, globular, pear- or spindle-shaped cell bodies composed of protoplasm with many fibres, nucleus, etc. From each neuron numerous short processes extend, the dendrites (ramified protoplasmatic offshoots) and one long fibre, the neurite or axon, which is a branched protuberance of the nerve cell. The branches of this axon are called collaterals.

A number of neurites, each surrounded by connective tissue, form the nerve fibres; they are grouped together in nerve bundles (like submarine cables) and create the actual nerve, divided into myelinated and non-myelinated nerves, depending on whether or not the neurite is covered by a thick viscid insulating layer, called myelin, composed of certain proteins.

Most nerve cell bodies are grouped together in segmentally arranged masses, known as ganglia. They are called post-ganglionic when the ganglion is located outside the central nervous system and pre-ganglionic when the cell bodies are located in the central nervous system. Points of contact between individual neurons are called synapses, the junction being generally formed by multiple points of contact. Many axon collaterals terminate in end-plates, swellings which are applied to the surface of both dendrites and the cell bodies of other neurons.

The actual physico-chemical mechanisms involved at the synapses and end-plates are not known with certainty. However, it is very likely that in many instances at least the transmission at a neuro-myal junction takes place through liberation of specific chemical mediators, the neuro-transmitters, such as acetylcholine, an active ester of choline (CH_2 $(OH) \cdot CH_2N$ $(CH_3)_3$ OH), and sympathin.* The transmission is affected by various physico-chemical conditions, e.g. fluctuations in the potassium and calcium content and acidity of the environment of the ganglia. The acetylcholine is rapidly destroyed through hydrolysis by an enzyme, cholinesterase, which is present in tissues and blood. Cholinesterase is inhibited by eserine.

Towards the central nerve organs in the body the nerves unite into larger bundles. Two main centres of nerve control are distinguished, the 'animal' or 'central' nervous system and the 'autonomic' or 'involuntary' nervous system.

The animal, or central, nervous system regulates our psychic processes and conducts consciously perceived phenomena controlled by our volition, such as movement, etc. It is composed of the spinal marrow, its continuation in the skull (medulla oblongata), the further extension into the brain, and the nerves connecting these central parts with the various organs (see Figs. 3.1, 3.4 and 3.5, pp. 59, 62, 70).

The autonomic or involuntary nervous system controls all unconscious physiological processes such as the heart, intestines etc. It is composed of two chains, the peripheral ganglia, situated on either side of the spinal column, the ganglia of each chain being connected by longitudinal fibres. The lateral ganglia send post-ganglionic fibres to the spinal nerves, which convey sympathetic post-ganglionic fibres to the blood vessels, sweat glands, smooth muscle, intestines, etc.

The autonomic nervous system has recently been classified into three (not two as in the past) main systems. These systems are known as the orthosympathetic, parasympathetic and peripheral nervous systems (see Fig. 3.4, p. 62).

The orthosympathetic nervous system, also known as the sympathetic nervous system proper, consists of the connector neurons, the fibres, ramifications and excitor neurons† of the thoracic region and the first two lumbar segments

*In recent years several new neurotransmitters have been discovered. In brain tissue three important neurotransmitters have been found: dopamine, serotonin and noradrenaline (Barondes, Jenden, Schlapfer, Woodson and others[47,48]).

†In the somatic spinal reflex arc three types of neurons are distinguished: the afferent or receptor neuron with its cell body in the posterior root ganglion; the connector neuron, the cell in the posterior horn of grey matter, which by means of its axon transmits the impulse to the anterior horn; and the excitor neuron, the anterior horn cell and its axon, which transmits the efferent impulses to a voluntary muscle.

of the spinal cord. Stimulation of this nervous system enables the organism to adapt itself quickly to sudden changes in the environment.

The parasympathetic nervous system consists of the connector neurons, their fibres and related excitor neurons located in the brain (called the cranial division) and in the second and third sacral segments of the spinal cord (sacral divisions). The principal function of this nervous system is the regulation and recovery of the disturbed physiological equilibrium of the organism as a result of environmental disturbing factors. Both nervous systems are rather antagonistic in their actions on organs which they both supply (e.g. heart, bronchi, pupil, etc.). Some organs are innervated by one division only: e.g. uterus, adrenal medulla and most arterioles by the orthosympathetic system only; glands of stomach and pancreas by the parasympathetic only.

The orthosympathetic fibres liberate peripherally a substance identical with adrenaline; parasympathetic fibres liberate acetylcholine. Both substances have different physiological effects.

Adrenaline causes dilation of the pupil of the eye and retraction of the lids; it accelerates and increases the force of the heartbeat; it increases the excitability of the heart; it increases the blood pressure; it causes constriction of arterioles and capillaries in the skin and of the vessels controlled by the splanchnic nerves (abdominal area), but causes dilation of vessels in heart and skeletal muscles; it relaxes the muscle wall of the bronchi but increases the excitability of skeletal muscles; it inhibits the intestinal wall; it converts glycogen into glycose in the liver and causes a rise in the blood sugar level.

This rise may be so great that sugar is excreted in the urine. Adrenaline also causes contraction of the capsule of the spleen and of the wall of the gall bladder and relaxation of the detrusor or contraction of the sphincter muscle (closing the urethra) of the urinary bladder. Also the clotting time of the blood is reduced by strong adrenaline secretion, due to indirect stimulation of the liver and this increases the risk of thrombosis and embolism in the body. Subcutaneous injection of adrenaline usually increases the red cell count, haematocrit value, haemoglobin concentration and the plasma protein concentration. The white cell and platelet count may likewise rise.

Acetylcholine, on the other hand, causes the pupils to contract, certain peripheral blood vessels to dilate and the blood pressure to drop (10 ng is enough to do this); it slows down the heartbeat (concentrations of 1×10^{-8} are sufficient); it stimulates lachrymal secretion and the production of gastric and pancreatic juices and increases movements of the oesophagus, stomach, intestines and bladder; it helps to transmit the nerve impulses to the muscle fibres (muscular contraction) and so forth.

The actions of the ortho- and parasympathetic systems are not necessarily opposed. Certain physiological phenomena can be either due to an orthosympathetic stimulation or it could be the result of a diminution of parasympathetic stimuli. However, it does not seem to be correct to deny the existence of these two separate autonomic systems, as has been suggested in recent years by certain physiologists. It is true that the antagonism between the two systems is less pronounced than may have been thought in the past, but anatomical, histological and histochemical studies have confirmed the existence of both systems e.g. by using the histochemical colour tests devised by Champy and Coujard[49] (for adrenergic fibres) or by Gerebtzoff[49] (for acetylcholine-esterase). French scientists,[49] more particularly, have pointed out that the result (stimulating or reducing effect) of ortho- or parasympathetic stimulation of an organ depends on various humoral conditions.

A third, and even more complex, nervous system is the peripheral nervous system. Its significance and function has always been a point of considerable controversy. Recent histological studies, notably by Meyling,[49,50] have demonstrated that this system probably consists of two parts: a peripheral intrinsic network of small autonomic ganglion cells (the interstitial cells of Cajal) and a synaptically built reflex system of afferent and pre- and post-ganglionic ortho- and parasympathetic nerve fibres, the end ramifications of which are connected with the terminal network of autonomic cells in an apparently synaptic way.

The peripheral nervous system seems to be also responsible for the tonus of cell fluids and their diffusion (e.g. in the artery walls), the interstitial cells acting as tension receptors. In view of these observations it is possible that the peripheral nervous system plays an important role in the causation of arteriosclerosis (for further details see Tromp[51]).

In Section 3.3 (p. 65) it was pointed out that the posterior part of the tuber cinereum of the hypothalamus regulates orthosympathetic stimulation, the anterior nuclei the parasympathetic reactions. Therefore it is evident that any climatic or meteorological stress condition affecting either the ortho- and parasympathetic nervous system could influence a very large number of physiological processes in the human body and might be pathogenic if the morbidity limit of an organ is reached shortly before stimulation impinges upon it.

3.4.2. The pituitary or hypophysis

The pituitary or hypophysis consists of two parts, a neural and a glandular division (Figs. 3.5 and 3.6, pp. 70, 71). The neural division is richly supplied by non-myelinated nerve fibres, which arise with supra-optic nuclei of the hypothalamus (p. 65). They form a track along the so-called pituitary stalk into the posterior parts of the hypophysis. A part of the glandular division of the pituitary is called the pars anterior, anterior lobe or adenophysis (Fig. 3.5, p. 70). The functions of this part of the pituitary can be summarized as follows: it controls the growth of the skeleton by secretion of a growth hormone; it regulates the activity of the gonads (both in males and females) by secretion of gonadotrophic hormones which are glycoproteins; it influences the growth of the mammary gland during pregnancy and controls the secretion of milk by means of a lactogenic hormone, prolactine; it controls the activity of the adrenal cortex by an adrenocorticotrophic hormone (ACTH), adrenotrophin; it regulates the secretory activity of the thyroid gland by means of the thyrotrophic hormone, thyrotrophin; it not only influences metabolism indirectly by thyrotrophin but also directly; the carbohydrate metabolism is influenced by the diabetogenic hormone which stimulates new sugar formation and raises the blood-sugar level; the fat metabolism is influenced by the ketogenic hormone which mobilizes depot fat and increases ketone body production; and tissue metabolism is influenced by the metabolic hormone, which stimulates the metabolism of the tissue cells directly via the thyroid. It is very likely that the three mentioned metabolic activities are actually due to the growth hormone.

Another part of the pituitary, the posterior lobe (see Fig. 3.6), has the following effects. Crude extracts of the posterior lobe of the pituitary, referred to as pituitrin, contain two active substances: vasopressin (which causes a slowing of the heart rate, usually a fall in blood pressure but sometimes a slight rise, constriction of the peripheral capillaries and decreasing urinary flow) and oxytocin, which stimulates contraction of the uterus and is responsible for the onset of parturition.

In view of the close connection between the hypothalamus and the pituitary (see Fig. 3.6) changes in our physical environment will affect the pituitary functions through the thermoregulatory system.

3.4.3. The rhinencephalon

The rhinencephalon is a third important brain centre, which can be seriously affected by the thermoregulatory centre as a result of changes in our thermal environment. The rhinencephalon is also known as the olfactory or emotional brain. There is a close nerve connection between the rhinencephalon and the hypothalamic-hypophysial system; as there is also a close integration between

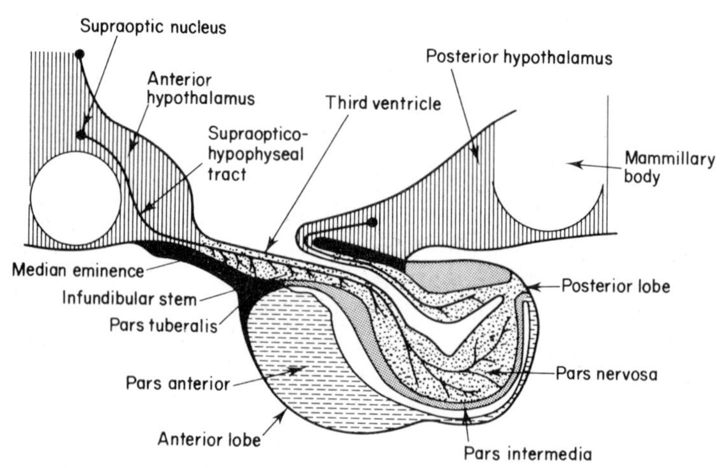

Fig. 3.5 Section through the hypothalamus and hypophysis (pituitary) of the cat (from: S. Wright, *Applied Physiology*, Oxford University Press, London, 1948, p. 260)

these three centres and the most differentiated parts of the cerebrum (or telencephalon covering seven-eighths of the upper part of the skull, see Fig. 3.1, p. 59), it is thus evident that weather and climate can seriously affect the psychological and behavioural processes in humans.

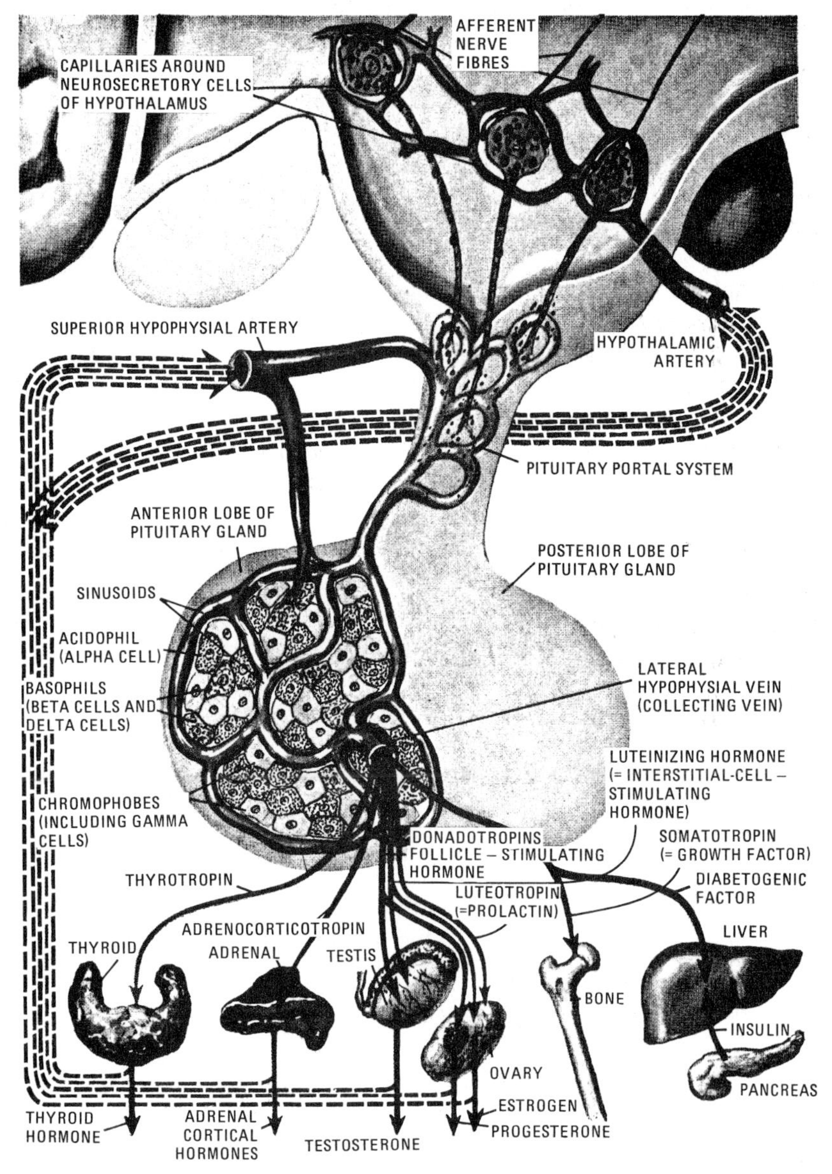

Fig. 3.6 Relations between hypothalamus and the anterior lobe of the pituitary and the principal secretory activities of the pituitary (from: Frank H. Netter, *The Ciba Collection of Medical Illustrations, Vol. 1, Nervous System,* 1958, p. 155)

On functional grounds the rhinencephalon is divided into different areas: The smell, gustatory and visceroceptive perception areas; areas mediating the unconscious instinctive-emotional life including the moods; areas concerned with the mediation of certain somato-motor, cardio-vasomotor, visceral and endocrinal activities, and finally areas which mediate conscious affective experiences (for further details see Prick[52]).

3.4.4. The epiphysis or pineal gland

The epiphysis or pineal gland is a fourth centre in the brain which, in addition to the hypothalamus, pituitary and rhinencephalon, plays an important role in the recording of weak external stimuli. The *epiphysis* or *pineal gland*, is also known as our photoperiodic brain (Fig. 3.1). The studies of Wurtman, Axelrod and Kelley[53] have shown the great significance of this gland in natural biological rhythms. The gland is located at the posterior lower part of the corpus callosum, which lies at the bottom of the cerebrum (Fig. 3.1). It is not a gland in the classical sense but rather a neuroendocrine transducer, regulating the timing of physiological events. As a result the pineal gland acts as a biological clock that informs the body about the light intensity of its environment and the hour of the day. The gland is rich in biologically active amines, as also is the hypothalamus. It is rich in noradrenalin, histamine and, in particular, serotonin (5-HT) which show striking diurnal changes in concentration. The serotonin content is highest at 1 p.m. and lowest at 11 p.m. A similar periodicity is clearly observed in the noradrenaline concentration in the pineal gland, with the maximum level occurring at night and the lowest during the day. Consequently the pineal gland is an important component of the endogenous time-measuring system. The release of the pineal hormone melatonin does not take place, as in most glands, in response to instructions received via the circulation, but is due to direct nerve impulses that are controlled by light stimuli reaching the pineal gland via the retina. The pineal gland is richly innervated by sympathetic nerve terminals. Through these sympathetic nerves the hypothalamus could stimulate the pineal gland after thermal stresses.

This brief summary indicates that hardly any physiological function escapes the effects of our thermoregulatory system, which explains the great variety of physiological effects after a meteorological stimulus.

3.5. MECHANISMS OF ACCLIMATIZATION AND HABITUATION

Around 1830 the term acclimatization was introduced to describe the adjustment of plants and animals to new environments. Different terms have been used to describe modifications of physiological functions in animals and man to meet the physical environment.

Four types of acclimatization have been described:

(a) Acclimatization sensustricto is defined as the complex of reversible changes of physiological response which increases the efficiency of individual organisms while they remain in a physical environment outside the neutral zone.

(b) Acclimation, according to Burton and Edholm,[54] denotes the long-term changes taking place during a life time. However, until a standard usage is accepted, it seems better to regard acclimation as a synonym for acclimatization, because the term is not commonly used.

(c) Habituation is defined as the alteration of reflexes, responses or percepts as a result of retained central nervous experience of specific stimuli. It has been studied particularly by Konorski,[55] Young,[56] Eccles[57] and Glaser.[58-62]

(d) Adaptation has two connotations:

(i) biologically it is used to generalize the concept of the evolutionary adjustment of species over many generations, as mutations are selected by environmental stresses;

(ii) physiologically, adaptation connotes the rapid changes of cellular function brought about by continued stimuli. The term applies mainly to nerves and receptors. Often adaptation is used, however, as a synonym for acclimatization. A distinction should be made between acclimatization to thermal stimuli, to high altitude and to weightlessness during space flights.

3.5.1. Acclimatization to thermal stimuli

Macfarlane[63-65] pointed out that the process of acclimatization comprises three aspects:

(a) detection by the body of changes in the surroundings, the skin and the nervous system;

(b) fast reaction by the body, by vascular, sweating or shivering responses through the nervous system;

(c) slower endocrine changes, modified cellular metabolism and adjustments of behaviour and judgement.

In hot climates, man decreases insulation and increases the rate of turnover of water and to a lesser degree, of salt. In cold regions, there is a greater turnover of chemical energy together with an increase of insulation. Other modifications depend upon these primary changes.

Acclimatization is mostly reached within one week, but some two or three weeks are required to approach a steady state. Reversal of the changes is slow, and acclimatization to heat survives two weeks in the cold. Some adaptations, particularly of behaviour and attitudes, are not achieved until years have passed, or they may require a new generation.

Acclimatization to heat is probably not a reflex reaction but is a complicated process of slow physiological adjustments.[58-62] The level of deep and superficial body temperature at which thermal balance is achieved depends on the preceding thermal environment.[58-62, 66] The effect of previous thermal stress may persist for very long periods.

Previous exposure to thermal stimuli facilitates adaptation at a later period of climatic stress.[61] Glaser and Whittow[60] demonstrated that pain or the rise in blood pressure which normally accompanies the immersion of a hand in icy water diminishes, or can be lost, after repeated immersions in the course of a number of days. People exposed to strong fluctuations of environmental temperature have a very efficient control of the skin temperature and can easily adapt to cold. The same is true for oxygen adaptation at high altitudes (Section 3.5.2). As this rule applies to various other adaptational processes, the primary cause must not be sought in one particular organ but in certain brain centres.

With a given stimulus higher sweat rates can be elicited in the summer than in winter during heat acclimatization. In other words, adaptation to heat stress is easier in summer than in winter. For cold stress it is the reverse. This may explain why the effect of the same meteorological factor or groups of factors is different in summer from what it is in winter. An influx of cold polar air or a cold front after a period of heat adaptation will have greater effect than after a previous cold period.

Experiments by Davis[67] have shown that repetition of stimuli (as experienced during acclimatization) may be accompanied by gradual changes in responses to those stimuli, a process described as habituation. It also explains the change in effect of drugs if applied at regular intervals. Konorski,[55] Young[56], Eccles[57] and Glaser[58-62] were able to show that habituation is probably caused by permanent ('plastic') changes of the nerve junctions as a result of previous stimuli. They could demonstrate that prolonged disuse of synapses leads to defectiveness of synaptic function, whereas repetitive stimulation increases the functional efficiency. The size of neurons and dendrites depends on the amount of afferent influences that falls upon them.

Several Canadian reports by Dugal and Thérien[68] suggest that application of vitamin C increases the resistance to cold stress and facilitates adaptation. It depends on the ascorbic acid level in the tissues, especially in the adrenals. Hypertrophy of the adrenal gland during cold stress is inhibited if vitamin C is given in large quantities.

Before complete acclimatization has taken place, the following physiological processes are observed in the case of long lasting heat stimuli: a rise in body temperature and in extreme cases a rise of basal metabolic rate, increase of respiratory ventilation, rise of blood pH and usually fall in pH of urine, increase in pulse and circulation rates, usually a fall in arterial pressure (unless the heat creates a psychological stress), increased sweating, dilatation of the arterioles and capillaries of the skin, increased blood volume, decreased haemoglobin content, usually a decrease in leucocytes and granulocytes, increase in blood sedimentation rates, decreased diuresis, increased chloride secretion, decreased 17-ketosteroid secretion (sudden heat waves may create a psychological stress condition with increased 17-ketosteroid secretion during the first two days), increased skin permeability and decreased capillary resistance. Cold stress has similar, but reverse effects.

3.5.2. Acclimatization to high altitude

The different short-term physiological effects of high altitude will be discussed in Section 5.3, p. 108.

In *Medical Biometeorology*[1] Hurtado has given a review of the acclimatization phenomena at high altitude, characterized by a reduced partial oxygen pressure. There is ample evidence that acclimatization to high altitudes is best investigated in healthy man born and raised in such environment. The constancy of the anoxic (shortage of oxygen) condition and probably hereditary or genetic influences are responsible for a high degree of adaptation and the study of subjects, living permanently at high altitude, provides an adequate comparative basis to understand and rate what happens in the newcomer through his early period of adaptation to the low pressure environment. Most of the data collected from permanent inhabitants of high altitudes have been obtained in the Peruvian Andean region, mainly in Morococha (Peru), which is a mining town located at an altitude of 4540 m with a mean barometric pressure of 446 mm Hg.

The Indian native resident in this place has a mean alveolar oxygen tension of about 50 mm Hg and his arterial oxygen saturation is around 80%. His high degree of acclimatization is indicated by the fact that this native Indian has been engaged in the hard physical work of mining operation for centuries. His rest periods are often employed in several hours of uninterrupted football playing.

Comparative studies of physical activity in residents at sea level and at high altitudes (Morococha) have shown a longer performance time in the latter place and also a higher efficiency (less oxygen consumption and lower pulse rate per kilogram metre of work performed).

Velásquez[69] has recently demonstrated in low pressure chambers a much higher ceiling tolerance in the Morococha natives than in men living at sea level. About half of the natives showed a useful consciousness time at 9000 m, and the mean value of this time was 88 ± 11 s at 12 000 m.

It is well known that in a man exposed to a low ambient pressure several physiological changes occur (Table 3.1 and p. 109). His pulmonary ventilation increases. This process partially compensates the drop in oxygen tension from inspired to alveolar air and, consequently, allows a higher acquisition of this gas by the circulating blood haemoglobin in the pulmonary vascular bed. At high altitudes the total pulmonary ventilation and the alveolar ventilation are about 22% and 28% higher than at sea level, and the difference is more significant if it is referred to body surface area, which is lower in the natives of Morococha.

At sea level, in resting condition, the pO_2 gradient falls about 8–10 mm Hg, when the gas passes from the alveolar air to the circulating blood through the alveolar membrane. At high altitudes the drop is practically nil. This results in an important economy of the pO_2 gradient. This adaptative process is due to a better rate of diffusion, influenced by dilated capillaries and also by enlarged alveoli. Velásquez[69] found a definite increase at high altitudes in the maximal diffusing capacity of the lungs. The ratio of ventilation to perfusion (Va/F) is also greater at high altitudes.

The polycythaemia (i.e. abnormal increase in red blood cells) of the high altitude residents is generally known. According to Hurtado[70] the Morococha native has an increased number of red blood cells and haemoglobin content of the circulating blood. It is due to a type of polycythaemia, with a greater cell volume and a normal or slightly reduced plasma volume. The increased red cell and haemoglobin contents result in a greater transport capacity for oxygen and help to reduce the fall in pO_2 gradient from arterial to venous blood. It has been demonstrated that polycythaemia is a consequence of a greater erythropoietic activity (i.e. formation of red blood cells due to reduced oxygen supply). The stimulating influence of the hypoxic factor acts through an increase in the erythropoietic factor (erythropoietin) present in the circulating plasma. According to Hurtado[70] there is evidence that the blood haemoglobin of the high altitude native shows a decreased affinity for oxygen, which favours the unloading of this gas to the tissues.

Permanent exposure to a low pressure environment modifies some of the chemical characteristics of the circulating blood. The lowering of the carbon dioxide tension in the arterial blood, brought about by hyperventilation (i.e. increased loss of carbon dioxide), is compensated by a proportional decrease of the bicarbonate ions so that the pH is kept within normal limits. In the electrolyte balance, it has been found that the low bicarbonate concentration is associated with a rise in the concentration of chloride ions. Experimental observations have shown that hyperventilation at high altitudes is apparently due to an increased sensitivity of the respiratory centre to the carbon dioxide chemical stimulus, and that this characteristic is not changed by oxygen administration, at least over a short period.

In spite of the compensatory processes which occur along the pO_2 gradient, there is still a decrease in the oxygen tension at capillary level at high altitudes. This requires the presence of adaptative mechanisms to facilitate the diffusion and utilization of oxygen in the tissues. There is important indirect evidence of this occurrence. It has been observed that lactate production in the active muscles, during all degrees of physical activity, is always lower in the high altitude natives compared with the sea level residents. This would indicate an adequate oxygen provision or utilization, or both.

Another important mechanism of high altitude acclimatization is the increased capillarity favouring the physical diffusion of oxygen from blood to the cells. Valdivia[71] has observed in native guinea pigs at high altitudes compared with the same animal species at sea level a definite increase in the ratio: number of capillaries/number of muscle fibres, per mm^2. The increase is about 35%.

3.5.3. Acclimatization to weightlessness

This will be discussed later in Section 5.9. In that section various laboratory studies will be described showing the effects of immobilization, achieved through bed rest and water immersion of the body. The lack of physical stress and strain on the muscular skeletal system in the weightless state would result in significant losses of calcium, nitrogen and related elements. However, certain adaptational processes do seem to occur as will be discussed.

REFERENCES

1. S. W. Tromp (Ed), *Medical Biometeorology*, Elsevier, Amsterdam, The Netherlands, 1963.
2. Lamberts and Veringa, private communication from Dept. Resp. Dis., Univ. Groningen, The Netherlands.
3. in W. Amelung and A. Evers, *Handbuch der Bäder- und Klimaheilkunde,* Schattauer Verlag, Stuttgart, West Germany, 1962.
4. G. H. Sidaway, *Spectrum* **33**, 3 (1967).
5. *idem* and G. F. Asprey, *Int. J. Biometeorol.* **12**, 321 (1968).
6. L. E. Murr, *ibid.* **10**, 135 (1966).
7. L. Schua, *Z. Vergl. Physiol.* **34**, 258 (1952).
8. E. Haine, H. L. König and H. Schmeer, *Int. J. Biometeorol.* **7**, 265 (1964).
9. J. A. Gengerelli and N. J. Holter, *Proc. Soc. Expl. Biol. (N.Y.)* **46**, 523 (1941).
10. S. R. Thompson, *Proc. Roy. Soc.* **82B**, 396 (1910).
11. R. Reiter, *Meteor. Rundsch.* **5**, 14 (1952).
12. G. Piccardi, *The Chemical Basis of Medical Climatology,* Thomas, Springfield, Illinois, USA, 1962.
13. C. Capel-Boute, *Proc. 9th Convegno della Salute,* Ferrara, 1962, p. 239.
14. J. Eichmeier and P. Büger, *Int. J. Biometeorol.* **13**, 239 (1969).
15. W. W. Fischer, G. E. Sturdy, M. E. Ryan and R. A. Pugh, *ibid.* **12**, 15 (1968).
16. G. H. Jonkers, *Z. Augenheilk.* **115**, 308 and 316 (1948).
17. J. F. Cuendet, *Opthalmologia* **117**, 296 (1949).
18. Deelder, private communication from Fisheries Res. Inst., Ijmuiden, The Netherlands, 1960.
19. H. Rohracher, *Mechanische Mikroachwingungen des Menschlichen Körpers*, Urban & Schwarzenberg, Vienna, Austria, 1949.
20. J. Benoit, *Colloq. Int. Centre Nat. Rech. Sci.* **172**, 121 (1970).
21. R. Miline, in *Pathophysiologia Diencephalica*, Springer Verlag, Vienna, Austria, 1958.
22. *idem*, *Compt. Rend. Ass. Anat.* **87**, 431 (1955).
23. *idem* and M. Ciglar, *ibid.* **91**, 1035 (1956).
24. F. Hollwich, *Ber. Tagg. Dtsch. Ges. Inn. Med.* **26**, 278 (1965).
25. *idem*, *Ann. Inst. Barraquer* **9**, 134 (1969).
26. F. H. J. Figge, *Science* **105**, 323 (1947).
27. J. Eugster, *Weltraumstrahlung*, Huber-Verlag, Bern, Switzerland, 1955.
28. S. G. Ong, *Scientia Sinica*, *Biophys Sect.* **2**, 263 (1964).
29. J. Barnothy and M. Forro, *Experientia (Basle)* **4**, 1 (1948).
30. H. Planel, F. Croute, R. Tixador and J. P. Soleihavoup, *C.R. Soc. Biol. (Paris)* **165**, 2189 (1971).
31. A. Rothen, *Proc. Nat. Acad. Sci. (Wash.)* **72**, 2462 (1975).
32. in *Progress in Human Biometeorology, Vol. 1, Part 1A*, S. W. Tromp (Ed), Swets & Zeitlinger, Amsterdam, The Netherlands, 1974, p. 112.

33. S. R. Briney and C. C. Wunder, *Amer. J. Physiol.* **202**, 461 (1962).
34. P. W. Ssawostin, *Planta* **11**, 699 (1930).
35. J. Lengyel, *Arch. Expl. Zellforsch.* **14**, 255 (1933).
36. J. M. Barnothy, *Nature (London)* **200**, 86 (1963).
37. D. E. Beischer, *Astronautics* **7**, 24 (1967).
38. F. A. Brown, Jnr., *Nature (London)* **209**, 533 (1966).
39. *idem* and C. S. Chow, *Biol. Bull.* **144**, 437 (1973).
40. S. W. Tromp, in ref. 1, p. 207.
41. *idem* in *Progress in Human Biometeorology, Vol. 1, Part 1B,* S. W. Tromp (Ed), Swets & Zeitlinger, Amsterdam, Netherlands, 1974, p. 413.
42. Y. Kuno, *Human Perspiration,* Thomas, Springfield, Illinois, 1956.
43. in H. Hensel, H. Precht and J. Christophersen, *Temperatur und Leben*, Springer Verlag, Berlin, 1955.
44. H. Hensel and Y. Zotterman, *Acta Physiol. Scand.* **22**, 96 (1951).
45. E. Dodt and Y. Zotterman, *ibid.* **26**, 345 (1952).
46. S. W. Tromp, *Int. J. Biometeorol.* **7**, 291 (1964).
47. in N. Ungerstedt, *Acta Physiol. Scand.* **80 suppl.**, 367, 35A (1971).
48. in K. Fuxe, *Triangle (Basle)* **17**, 1 (1978).
49. in H. A. Meyling, *J. Comp. Neurol.* **99**, 495 (1953).
50. *idem*, *Acta Neuroveg. (Vienna)* Suppl. 6 (1953).
51. S. W. Tromp, in ref. 1, p. 194.
52. J. J. G. Prick, *ibid.*, p. 302.
53. R. J. Wurtman, J. Axelrod and D. E. Kelly, *The Pineal*, Academic Press, London, 1968.
54. A. C. Burton and O. G. Edholm, *Man in a Cold Environment*, Arnold, London, 1955.
55. J. Konorski, *Conditioned Reflexes and Neuron Organisation*, Cambridge University Press, London, 1948.
56. J. Z. Young, *Proc. Roy. Soc. B* **139**, 18 (1951).
57. J. C. Eccles, *The Neurophysiological Basis of Mind*, Clarendon Press, Oxford, UK, 1953.
58. E. M. Glaser, *J. Physiol. (London)* **110**, 330 (1950).
59. *idem*, *Lond. Hosp. Gaz.* **61**, Suppl. 2 (1958).
60. *idem* and G. C. Whittow, *J. Physiol. (London)* **112**, 43 (1953).
61. *idem*, *The Physiological Basis of Habituation*, Oxford University Press, London, UK, 1966.
62. *idem* in *Pharmacological Analysis of Central Nervous Action*, W. D. M. Paton (Ed), Oxford University Press, London, UK, 1962.
63. W. V. Macfarlane, *Arch. Sci. Rev.* **1**, 1 (1958).
64. *idem* in ref. 1, p. 372.
65. *idem* in ref. 41, p. 468.
66. J. Wilder, *Z. Ges. Neurol. Psychiat.* **137**, 317 (1931).
67. R. C. Davis, *J. Expl. Psychol.* **17**, 504 (1934).
68. L. P. Dugal and M. Thérien, *Canad. J. Res.* **25 E 3**, 111 (1947).
69. T. Velásquez, *J. Appl. Physiol.* **14**, 357 (1959).
70. A. Hurtado, *Milit. Med.* **117**, 272 (1955).
71. E. Valdivia, *Amer. J. Physiol.* **194**, 585 (1958).

CHAPTER 4

INTERACTION BETWEEN EXTERNAL WEATHER CONDITIONS AND MICROCLIMATES IN BUILDINGS

Most people spend at least half their lives in the artificial climates which exist within the walls of their houses, office buildings and other workplaces. Man's health depends therefore as much on the indoor microclimate as on the external weather conditions.

Despite many similarities, on the whole the climate inside buildings is different from that prevailing outside, even in the case of buildings in which no mechanical equipment is provided to modify the environment. A good example is the uncomfortable experience of persons sitting in a warm room near an outside wall of a building during cold weather. Athough the temperature of the air in the room near the wall is more than 20°C, the outside wall may be only 10°C. As a result the human body sitting near the wall will radiate towards the wall and will experience a cold stimulation. Still most people are inclined to believe that in a well-built modern house sudden changes in weather conditions outside do not affect the internal microclimate. This, however, is not true and many research workers have studied the inter-relationship between internal microclimate and external weather conditions. The following studies may demonstrate this inter-relationship:

4.1. STUDIES ON THE ELECTRICAL CONDUCTIVITY OF THE AIR

Schilling and Carson[1] carried out simultaneous measurements of the electrical conductivity of air inside and outside a building with two identical Gerdien tubes. The observations were made in a closed room on the 4th floor of a newly constructed campus building at the University of California (Los Angeles, USA). The simultaneous recordings showed the following results: the general trend of the conductivity curve was the same in the building and on the roof; both internal and external conductivity curves showed a diurnal variation; the conductivity, measured inside, was usually a little higher than that outside;

occupancy of the room had a definite effect on the large ion concentration of the room but not on its air conductivity; changes in conductivity were often observed in a room, hours in advance of approaching meteorological disturbances such as weather fronts.

4.2. STUDIES ON ELECTRIC SPACE CHARGES

During thunderstorms or rain, considerable space charges are recorded in the atmosphere. According to Mühleisen,[2] values of more than 2000 e cm^{-3} have been observed. The electric charges are usually attached to water droplets of only a few μm in diameter. Simultaneous recordings by Mühleisen in the open air and in a closed hall of a building have shown the same fluctuations in space charges inside and outside during changing weather conditions. Mühleisen[2] assumes that the high electric space charges observed in buildings during thunderstorms did not penetrate through the walls but were due to very small charged particles which penetrate into buildings through tiny cracks in doors and windows. Here again a close relationship between the inner and outer environment was demonstrated.

4.3. STUDIES ON WALL TEMPERATURE AND AIR MOVEMENT

It was mentioned earlier that, during a sudden influx of cold weather, particularly if it is accompanied by rain and great atmospheric turbulence, the inner surface temperatures of the outer walls of completely closed rooms or buildings may drop suddenly several °C. This is due to the strong cooling of the wet walls in the open air, particularly as a result of considerable air movement and evaporation of moisture on the walls. The suddenly increased radiation loss of the body of a person sitting near such a wall will be experienced physiologically by this person. In other words, the change in weather conditions outdoors, even if the air temperature in the closed room is kept very high, will be noticed by people living in such a room. Furthermore, considerable air movement is perceptible near windows and doors, even in modern buildings, which creates a chilling effect on the body.

4.4. STUDIES ON HUMIDITY

The relative humidity in heated buildings is normally lower than that outside, but the absolute humidity is usually higher. The decrease in relative humidity which results from the raised internal temperature, is offset to some extent by the moisture produced internally by the occupants themselves and by their activities like cooking, bathing, washing, etc. If the absolute humidity indoors becomes too high, severe condensation is likely to take place, particularly on the colder surfaces of the room. Such condensation may often be reduced by increasing the ventilation rate. Very low relative humidities which tend to occur in winter in heated buildings with cold outside air masses are undesirable indoors. They may cause an unpleasant dryness of the throat and contribute to the multiplication of certain viruses and bacteria (e.g. the influenza virus, Section 6.7). Very high humidities are liable to lead to dampness of clothes, etc. fungal attacks on walls and so on. It is normal in good building practice to control the relative humidity around 50%. In summer the absolute humidities indoors are usually only slightly greater than outdoors, as considerably higher ventilation rates occur then than in winter.

4.5. MEASUREMENT OF AIR EXCHANGE

Von Pettenkofer[3] pointed out more than a century ago that a regular exchange takes place between the air in an inhabited room rich in carbon dioxide, and the free atmosphere. In his view this exchange takes place through the porous walls of the rooms (wall breathing). Others are more inclined to believe in an exchange through cracks near windows and doors, etc. Hottinger,[4] Courvoisier,[5] Küster,[6] Georgii[7] and others observed, in a series of experimental studies, that with wind speeds of 6–7 m s^{-1}, the air of a room is, as a rule, completely replaced by fresh air within about one hour, a process known as autonomous air exchange. With wind speeds of 2–3 m s^{-1}, only about 40% of the room air will be replaced. This process depends mainly on three factors:

(a) the pressure difference in the air on the windward and the lee side of a building;
(b) the difference in temperature between room and outer atmosphere which creates small pressure differences;
(c) diffusion of gases or trace substances with different concentrations inside and outside buildings. For further details see Tromp.[8]

Further experiments have shown that the smaller the rooms are, the greater, usually, is total air exchange. Wooden floors permit greater exchange than linoleum covered floors, etc. Buildings along streets with heavy traffic will suffer more from minor cracking and, as a result, the processes mentioned above may be more pronounced. It is evident that the position of a building in relation to the dominating wind direction is likewise important.

The foregoing summary clearly indicates the close relationship between the internal and external environment of buidings. Most people will spend at least a few hours of the day in the external environment, and will therefore be subjected to the influence of the daily changes in the weather. But, even if a person stays at home, he always experiences physiologically the changes in the outer environment. This is a most important fact which must be kept in mind.

Finally interesting observations were made by Landsberg[9] in bedrooms. During 100 nights (September–December 1937) he studied the relationship between outside temperature and bed temperature. The former fluctuated between +17 and −14°C. As during 75% of the days the difference between the temperature of the room (with an open window) and the bed temperature

TABLE 4.1
Human comfort indices (compiled by S. W. Tromp)

Name of index	Introduced by	Definition	Factors included	Comments
Effective temperature index (ETI)	Jaglou *et al.*[13,18]	Temperature of still air saturated with water vapour in which subjects experience an equivalent sensation of warmth.	Air temperature, humidity, air velocity, clothing (no radiation).	In original ETI no radiation heat included. It can be included by substituting globe thermometer temperature for air temperature.
Radiation–convection-temperature index (RCI)	Vernon[19]	Globe temperature indicating combined effect of radiation and convection of human body.		Only to be used in closed rooms with little air movement.
Resultant temperature index (RTI)	Missenard[20]	Improved ETI	Air temperature, humidity, radiation, air velocity.	
Predicted 4-hour sweat rate index (P_4SRI)	McArdle[21]	Sweat rate resulting from a 4 h exposure to given conditions.	Metabolic level, two types of clothing (shorts and overalls), air temperature, air velocity.	
Excessive heat stress index (EHI) or discomfort index	Thom[22] and Bosen.	Temperature index to determine maximum discomfort permitted in government offices.	$THI = 0.4\,(T_d + T_w) + 15$ $THI = 0.55\,T_d + 0.2\,T_{dp} + 17.5$ $THI = T_d - (0.55 - 0.55\,RH) \times (T_d - 58)$.	T_d = dry-bulb (air) temperature in °F T_w = wet-bulb temperature in °F T_{dp} = dew-point temperature in °F.
Heat stress index (HSI)	Belding and Hatch[11].	Effect of heat stress on man exposed to certain thermal environment, heat produced by metabolism for various degrees of activity and evaporative capacity of the environment.	Metabolic level, radiation, convection, vapour pressure, air velocity.	Heat stress = $\frac{\text{Required evap.}}{\text{Max.evap.cap.}} \times 100$

TABLE 4.1—contd

Name of index	Introduced by	Definition	Factors included	Comments
Relative strain index (RSI)	Lee and Henschel[14] (incorporating equations of Burton and Edholm, 1955, of heat transfer between a clothed body and the environment).	Ratio between total heat load (sum of metabolism and heat exchange with environment by convection and radiation) and evaporative capacity of the air.	Same as HSI	
Equivalent temperature index (TEqI)	Dufton[23].	Surface temperature of an instrument (eupatheostat) corresponding with that of the human body during comfortable conditions.	$T_{Eq} = 0.522\ T_A + 0.478 T_S - 0.01474\ \sqrt{V}(100 - T_A)$.	Effective temp. + radiant heat, but humidity neglected. T_A = temp. of the air in °F (measured with a dry-bulb thermometer); T_S = med. temp. of the surroundings in °F; V = velocity of air in ft min^{-1}.
Thermal stress index (TSI)	Givoni[10,24] and Bernez-Niz[25].	Total thermal stress on the body (metabolic and environmental) and sweat rate required to maintain thermal equilibrium.	Air temperature, vapour pressure, air velocity, radiation, metabolic rate, clothing (semi-nude, light summer clothing, industrial or military overalls).	$S = [(M-W) \pm (C \pm R)]\ (1/f)$ S = required sweat rate in equivalent kcal h^{-1} M = metabolic rate, kcal h^{-1} W = metabolic rate transformed into mechanical work, kcal h^{-1} C = convective heat exchange, kcal h^{-1} R = radiant heat exchange, kcal h^{-1} f = cooling efficiency of sweating, dimensionless.

was 7.6–12.5 °C, a considerable amount of energy would be required by the thermoregulation centres of the body to keep the bed temperature at a constant level. This explains various physiological experiences of rheumatics during cold and humid nights. It also demonstrates that for many people a warm bed is not a

sufficient protection against meteorotropic influences and the use of electric blankets is highly recommended.

4.6. COMFORT INDICES

According to Givoni[10] the thermal comfort of man and his physiological responses to thermal stress depends on his metabolic heat production, the level of the environmental factors (air temperature, humidity, air velocity and the radiant temperature) and the type of clothing he is wearing. It has been recognized that none of these factors can be used as a single criterion for the evaluation of the thermal stress operating on the body or of its expected responses to the stress. Not only do all these factors affect the body simultaneously, but the effect of any of them depends on the level of the others.

In order to estimate the combined effect of the various physical stimuli of the meteorological environment many research workers[10-15] have tried to introduce formulae which could predict in a mathematical form the thermal stresses and physiological responses of the human body in a given environment. Such formulae are known as 'biometeorological indices' and they are divided into two groups, human comfort indices and miscellaneous biometeorological indices.

4.6.1. Human comfort indices

In Tables 4.1 and 4.2 the nine most important comfort indices and their reliability have been summarized. For further details the reader is referred to standard texts.[16,17]

TABLE 4.2
Reliability of human comfort indices

ETI:	least reliable in predicting the expected physiological and sensory responses, both under comfortable conditions and under heat stress.
RTI:	satisfactory in predicting responses of people at rest or engaged in sedentary activity.
P_4SRI:	satisfactory, under light to heavy heat stress conditions, in predicting the sweat rate response. It is not suitable around the comfort zone and under cold exposure.
HSI:	suitable for analysing the relative contribution of the various factors resulting in thermal stress, but not for predicting quantitative physiological responses to the stress.
RSI:	seems suitable for analysing the relative contribution of the various environmental factors to the thermal stress of sedentary subjects under different clothing conditions. The available information does not enable evaluation of the RSI with respect to working people, or for prediction of the quantitative physiological response to heat stress.
TSI:	suitable for analysing the individual contributions of metabolic and environmental factors and for prediction of the sweat rate expected in the case of resting or working people. It is reliable in the range of conditions between comfort and severe stress, providing thermal equilibrium can be maintained (stabilized rectal temperature and pulse rate). Beyond this limit the index does not apply.

TABLE 4.3
Miscellaneous biometeorological indices

(a) *Cooling or windchill indices* of Bedford[26]-Siple[27]-Court[28]-Thomas-Boyd[29]. (This formula applies to dry, naked skin in shadow)

$H = \Delta t \times (9.0 + 10.9 \sqrt{V} - V)$ where

- H = cooling or heat loss in kcal m^{-2} h^{-1}
- Δt = difference in temperature between subject and environment in °C
- V = windspeed in m s^{-1}

In Polar areas the formula of Wilson[30,34] has been used.

$H = (\sqrt{100\,V} + 10.45 - V)(33 - t)$ where

- H = heat loss in kcal m^{-2} min^{-1}
- V = windspeed in m s^{-1}
- t = air temperature in °C.

$H = 50$ indicates excessive heat, 2500 enormous cold

(b) *Total temperature index of Boerema*[32]

According to Boerema the total temperature curves, based on evening and morning temperatures of 200 patients in surgical and internal clinics and also the curves of their pulse frequencies, show similar daily fluctuations in independent clinics during the period middle May to middle November.

Post operative complications (e.g. post operative pneumonia and thrombo-embolism) are most likely to occur, if a high peak in the temperature curve is followed by a deep fall.

(c) *Vegetative or autonomic nervous system index of Kerdö*[33]

V.I. $= (1 - \frac{d}{p})\,100$ in which

- d = diastolic pressure in mm
- p = pulse rate per min

During the so-called 'ergotropic phase' (i.e. domination of orthosympathetic processes, increased metabolism, increased oxygen consumption, increased bloodflow and acidosis) d is lower, $\frac{d}{p}$ is less than 1, and V.I. is positive.

In the 'trophotropic phase' (domination of parasympathetic stimulation) $\frac{d}{p}$ is greater than 1, and V.I. is negative. V.I. fluctuates between 0 ± 15 (between 8 and 10 a.m.).

(d) *Pulse/breathing index of Hildebrandt*[34]

During perfect autonomic balance $\frac{\text{Pulse}}{\text{breath}} = 4$. Values greater than 4 indicate labile autonomic conditions. Lower values suggest very high stability.

Subjects with high values experience early and strong reactions to balneo- or thalasso-therapy.

(e) *Climatic-tourist indices of Rivolier*[35]

Monthly index $I = \frac{S + T - 5D}{5}$ where

- S = number of hours of solar insolation during one month
- T = average monthly temperature in °C
- D = number of hours of rainfall during one month

Weekly index $= \frac{30s + 7t - 150d}{35}$ where

- s = number of hours of solar insolation during one week
- t = average weekly temperature in °C
- d = number of hours of rainfall during one week.

TABLE 4.3—contd

(f) *Aggressive biometeorological index of Rivolier*[35]

Daily index ABI $= K_1 + \Delta T + V + K_2 + K_3 - 2I$ in which

K_1 = humidity coefficient = (rel. humidity above 50% + mm rainfall + duration of rainfall in hours) divided by 5

ΔT = number of °C below 12°C and above 25°C

V = wind velocity in m s^{-1}

K_2 = each sudden pressure change is counted as 5. The total represents the K_2 coefficient

K_3 = each abnormal meteorological phenomenon (fog, thunderstorm etc.) is counted as 5 and totalled

I = number of hours of daily insolation

4.6.2. Miscellaneous biometeorological indices

In Table 4.3 six indices are summarized, which have been proposed to describe certain physiological conditions as a result of meteorological stimuli.

REFERENCES

1. G. F. Schilling and J. Carson, *Arch. Meteor, Geophys. Bioklim.* **5**, 191 (1954).
2. R. Mühleisen, *Int. J. Bioclimatol. Biometeor* **2**, 1D (1958).
3. M. von Pettenkofer, *Uber den Luftwechsel in Wohngebäuden*, Cotta Verlag, Munich, 1858.
4. M. Hottinger, *Lüftungs und Klimaanlagen*, Springer Verlag, Berlin, 1940.
5. P. Courvoisier, *Arch. Meteor, Geophys. Bioklim. Ser. B* **2**, 162 (1951).
6. E. Küster, *Z. Hyg.* **110**, 324 (1928).
7. H. W. Georgii, *Arch. Meteor. Geophys. Bioklim.* **5**, 191 (1954).
8. S. W. Tromp, *Medical Biometeorology*, Elsevier, Amsterdam, 1963, pp. 84–85.
9. H. E. Landsberg, *Arch. Meteor. Geophys. Bioklim.* **12B**, 159 (1938).
10. B. Givoni, *Estimation of the Effect of Climate on Man: Development of a new Thermal Index*, UNESCO Res. Rep., U.S. Dept. Hlth. Educ. Welf., Haifa, 1963.
11. H. S. Belding and T. F. Hatch, *Heating Piping Air Condit.* **27**, 129 (1955).
12. P. O. Fanger, *Thermal Comfort*, Dan. Techn. Press, Copenhagen, 1970.
13. F. C. Houghten and C. P. Jaglou, *ASHVE Trans.* **29**, 163 (1923).
14. P. H. K. Lee and A. Henschel, *Evaluation of Thermal Environment in Shelters*, Rep. U.S. Dept. Hlth. Educ. Welf., P.H.S., Tr-8, 1963.
15. W. H. Terjung, *Geograph. Rev.* **60**, 31 (1970).
16. Ref. 8, pp. 40–45.
17. S. W. Tromp (Ed), *Progress in Human Biometeorology, Part 1A*, Swets & Zeitlinger, Amsterdam, 1974, pp. 127–193.
18. C. P. Jaglou and W. E. Miller, *ASHVE Trans.* **31**, 89 (1925).
19. H. M. Vernon, *J. Physiol.* **70**, 15 (1930).
20. H. Missenard, *Chaleur et Industrie*, 1948.
21. McArdle *et al.*, *Mem. Med. Res. Counc. (London)*, Rep. RNP 47/391, 1947.
22. E. C. Thom, Air Condit. Heat. Ventil., June, 1957.
23. A. F. Dufton, *Bldg. Res. Techn. Paper 13*, 1932.
24. B. Givoni, *Man, Climate and Architecture*, Elsevier, Amsterdam, 1969.
25. Bernez-Niz in ref. 10.
26. T. Bedford, *Basic Principles of Ventilation and Heating*, Lewis, London, 1948.
27. P. A. Siple, in Proc. Amer. Phil. Soc. **89**, 177 (1945).

28. A. Court, *Bull. Amer. Meteor. Soc.* **29**, 487 (1948).
29. M. K. Thomas and D. W. Boyd, *Canad. Geograph.* **10**, 29 (1957).
30. O. Wilson, *Metabolism,* **5**, 543 (1962).
31. *ibid.* in *Biometeorology I,* Pergamon, Oxford, 1962, p. 411.
32. I. Boerema, *Mededelingen Chirug. Klin. v. Prof. Gerland*, Groningen, Netherlands, No. 2, 1941.
33. L. Kerdö, in *Biometeorology II*, Pergamon, Oxford, 1967, p. 147.
34. G. Hildebrandt, *Arch. Phys. Ther. (Leipzig),* **14**, 39 (1962).
35. J. Rivolier, *Int. J. Biometeorol.* **19**, 12 (1975).

CHAPTER 5

EFFECTS OF METEOROLOGICAL STIMULI ON BASIC PHYSIOLOGICAL PROCESSES IN HUMANS

5.1. EFFECTS OF THERMAL STIMULI

5.1.1. General effects of heat

General effects of heat were briefly discussed in parts of Sections 3.1, 3.3 and 3.5. During exposure to heat, increased sweating rates increase the losses of components normally found in sweat. Some of these losses are quite considerable, as in the case of sodium, potassium, amino acids and nitrogen. Although the concentrations of substances in sweat decrease with acclimatization, the increased rate of hypotonic sweat after acclimatization is such that the total amounts lost remain relatively unchanged.

Various factors influence the response to heat:

(a) *Ethnic factors.* It has been shown that differences between ethnic groups such as Caucasians and Bantu are not very great; much of the reported differences can be attributed to the fact that living and working in a hot climate causes a certain measure of acclimatization.

(b) *Sex.* Studies on sex differences in response to heat and heat acclimatization are few and show large disagreements and uncertainties. The only apparently consistent findings are that women sweat less than males, that their sweating response is suppressed more readily than that of males and that they respond to heat with a higher body temperature. Some of the reasons for these differences are due to the fact that women probably prefer to avoid sweating more than men, that normally they wear lighter clothing and take less part in athletic activities demanding a sustained high level of energy expenditure. Physiologically they have a lower basal metabolic rate and can therefore more readily maintain thermoequilibrium by convective and radiant exchanges. Anatomically women usually have a thicker layer of subcutaneous fat than men. This has an important advantage in providing a much bigger heat sink in the peripheral shell.

(c) *Age*. The decrease in thermoregulatory efficiency and other geriatric changes in respect to heat are discussed in Section 6.13.4, p. 192.
(d) *Training*. Highly trained long-distance runners appear to be well acclimatized to work in the heat without previous exposure, their physical conditioning improves their capacity to acclimatize to a severe heat stress. Although physical training may improve performance under conditions of heat, it certainly cannot replace acclimatization.
(e) *Hydration*. During work in the heat a progressive dehydration takes place as a result of sweat production exceeding voluntary water replacement (see p. 229). This dehydration produces impairment of performance which is reflected in specific physiological functions. Prevention of this dehydration by prior or concurrent replacement of anticipated sweat loss will reduce these impairments. Over-hydration prior to work in the heat also significantly reduces the indices of physiological strain. Greenleaf[1] has found that the volume of water consumed is proportional to the total severity of experimental conditions and that the rate of water consumption following exposure is independent of the level of hypohydration. He concludes that thirst is associated more with the severity of the total stress than with the degree of water deficiency. In another study it was shown that acclimatized men drink more than unacclimatized men, in response to an elevated serum osmotic pressure caused by the secretion of larger volumes of hypotonic sweat after acclimatization.

5.1.2. General effects of cold

The maintenance of body temperature during cold stress can be achieved either by increasing heat production or by reducing heat loss from the body.

When the body tends towards a negative heat balance, at air temperatures below 20 °C (clothed, at rest) or 28 °C (naked), compensatory responses occur. A non-acclimatized nude subject responds to a lower temperature by four principal mechanisms: peripheral vasoconstriction, piloerection (only important in arctic animals), shivering, and non-shivering thermogenesis. Skin temperature is recorded by the thermoreceptors, the Krause End Bulbs, in the outer-most layer of the skin (p. 65); these discharge as the temperature falls. Nerve endings in the skin cause a constriction of the blood vessels in the skin of the toes, feet, hands, nose and ears roughly in that order. The blood pressure in the veins rises, and venous volume decreases gradually during acclimatization to cold, less total vasoconstriction occurs and more arterio-venous anastomoses (i.e. connections between two tubular organs) remain open to allow the extremities to feel warmer.

Vasoconstriction increases the insulation offered by the thermal shell of the body by only a small amount. As this amount is small as compared with the rate of heat loss, other responses are induced. The most important of these is increased heat production (primarily from fats) by one of the following mechanisms:

(a) *Increased muscle action (tone or shivering).* The skin temperature and, to a lesser extent, the heat content of the body and blood reaching the hypothalamus determine the onset of shivering. Europeans can become accustomed to sleeping in the cold, while shivering and thus producing 50% more heat than in the normal state. In some Australian Aboriginals the skin temperature may fall as much as in Europeans while sleeping without shivering and without increasing oxygen consumption. Much of this response appears to be habituation to low skin temperatures with reduced reflex shivering.

(b) *Non-shivering thermogenesis.* This process results from increased liver metabolism, combined with thyroid and adrenal hormone action. Initially there is an increased output and turnover of thyroxin and adrenal cortical hormones. After 3–4 weeks, in rats, the required amount of these adrenal hormones is diminished.

(c) *Voluntary exercise.* This is the third mechanism to increase heat production through increased muscle metabolism.

The most striking features of even a moderate cold stress are the increased urinary and 17-ketosteroid output, the increased diastolic blood pressure, the drop in the erythrocyte sedimentation rate and fibrinogen content of the blood, and the rise in haemoglobin level of the blood (see Table 3.1, p. 55).

5.1.3. Thermal effects on the autonomic nervous system

Differences in the reactivity of the orthosympathetic and parasympathetic nervous systems have been studied in relation to weather stress by Becker, *et al.*[2] Pronounced differences in excitability of both nervous systems were found as revealed by electrophoretic adrenalin and acetylcholine skin tests of the thorax of the same person under different weather conditions. There were differences in the size of the vesicles on the skin produced by adrenalin and in redness of the skin produced by acetylcholine. The colour changes in the skin were strongest during the passage of fronts and after sudden changes in air mass; the more the weather changed, the stronger was the colour reaction. The orthosympathetic reaction was usually stronger 3–5 h after the passage of the front. The parasympathetic reaction often occurred 5–6 h before the passage of the weather front or air-mass influx. Increased stimulation lasted in several cases for 16–30 h after the passage of the front. It was also found that, in persons with a particularly unstable sympathetic nervous system, the adrenalin reaction may precede the acetylcholine colouring reaction.

5.1.4. Thermal effects on endocrinal functions

(a) *Pituitary.* In view of the neural and neurohumoral connections between the pituitary and the hypothalamus (see Section 3.3.2), there is an intimate inter-relationship between the meteorological environment and the functioning of the pituitary and of other endocrine glands. The effects observed in the pituitary are due to both thermal and radiation stresses. They explain the increase in growth hormone production in children in spring; of gonadotrophic hormones affecting the seasonal fertility of man and animals; thyrotrophin production of the anterior lobe of the pituitary fluctuates due to cold, heat, and/or light stress and affects the secretion of thyroxine by the thyroid and the general activity of this gland.

Considerable daily fluctuations in urinary output (diuresis) are observed despite the same fluid consumption. Diuresis is high during the influx of cold polar air masses or other cooling effects in the atmosphere. This seems to be due, at least partly, to fluctuations in the liberation of the antidiuretic hormone in the posterior lobe of the pituitary.

(b) *Thyroid gland.* Cold produces intensive activity of the thyroid gland, which is manifested by intense congestion of the capillaries and the disappearance of the colloid from the intra-aveoli (see Figs. 5.1, 5.2). It may lead to hyperthyroidism. Heat stress induces inactivity of the gland, with colloid accumulation in the alveoli etc. All these processes seriously affect the metabolic processes in the body.

(c) *Parathyroid gland.* Cold stress creates a parathyroid deficiency and fall in the calcium level of the blood serum. The parathyroid seems also to control the magnesium level. With temperatures below 5.5 °C, scarce sunshine, high rainfall and strong winds, a magnesium deficiency is created in grass causing 'grass-tetany' disease in cows.

(d) *Adrenal gland.* Its activity is controlled by adrenotrophin (ACTH), a hormone secreted by the pituitary. Weather changes affect the adrenal gland through the hypothalamic-pituitary tract. Increased gland activity decreases the ascorbic acid and cholesterol level of the gland. Cold stress causes a hypertrophy of the adrenal cortex which is inhibited if large quantities of vitamin C are injected. Both the influx of a cold weather front and of cold polar air increases the 17-ketosteroid (an important adrenal hormone) excretion in 24 h urinary samples by up to 100%. After a series of fronts, following one after another, the effects gets smaller and smaller. In winter the average excretion is higher than in summer. As both the adrenal cortex and the pituitary are closely connected with the control of the capillary resistance and the permeability of cell membranes it is not surprising that cold stress increases the capillary resistance.

(e) *Thymus gland.* As meteorological stimuli affect the pituitary and the production of adrenalin and corticosteroids by the adrenal gland, and as each of these processes influences the lymphocyte production and other activities of the thymus gland (e.g. the thymus plays a major role in a number of auto-immune diseases in man and animals and is responsible for the rejection of genetically non-identical tissue), it can be expected that the various processes controlled by the thymus are affected by weather and climate.

(f) *Pancreas.* Parasympathetic stimulation of the pancreas, and of the

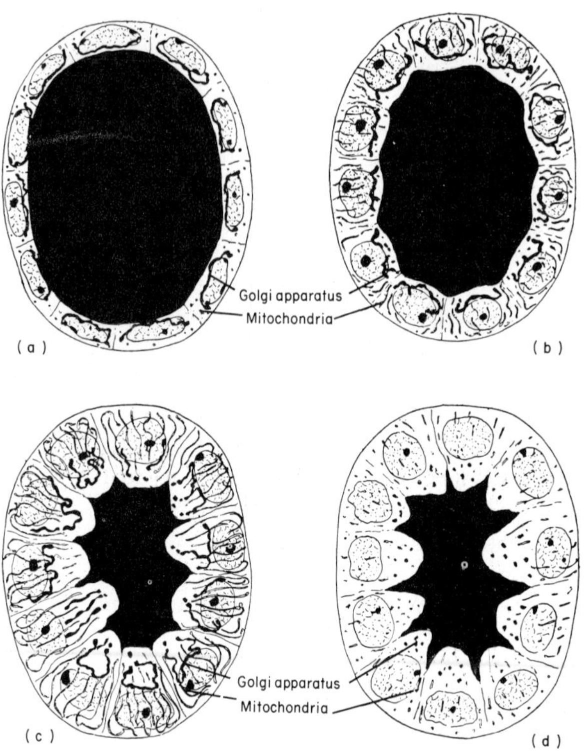

Fig. 5.1 Influence of heat and cold stress on the thyroid gland of the mouse (after: W. Cramer and R. J. Ludford: On cellular activity and cellular structure as studied by the thyroid gland, *J. Physiol.*, 61, 398–408, 1926) (a) resting, thyroid fed, exposure to heat; (b) moderate degree of activity; fasting; (c) extreme degree of activity, exposure to cold, action of β-tetrahydronaphthylamine; (d) after extreme degree of activity, following exposure to cold or action of β-tetrahydronaphthylamine

'Islands of Langerhans', is followed by a reduction of the glucose level due to increased insulin production; however, if the level falls too much then orthosympathetic stimulation with adrenalin secretion will cause an outpouring of glucose from the liver. It is evident that all meteorological stimuli causing a dysfunction of the hypothalamic-pituitary system will affect the pancreas and the various processes controlled by this organ. Around 1960 it was demonstrated in Munich[3] that during the influx of humid, warm air masses the blood sugar level of unfed animals was lowered to a greater extent than during cold-humid or cold-dry weather conditions (see also Section 6.13.2).

5.1.5. Thermal effects on liver, spleen and gall bladder

(a) *Liver.* Cold stress increases the conversion of glycogen in the liver, the formation of liver enzymes (SGOT, SGPT and SLDH), the respiration of liver cells and also changes the liver acetate metabolism.

(b) *Spleen.* Meteorological stimuli affecting the orthosympathetic nervous system (e.g. cooling affecting the adrenal gland and adrenalin excretion) may increase the sudden output of erythrocytes, leucocytes and thrombocytes. This explains the short term changes in the haemoglobin content of the blood after abrupt weather changes.

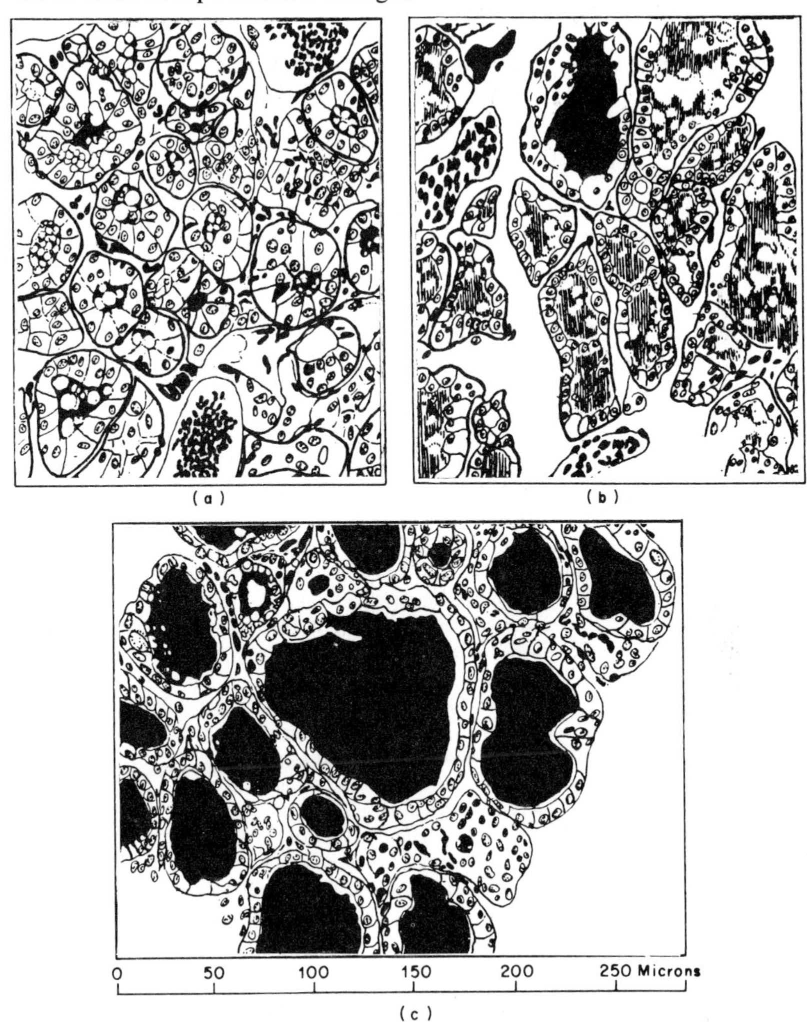

Fig. 5.2 Influence of cold (following heat) and heat (following cold) on the thyroid of the mouse (after: W. Cramer and R. J. Ludford: see Fig. 5.1) (a) the thyroid was kept at ordinary room temperature; (b) mice were kept in glass jars placed in an incubator at 37 °C at 10 am and removed at 5 pm. From 5 pm till 5 am they were kept in a room with open window, during cool weather, in March (immediate effect of cold, following heat); (c) after this routine took place for several days, the mice after being back in the warm incubator for 6 h were killed and the thyroid was studied (immediate effect of heat following cold)

(c) *Gall bladder.* Disorder of the metabolism may cause high cholesterol excretion in bile (e.g. during thyroid deficiency), together with diminishing bile salt secretion causing cholesterol precipitation and the formation of cholestrol gall-stones. Sudden strong stimulation of the orthosympathetic nervous system may cause gall-stone attacks.

5.1.6. Thermal effects on various components, and physico-chemical properties, of the blood

The influence of thermal stresses on the physico-chemical state of the blood is particularly evident not only in the case of the blood sedimentation rate (BSR), the haemoglobin (Hb), albumin, γ-globulin and fibrinogen levels of the blood and the blood-clotting time, but also in other components of the human blood (see Table 5.1).

TABLE 5.1
The effect of meteorological stimuli on basic physiological processes in man

Physiological parameter	Observed Biometeorological Relationships[6]
Haemoglobin	After heat stress, increased plasma volume, decreased Hb; after cold stress, increased Hb. Hb lower in summer than in winter. Minimum around June.
Leucocytes	High in winter, maximum around December; minimum in August.
Thrombocytes	In period 1958–1962 in The Netherlands: maximum March–April; minimum in August.
Eosinophils	Increase from winter to spring. In 1958 in The Netherlands, low from May–September (minimum July–August); high from November–April (maximum in March).
Prothrombin	Minimum in children in Hungary between September–December. In adults minimum in winter and spring. Fluctuation in prothrombin index with air mass and barometric pressure changes.
Fibrinogen	Low after atmospheric cooling; usually lower in winter than in summer.
Serum proteins	Total serum protein usually decreases from winter to summer from 8.5–7.5 g/100 ml serum. Albumin level high in summer low in winter; γ-globulin usually high in winter, low in summer.
Fibrinolytic properties of serum	Blood-clotting time very low shortly before passage of cold front. After cold front passage strong fibrinolysis. Fall in fibrinogen after strong cooling.
Blood volume	Increases with heat stress, decreases with cold (cold fronts, polar air). Lower in winter than in summer.
Packed red cell volume	Maximum: February–March; minimum: July and August in Japan.
Specific-gravity red blood cells	In Japan usually rising in summer.
Oxygen capacity of blood	Increases from January–May (in children under 6 years); from February–August (in older children).
Carbon dioxide capacity of blood	Maximum absorption of blood around 21 December; minimum around 21 June.
Serum calcium	Minimum in February–March (8.5 mg/100 ml); maximum in August (11 mg/100 ml).

TABLE 5.1—contd

Physiological parameter	Observed Biometeorological Relationships
Serum phosphate	Minimum in February; maximum in summer and autumn (average level 3–5 mg/100 ml), due to increased vitamin D synthesis in the skin by UV light.
Serum magnesium	In Japan, minimum in February (2.12 mg); maximum in December (2.85 mg).
Serum iodine	Minimum in winter (December–April) (8.35 mg%); maximum in summer (July–August) (12.85 mg%).
Serum copper	In cows in Holland increasing from 76 μg% in September to 104 μg% in May: decreasing again in summer.
Serum ascorbic acid	Low in winter, high in summer. Rats living under rhythmical light-dark changes have maximum values during the light periods and minimum values during darkness; ascorbic acid secretion always precedes the secretion of corticosteroids.
Erythrocyte sedimentation rate (ESR)	Low in winter, higher in summer. Short-term fluctuations: low ESR values after influx of cold air or general cooling.
Blood coagulation	Blood clotting time particularly low shortly before cold front passage. Clotting time increases after front passage.
Bleeding after treatment with anticoagulants	Maximum in January–February; minimum in July.
17-ketosteroid excretion in urine	Increase after cold stress, e.g. cold fronts, polar air masses; decrease with increasing temperature. During very strong heat stress, also increasing. Corticosteroid secretion of mice reaches maximum values 2 h before the beginning of the dark period, decreasing during the following dark period. Corticosteroid secretion preceded by ascorbic acid secretion.
Diuresis	Decreasing after heat stress; increasing during fall in temperature (cold front, polar air influxes).
Metabolism	In children maximum during autumn. General metabolism decreasing sharply during winter. In diabetics, during late spring, summer and early autumn, higher metabolism. In winter low metabolism, high blood-sugar content, high insulin consumption.
Diastolic blood pressure	High in winter months (particularly February); low in summer.
Thyroid function	Increased thyrotrophin production and general increase in activity after cold stress followed by temporary hyperthyroidism. After continued cold stress, effect becomes less.
Permeability of tissue	After cold fronts, decreasing; after warm fronts or föhn[a] winds, increasing.
Capillary resistance	Increasing after cold fronts, decreasing after warm fronts.
Skin capillary test	Capillary structure of the skin changes rapidly in spring; excitability of peripheral nerves in skin increases as shown by dermographic tests.
Adrenaline and acetylcholine skin tests	Acetylcholine test shows the influence of cold front passages (8 h and more) earlier than adrenalin tests.
Direct current resistance of skin	Very high during falling barometric pressure and föhn.
Muscle metabolism	Phosphoric acid and glycogen content of muscles of rabbits high with cold fronts, low with föhn winds and warm fronts.
Muscle strength	Changing during different weather conditions.

[a]For the definition of the föhn see p. 190.

TABLE 5.1—contd

Physiological parameter	Observed Biometeorological Relationships
Gastric acidity	Hyperacidity high in winter, low in summer, anacidity high in summer.
Growth and weight of children	Slow growth in winter, rapidly increasing in spring; height increase maximum in March–May and November–January in Stockholm. Weight increase in Stockholm maximum in September–November.
Birth weight	Greatest in June–July; smallest December–March.
Birth frequency	Highest number of conceptions in June (legitimate children) or May (illegitimate); stillborn children maximal in January, neonatal deaths maximum in February.
Mortality	In western Europe maximum in December–January; minimum in July.
Acclimatization	More difficult with respect to cold stress in summer than in winter; acclimatization to heat-stress is more difficult in winter.

(a) *Effects on blood sedimentation rate.** Analysis of the blood sedimentation rate (BSR) data per donor day (i.e. a day on which many blood donors visit a donor centre) has shown[4,5] that in western Europe the 1 and 2 mm values together (determined with the Westergren method), but in particular the 2 mm values after the first hour, are by far the most common blood sedimentation rates recorded in healthy male donors. However, it was observed that the percentages of donors with 1 to 2 mm BSR values varied considerably from one donor day to another. The observed BSR data showed short and long term changes. The correlation with weather changes revealed some striking relationships in the case of short-term changes.

(i) Short term changes: very low BSR values (i.e. high percentages of donors, often over 80%, with BSR 1 to 2 mm/1st hour) were observed, when the donor days were preceded by one or more days with low or strongly falling temperature and increasing atmospheric turbulence, in other words during strong cooling conditions.

On the other hand very high BSR values (i.e. very low percentages of donors, often less than 40%, with BSR 1 to 2 mm/1st hour) occurred if the donor day was preceded by one or more warm days with little wind.

(ii) Long term changes: both changes during the year (seasonal) and changes from year to year have been observed:

(a) Highest BSR values 1 to 2 mm/1st hour occur during summer, lowest values in winter or early spring;

(b) Yearly averages show irregular, long term changes, both in the northern and southern hemisphere. The BSR values decrease

*Blood is collected and a special solution is added to prevent coagulation. The mixture is sucked up in a 300 mm long pipette. After one hour the number of mm that the erythrocyte column has fallen is registered. This represents the BSR in mm. In healthy males it is usually below 5 mm/1st hour, in healthy females 5–10 mm, during most diseases over 15 mm/1st hour. Therefore BSR is a good indication of the general health of a person. Instead of BSR, ESR (erythrocyte sedimentation rate) is also used.

from the equator to the northern countries and with increasing altitude. The changes in BSR pattern with latitude and altitude may be partly due to differences in the infection rate of the population, but for a greater part it seems to be due to the average lower temperatures and decreased solar radiation (infrared radiation accelerates BSR, ultraviolet reduces the precipitation speed). These very long term BSR changes (both in the north and south hemisphere) cannot be correlated with any known meteorological parameter or other terrestrial factors and it is therefore not excluded that extra-terrestrial factors may be involved.

(b) *Effects on the haemoglobin (Hb) content of the blood.** The Hb percentages of healthy male blood donors in Western Europe show both short and long term changes.[7]

(i) Short term changes: if a donor day is preceded by one or more days with very cold weather (particularly during falling temperature accompanied by rising wind speed) very high Hb values are observed. With warm weather preceding the donor days the average Hb values are low. If a cold period is immediately preceded by a very warm period it usually takes a few days before the increase in Hb, as a result of the cooling, can be observed.

(ii) Long term changes: the seasonal fluctuations of the average monthly percentages of donors with high Hb values show, in western Europe, that the average winter and autumn Hb values are higher than in summer or spring. The short term Hb changes may be due to an increased adrenal activity during cooling which may affect the spleen or other blood producing centres in such a way that a rapid expulsion of blood cells into the blood stream occurs. It is also possible that rapid changes in blood volume must be considered in order to explain these short term changes.

The seasonal changes are probably related to changes in blood volume. The known increase in blood volume during summer, due to a rise in the environmental temperature, causes a relative decrease in the average Hb content of the blood. In winter the reverse can be observed. Similar observations were made during long term climatic chamber studies.

The yearly changes are more difficult to explain. It is known that a high protein intake combined with sufficient gastric acid and bile secretion increases the haemoglobin level of blood. Undoubtedly for the first five or 10 years after the war one could have expected in western Europe such a dietary effect. However, it seems difficult to explain the periods of decrease and increase since 1965 on the basis of

*Haemoglobin is the oxygen-carrying pigment of human blood. A considerable shortage of the haemoglobin in blood causes an anaemia.

dietary factors alone. Perhaps the long term weather changes during the last five or 10 years are responsible.

(c) *Effects on the albumin* and γ-globulin* level of the blood.* Periods of high, or low average albumin values of the serum of blood donors are not closely related to season, periods of infectious diseases or changes in food conditions during the year. A statistical analysis of the albumin and γ-globulin values collected at Leiden, The Netherlands,[8] shows a clearly inverse relationship. Long or short term periods with high albumin values are accompanied by similar periods of low γ-globulin values and vice versa. Periods with low albumin values are related to cold periods, high albumin values to periods with temperatures above normal. Periods of low γ-globulin levels (and of probably related anti-bodies), which occur after a stay in a warm environment, may coincide with reduced general resistance of the population against infections, and may explain why after holidays in warm sunny climates people often suffer more from infectious troubles after they return to their colder climate at home.

(d) *Effects on the fibrinogen content of the blood.* There is a clear relationship between BSR and the fibrinogen content of the blood (fibrinogen is produced by the liver). With rising fibrinogen values the BSR rises and the diastolic blood pressure of donors also shows a similar tendency. The fibrinogen content does not show a relationship with the haemoglobin level. However, the fibrinogen content increases with increased age of the subjects. It has also been observed that very low fibrinogen values are found during days with considerable atmospheric cooling, whereas during warm weather conditions high fibrinogen values are common.

(e) *Effects on blood clotting time.* In western Europe the blood clotting time is shortened during warm periods (i.e. periods with high BSR and fibrinogen values) and lengthened during cold periods. However, also during relatively warm winters a shortening of blood clotting time can be expected. These meteorotropic changes in fibrinogen and blood clotting time may have considerable effects in the case of heart diseases.

5.1.7. Thermal effects on the electrolyte balance of urine

Under conditions of constant fluid intake the urinary output is closely related to the daily and seasonal cooling conditions in the atmosphere and to the microclimate in houses and buildings. Even a small cooling causes, in well-thermoregulated subjects, a rise in urinary output within a few hours. In subjects with an inefficient thermoregulation a delay of 12 h or more may occur.[9]

Chloride, urea and sodium excretion decreases during cooling. Potassium excretion does not show a clear weather relationship because the excretion depends considerably on the type of ingested food.

*Simple proteins produced by the liver; γ-globulins are concerned with the phenomenon of immunity.

5.1.8. Thermal effects on blood pressure

Both short and long term changes have been observed in healthy male blood donors.[10] In a large constant group of male donors in western Europe, of 20–40 years of age, the majority of donors has a diastolic blood pressure (DBP) of 80 mm Hg. The percentage of donors with DBP above 80 mm (in a constant group of males) increases suddenly if the donor day is preceded by one or two days of strong atmospheric cooling (due to falling temperature and/or rising wind-speed). During warm periods the DBP falls, and the percentage of donors with DBP above 80 mm decreases. In western Europe during the usually coldest months of the year (January–February) a steep rise in the percentage of donors with a DBP over 80 mm is observed. Studies in Bavaria[11] with hypertensive subjects aged 29–55 indicate an initial rise in DBP during the first six days of their stay at 2000 m altitude. The DBP normalizes during the second week. After the subjects returned to sea-level, after a stay of four weeks at high altitude, their DBP values fell below their initial values. This condition lasted for at least four months.

5.2. EFFECTS OF SOLAR RADIATION*

The biological effects of solar radiation depend on its wavelength (λ). For practical purposes the wavelength range is divided into three main regions: infrared, visible light and ultraviolet.

The infrared region is further divided into 'far' or non-penetrating infrared (λ over 3 μm) and the penetrating infrared (λ below 3 μm) is subdivided into intermediate infrared ($\lambda = 1.5$–3 μm) and near infrared ($\lambda = 0.75$–1.5 μm).

The skin temperature is most elevated by the penetrating infrared because part of the visible and non-penetrating rays are reflected by the skin. Maximal penetration in the skin takes place with wavelength of 0.75 μm (750 nm) 99% of this radiation is absorbed after penetrating a few millimetres.

Radiation of wavelength less than 320 nm does not penetrate more than 0.1 mm. The visible light comprises wavelengths between 380 and 750 nm. It is this part of the solar spectrum on which our visual concept of the environment and plant life depends.

The ultraviolet is divided into ultraviolet A (315–380 nm), ultraviolet B or Dornoradiation (290–315 nm) and ultraviolet C (below 290 nm).

5.2.1. Biological effects of infrared radiation

(a) *Effects on the human skin.* Infrared radiation produces an erythema (reddening) of the skin. However, in contrast to ultraviolet erythema, infrared or heat erythema becomes manifest immediately after exposure and disappears within a few minutes of irradiation. Usually, it is not followed by pigmentation, but repeated exposures to infrared may cause a heat pigmentation, known as *cutis marmorata*.

*The wavelength of solar radiation can be expressed in Ångstrom units (1 Å = 10^{-8} cm) or mμ (1 mμ = 10Å) or nm (1 nm = 1 mμ).

The effects of infrared on the skin are not different from those produced by heat conduction. During slightly cloudy hazy weather humid air causes a strong scattering of sunlight. The dispersed and reflected rays have wavelengths in the infrared part of the spectrum and penetrate more deeply into the skin. This explains the well-known 'burning' feeling of sunlight on such humid days.

(b) *Effects on the eye.* Most of the infrared radiation is absorbed by the conjunctiva (the membrane lining the eyelids and covering the eyeballs) and cornea (the transparent anterior part of the eye); therefore these are the parts primarily affected. Irradiation produces a painful inflammation of the conjunctiva and cornea, clinically designated as photo-opthalmia. The changes are similar to those produced by ultraviolet radiation but occur more rapidly.

The formation of cataracts (opacity of the lens of the eye) by chronic exposure to infrared radiation is established beyond doubt. Since infrared radiation of 1 μm wavelength penetrates to the retina, the occurrence of retinal lesions is not surprising. They may be produced by careless observation of the sun during an eclipse with the eye not protected by darkened glasses.

5.2.2. Biological effects of visible light

Both psychological and physiological effects have been reported. Very little is known about psychological and physiological effects of spectral colours of visible light on the human and mammalian body. The human skin seems to possess a rudimentary sensory organ for spectral colours. An indirect effect of red light perceived by the eye and mediated via the pituitary gland to the gonad has been demonstrated experimentally in ferrets.[12] The emotional effect of spectral colours on man was first pointed out by Goethe, in his treatise *Farbenlehre.* For modern public buildings (hospitals, office rooms, etc.) the emotional effects of spectral colours are often used to good advantage. Apart from psychological effects, Benoit, Assenmacher and others in France,[12,13] Miline in Yugoslavia[14-16] and Hollwich in Germany[17,18] were able to demonstrate a great number of important physiological effects of light, both in man and animals (see pp. 267, 268).

Benoit[12] and Assenmacher[13] demonstrated that stimulation of the optic nerve by visible sunlight (380–780 nm) affects the secretion of gonadrotrophic hormones. The light impulses reach the visual centre of the cerebral cortex which, in its turn, probably stimulates the supraoptic area of the anterior nuclei of the hypothalamus through the thalamus (see p. 59). As a result of neurohumoral stimulation, the anterior lobe of the pituitary is affected. In ducks it was found that 15 h of light stimulation per day, at temperatures of 8–15 °C, increased the size of the testicles within 2 d. Removal of the anterior lobe of the pituitary prevents this phenomenon. Benoit[12] also found that direct irradiation of the skull (even after removal of the eyes and eye nerves) with wavelengths above 600 nm (orange-red and infrared) stimulates the hypothalamus in various mammals (e.g. rats, rabbits, and monkeys). It is thus evident that the seasonal variation in sunlight, in fact even long periods of heavy fog, must affect the functioning of the pituitary and the production of gonadotrophic hormones.

Further experiments with monochromatic light demonstrated the following:

(a) direct stimulation of the hypothalamus through the skull is possible with all wavelengths from indigo to red; the retina reacts only to orange and red;

(b) yellow radiation has no optical effect, although the eyes of birds (e.g. pupil diameter) have a maximum sensitivity to this wavelength. This suggests that the effects of light waves, in case of visual and hypothalamic stimulation, are different.

Studies by Miline[14-16] showed permanent changes in the supra-optic and paraventricular nuclei of the hypothalamus of 30 rats (see Fig. 3.1) after one month (May) of continuous light stimulation (natural daylight during the day and electric light at night). Hypertrophic nuclei and capillary congestion were observed. After 30 days of complete darkness, regressive structural changes were noticed.

Ganong *et al.*[19] studied the penetration of light into the brain of mammals in view of the pronounced effects of light on gonadotropin secretion in birds and mammals. These effects are partly mediated via the retina and optic nerves, but part seems to penetrate the skull where it seems to act directly on receptors in the region of the third ventricle. This assumption was confirmed in mammals.[19]

Hollwich[17,18] also studied the physiological effects of light penetrating the eye of both animals and human subjects. In the latter case the physiological changes were studied in totally blind, perceptive blind (subjects can distinguish between light and dark) and socially blind subjects (the latter cannot read but can move around without any assistance). In the case of operated cataract patients, who regained eyesight, a reversal of the observed changes were noted. His major findings can be summarized as follows:

(a) *Animal experiments.* Rabbits were kept in constant darkness for 38 days. The circadian rhythms in urinary metabolites (uric acid, creatinine, potassium, sodium chloride, phosphate, bilirubin and urobilinogen) were disturbed. The amplitude and average excretion of metabolites decreased with the length of the period of darkness. After re-establishing the normal day and night conditions the physiological functions were normalized within three days.

(b) *Human experiments.* Similar experiments in blind subjects showed also a disturbance of the circadian rhythms of the following urinary parameters: 17-ketosteroids-, 17-OH-corticosteroids, urea-N, bilirubin, calcium, phosphate etc. The rhythm amplitude and average excretion were reduced.

With practically blind patients before and after cataract operations a number of changes were observed after the operation. While the subjects had very little perception of light, a diminished excretion was observed of corticosteroids, chlorides, uric acids, phosphate, potassium, urea-N and bilirubin; but sodium excretion increased. After the operation the excretion of metabolic products became normal.

In normal subjects the cortisol (17-hydroxycorticosterone) plasma level rises from 6 to 8 a.m. to about 15 μg ml^{-1} and reaches a minimum at night (about 6 μg). In blind subjects hardly any fluctuations are observed. The average level is about 6 μg ml^{-1} against 9 μg in normals, but after a cataract operation the cortisol rhythm is restored.

The circadian rhythm of thrombocytes in blind patients was compared with that of subjects with normal ocular light perception. In normal persons there is a marked rhythm with a minimum between 11 and 12 a.m. and maximum around 8 p.m. In blind subjects the rhythm is disturbed (only a small minimum around 8 p.m.), the amplitude is lowered and the absolute number of thrombocytes in the blood is reduced.

The daily eosinophil rhythm of normally sighted persons shows a maximum early in the morning after a minimum in adrenal activity and in cortisol level at night. Between 11 a.m. and 4 p.m. this reaches a minimum. According to studies by Radnot and Wallner,[20] and Hollwich and Dieckhues[21] 20–30% of the late morning decrease is due to light. In blind subjects this reduction does not occur and the whole rhythm has disappeared, but returns after a successful cataract operation.

Other changes observed in blind persons are the increased nightly diuresis (reduced in normals), an effect also observed in normal subjects during polar nights; a low fasting glucose level of 85 mg% at 8 a.m. against 98 mg% in normals and after cataract operations; higher average fasting values of cholesterol and free fatty acids 305 mg% and 661 mg% respectively, as compared with 255 and 530 mg% in normally sighted persons. Also changes in blood clotting time and liver enzymes have been reported.

Experiments by Tromp and Bouma[22] showed the following additional changes in blind children: in 24-h urinary samples the 17-ketosteroid excretion was considerably below normal; the amount of excreted hexosamines was below normal; urea and sodium values were very high as compared with normals; all blind children have a poor thermoregulation efficiency.

5.2.3. Biological effects of ultraviolet radiation

(a) *Local effects on human skin, on the hair and on the eyes* have been fully described by Ellinger.[23]

(i) Effects on the skin: exposure of the human skin to sunlight causes a reddening of the skin (due to increased blood flow) the so-called photo-erythema. It is characteristic of this phenomenon that it does not occur instantaneously, but that it appears after a certain amount of time has elapsed, a so-called 'latent period'. The length of the latent period is inversely proportional to the intensity of the radiation and may be as short as 1 h at high intensities. Exposure to moderate intensities, as they prevail at sea level, produces a photo-erythema with a latent period of from 4–7 h usually. The photo-erythema increases in intensity and usually subsides within 18–24 h. As the erythema starts to fade, the skin acquires a brownish colouration, a phenomenon designated as photo-pigmentation, which may last for months or even years. This colour is due to oxidation of the melanin pigment in the epidermis (outermost layer) of the skin. The *stratum corneum* (outer, horny, layer of the epidermis) is thickened. Both processes reduce the effect of further irradiation. If exposure to sunlight takes place at higher intensities, as they prevail in high altitudes, or in the course of excessive sunbathing, the result is a painful redenning of the skin with swelling (oedema), a phenomenon which is called a 'second degree' reaction. Still more intensive exposure to ultraviolet radiation may lead to the formation of fluid-containing blisters, the so-called 'third degree' skin reaction. If no infection of these blisters takes place they heal without *sequelae*. Second or third degree skin reactions are usually also followed by pigmentation, which, however, is more intense than normal photo-pigmentation. As a result of irradiation with ultraviolet radiation, a dilatation of the skin capillaries takes place, which accounts for the phenomenon of photo-erythema. At the same time a progressive pigment migration to the uppermost layers of the skin, the *stratum corneum*, takes place. This is due to the development of dendritic cells into which the pigment is transferred from its original location somewhat deeper in the skin (*stratum pigmentosum*) by intracellular deposition. Besides this pigmentation, subsequent to the production of an erythema (secondary tanning effect), there exists also a direct tanning effect due to an increase in pigment granules in the basal cells of the skin. A further result of exposure of the skin to sunlight consists of an increased cornification. This represents a compensatory hypertrophy of the skin to that seen after mechanical skin irritation and well known as callus formation. Increased skin cornification, together with pigment formation, accounts for the phenomenon of photo-adaptation, a condition in which the skin is rendered more resistant to ultraviolet radiation, i.e. a second exposure to the same or even higher intensity than the one which produced these phenomena, is tolerated without notable erythemic response.

Not all ultraviolet rays are equally effective in producing erythema and pigmentation. It appears that there exist two wavelength maxima for the erythemic response of the human skin: a first one at 380 nm and a second one between 303 and 297 nm. A specific absorption of these rays in the skin has been recognized as the reason for this specific action. Ultraviolet radiation with longer wavelengths is more effective in tan production than that of shorter wavelengths. It is for this reason that antisunburn media which filter out the shorter wavelengths, and in this manner prevent painful erythemic responses, permit nevertheless suntanning (photo-pigmentation).

The deeper physiological action mechanisms of ultraviolet radiation on the skin consist of the formation of vitamin D from ergosterol (a steroid, occurring in human tissues, which becomes a potent antirachitic substance after ultraviolet irradiation), the photochemical formation of histamine from the amino acid histidine, and/or the release of pre-existing histamine from ultraviolet-damaged skin cells.

The sensitivity of living cells to solar radiation is called photosensitivity. It varies not only with the wavelength of the radiation, but also with the physiological and pathological conditions of the human body. Fundamental studies in this field have been performed by many scientists. From the accumulated evidence it appears that the following factors determine photo-sensitivity:

(a) Colour of the hair, age of the individual, sex and season: blond persons, and particularly red-blonds, are in general more photosensitive than persons with dark hair. Young children are less photo-sensitive than adults (age range 20–50 years). During the involutionary period of life photo-sensitivity declines. Males appear to be more photo-sensitive than females. Finally, photosensitivity is higher in spring and early fall than in winter and summer;
(b) The body section: photo-sensitivity decreases in the order breast-abdomen-back-cheeks-upper and lower extremities;
(c) Constitutional factors: photo-sensitivity increases during the first day of the menstrual cycle of women and also during pregnancy.
(d) Effects on the hair: the bleaching effect of sunlight is a matter of common knowledge, which may be accompanied by damage to the skin surrounding the hair root.
(e) Effects on the eye: as a result of exposure of the eye to ultraviolet radiation, more or less severe inflammation of the eyelids and *conjunctivae* may occur. These symptoms may be very painful, but usually clear up without serious consequences.

(b) *General effects on the human body.* As a result of exposure of the human skin to ultraviolet radiation six important general physiological effects are observed: these are effects on the gastrointestinal tract; the circulatory system; the general metabolism; the hypothalamus; the pituitary gland and the endocrinal functions (thyroid, parathyroid, adrenal, pineal gland and testis). Part of these effects seems to be the result of histamine or histamine-like substances, formed as the result of exposure of the skin, which have entered the blood circulation.

(i) Effects on the gastrointestinal tract: as the result of exposure to erythema-producing doses of ultraviolet light, increased gastric secretion has been observed in normal subjects as well as in patients. Severe over-exposure to ultraviolet light may produce inflammatory changes of the stomach (gastritis).

(ii) Effect on the circulatory system: this consists of a fall of the blood pressure. In healthy young adults the fall may amount to 6–8 mm Hg. It is related to the dilatation of the capillaries of the skin due to the liberation, or new formation, of histamine. The effects of ultraviolet radiation on blood components are an increase in the number of erythrocytes and the haemoglobin value, as well as an initial increase in polymorphonuclear white cells; increases in calcium, potassium, sodium and phosphorus contents have also been reported.

(iii) Effects on the metabolism: a reduction of the respiratory quotient has been observed after acute exposure of man, but the basal metabolic rate remained unchanged. Repeated exposures may, however, cause a reduction of basal metabolism. Mineral metabolic changes are reflected in observations on blood chemical changes. Protein metabolism appears to be increased, as manifested by increased urinary nitrogen excretion.

(iv) Effects on the hypothalamus: histophysiological experiments, particularly by Miline[14-16] in Yugoslavia, demonstrated the photo-sensitivity of the hypothalamus either by a nervous pathway of retinal origin or by a direct transorbital route. In infantile rabbits, exposed to successive periods of solar radiation, some changes in the structure of the supra-optic nucleus have been observed. In rabbits, sacrificed after two and three weeks, changes of a progressive character can be noticed, such as hypertrophy of the eccentrically located nuclei and nucleoli and the presence of fine granulated hypochromatic neurosecretory material, which is less than in the control group. Processes of vacuolization can be noticed in some pericaryons. Neurosecretory fibres are thinner, capillaries are enlarged. Nuclei of the satellite neuroglia cells are also found to be hypertrophic.

(v) Effects on the pituitary gland: studies by Miline[14-16] have shown that, in the distal part of the adenohypophysis of infantile rabbits, a number of important processes take place as a result of solar radiation. Intense degranulation of the ACTH cells; an increase in number of somatotrophic cells; gradual accumulation of the secretory granules in the cells secreting follicle-stimulating hormone and luteinizing hormone; intense polymorphism of the cells responsible for the secretion of the thyrotropic hormone. The *pars intermedia* of the hypophysis is enlarged due to hypertrophy and hyperplasia of the intermediocytes. The most numerously represented are the light glandular cells, very large, with a hypertropic nucleus and a prominent nucleolus. In most of these cells the nucleus contains a big vacuole with pyroninophil material (RNA). Cytoplasm of the dark cells is found PAS positive. Penetration of neurohypophyseal fibres are less manifested than in control animals. Activity of the alkaline phosphatase in the cell wall is found to be more positive. The reactive changes described suggest the hyperactivity of this part of the hypophysis, i.e. of intensified secretion of the melano-stimulating hormone (MSH).

(vi) Effects on the thyroid gland: the effects of ultraviolet irradiation upon the thyroid gland have been studied by Dempsey,[24] Higgins,[25] Miline[14-16] and many others.[6] The thyroid gland in pigeons exposed to the radiation from a quartz lamp shows a hypofunctional stage, and the lack of ultraviolet rays determine the hyperactivity of this gland in rats. Rabbits exposed to solar radiation in Yugoslavia in August, for seven days, showed the following reactive changes of a progressive character in the thyroid: an abundance of small dimension follicles (neogenesis of micro-follicles) with highly prismatic epithelium, hypertrophy of the thyrocytes nuclei, lack of colloids in numerous follicles. In the structure of the follicular epithelium surrounded by hyperemical capillaries more light cells can be found than in the control groups. These changes in the structure of the thyroid gland are more evident in rabbits sacrificed after two or three weeks of solar radiation. Apart from stimulated excretory secretory activity of the thyrocytes, they are characterized by a hyperplasia of parafollicular, or C, cells which are distributed separately or in groups. Their cytoplasm is richer in argorophil granules, the Golgi area is enlarged and nuclei are hypertrophic and nucleoli more numerous.

(vii) Effects on the parathyroid gland: it has been found that the calcium content of the blood serum is increased after ultraviolet irradiation; in milk-cows exposed to ultraviolet radiation milk acquires antirachitic properties. Pig production was increased by 27% if the animals were exposed to ultraviolet radiation during and after the period of lactation. The enlargement of the parathyroid develops under the influence of ultraviolet rays. In rats the glands are enlarged by 28% in summer as compared with their size in winter. In infantile and adult rabbits exposed to the systematic influence of solar radiation for three weeks, several structural changes took place in the parathyroid glands. The most important changes described by Miline[14-16] are: hyperplasia of the parathyrocytes, microbulation and capillary congestion. The light type parathyrocytes of the hypertrophic nucleus and nucleolus, with outstanding perinuclear basophilia, and with more extended Golgi apparati predominate. A greater concentration of nucleic acids has been found in nuclei. Alkaline phosphatases are more abundant in the walls of the capillaries.

(viii) Effects on the adrenal gland: exposure of rabbits to ultraviolet irradiation causes adrenal enlargement, changes in the content of lipids and plasmalogens, a decrease of lymphocytes etc.[6,26] In experimental animals, under influence of ultraviolet radiation, stimulation of the sympathico-adrenal system has been observed (increase of the catecholamines), together with hypertrophy of the adrenal cortex and an increase in the concentration of vitamin C. If infantile rabbits are exposed to systematic progressive insolation in July, and August, reactive histophysiological changes of the adrenals may occur. Hyperplasia of all three layers of the adrenal cortex was observed in rabbits sacrificed after seven days. In the border area of the glomerular and fasciculated zone the same has been found and also the presence of 'lumina', typical formations in the adrenal cortex in hyperactive stages. In rabbits sacrificed after three weeks of solar radiation, the most characteristic changes in the adrenal structure are the following: neogenesis of the subcapsular adenomes mainly formed of the constructive elements of the glomerular zone; the frequent occurrence of 'lumina', hyperplasia of the *zona fasciculata*; profound invagination of the ramifications of the reticular zone in the medulla; accumulation of vitamin C; and hyperplasia of the medullocytes.

(ix) Effects on the pineal gland: under the influence of light some structural changes of a depressive character occur whereas in darkness contrary changes take place. Studies on the influence of solar radiation on the histophysiological characteristics of the pineal gland show an opposite effect as compared to the effects on the hypophysis. In infantile rabbits, sacrificed after three weeks of solar irradiation, a number of structural changes were found: pineocytes have small dimensions and are seldom found; their nuclei are less voluminous, with less nucleoli; the number of intranuclear phloxinophil granules is small. Reactive changes in the structure of this organ, which is the source of melatonin secretion, are opposite to the histophysiological reactions of the *pars intermedia* of the hypophysis which is the site of the melanostimulating hormone secretion.

(x) Effects on the testis: solar irradiation in infantile rabbits accelerates the maturation of the seminal epithelium; it causes hypertrophy of interstitial cells and congestion of capillaries. These results suggest gonadotrophic effects in infantile animals.

5.2.4. Influence of photo-dynamic substances

The presence of substances usually exhibiting the physical phenomenon of fluorescence, has been recognized as the cause of light sensitization. By this an abnormal reaction of the skin, both from the qualitative as well as the quantitative point of view, is understood, i.e. sunlight with wavelengths which ordinarily may not produce an erythema or pigmentation of the skin may do so.

Usually ineffective sub-threshold radiation doses may produce these skin effects.

The first observations of this effect were made by Raab[27] who described how the addition of a dilute solution of the fluorescent dye acridine to a colony of *Paramecia* killed these more readily if exposed to sunlight than if kept in the dark. In later studies it was shown that a similar effect could be obtained with numerous other fluorescent dyes and also if such dyes were injected into mammals (mice). For the phenomenon of sensitization to sunlight by fluorescent substances of various origins (plant materials, metabolites, etc.), Von Tappeiner[28] coined the term photo-dynamic action. Substances which produce a photo-dynamic action are called photo-sensitizing agents. A list of photodynamic substances and their actions was published by Blum.[29] In recent years a new kind of photosensitizing agent has been discovered in the form of certain sulphonamide drugs often applied in modern treatments. Due to the influence of light the sulphonamide may be oxidized into a substance (e.g. *p*-hydroxyaminobenzolsulphonic acid amide) which acts as an antigen. With respect to their role in producing skin diseases, photo-sensitizing substances are classified as either endogenous or exogenous agents. According to Haxthausen and Hausmann[30] endogenous agents are substances which may be formed within the body as the result of a metabolic anomaly, e.g. certain porphyrines, or substances which enter the body as drugs administered for theurapeutic purposes, such as luminal. Among the exogenous agents are certain cosmetics.

Photo-sensitization should be distinguished from photo-allergy,[31] a relatively rare condition in which the process of photo-sensitization can be transmitted passively from one subject to another by serum injections (for further details on photo-sensitization see Reference 6, pp. 801–808).

5.3. BIOLOGICAL EFFECTS OF CHANGES IN ATMOSPHERIC PRESSURE

The biological effects of changes of more than 125 mm Hg in atmospheric pressure, which are observed if subjects move from sea level to a mountainous area above 1500 m altitude (635 mm Hg), have been extensively studied in high altitude biometeorology[32] and have been discussed briefly in Section 3.5.2. However, considerably less is known about possible effects of minor daily changes in atmospheric pressure, e.g. a fall in barometric pressure from 770 to 720 mm in two or three days which one can observe during the approach of very active deep low pressure areas. However, as such strong barometric changes are always accompanied by strong air turbulence and by changes in temperature and often by rainfall, hail etc. it is difficult to determine the effect of the barometric fall alone.

5.3.1. Effects of high altitude (changes in atmospheric pressure of more than 125 mm Hg)

5.3.1.1. Effects on basic physiological mechanisms

Studies reported by several high altitude research stations (e.g. Jungfrau Joch in Switzerland, high altitude stations in Peru, California, Rocky Mountains etc.) and studies in low pressure climatic chambers have shown a number of clinically important physiological changes after a stay at high altitude above 1500 m (below 635 mm Hg), both in healthy subjects and particularly in subjects suffering from respiratory and vascular disorders. Different observations indicate that high altitude effects can only be observed if the subjects stay above the critical boundary of 1500 m altitude. This statement is supported by a number of observations:

(a) In asthmatics, who are wheezy at sea level, the complaints decrease or disappear only above the 1500 m level equivalent pressure in a low pressure chamber, but not earlier. The subjects do not know the exact simulated altitude, so suggestion is excluded.
(b) Patients suffering from slight sinusitis (inflammation of the maxillary sinus near the nose) experience serious pains in this area as soon as the 1500 m level is surpassed.
(c) Subjects suffering from diaphragmatic hernia (hernia of the muscle between the thoracic and abdominal cavities) have serious stomach complaints as soon as they surpass the 1500 m level but not earlier.
(d) An increased peripheral blood flow occurs in most subjects above 1500 m, particularly in patients suffering from vascular disorders. It shows up in a rise in temperature of the palm of the hands and feet only above 1500 m altitude.
(e) A number of biochemical changes were observed in the urine of asthmatics above the 1500 m altitude level, but not below this level; for example, increased excretion of 17-ketosteroids, increased pH (alkalosis) and decrease in hexosamine excretion (see Section 6.9).
(f) The improvement of the thermoregulation efficiency, determined with the water bath test, can only be observed above 1500 m altitude.

The observed high altitude effects differ depending on the duration of the stay above 1500 m altitude.

Effects after a one-hour stay at an altitude above 1500 m. The following changes can be observed: increased lung ventilation and vital capacity; increased peripheral blood flow, which shows up as a rise in temperature of the palm of the hand and the nasal mucosa, often a reddish colouring of the face and an increased blood flow in the brain capillaries, particularly in the hypothalamus (confirmed, during animal experiments, by Betz[33]); rise in urinary pH from e.g. 5.1–6.9 (the better the normal respiratory function of the subject, the greater the rise; in subjects suffering from emphysema the effect is very small); slight changes in the blood sugar level and a decrease in the hexosamin content of the excreted urine.

Effects after a repeated stay of one hour a day for a long period, at altitudes above 1500 m

(i) Considerable improvement of the respiratory function.
(ii) Increased sensitivity of the autonomic nervous system: the sensitivity

to various stimuli increases considerably, at least up to 2500 m. Above 3000 m the opposite phenomena may be observed. Repeated high altitude treatments of one or two hours a day cause a better balance between the ortho- and para-sympathetic system of the body and may improve the condition of a subject suffering from a disturbed autonomic nervous system, e.g. due to serious mental stresses, long lasting infections etc.

(iii) Stimulation of the hormonal function of the adrenal gland: this is observed particularly in asthmatic patients, usually suffering from a strongly reduced functioning of the adrenal gland which shows up, for example, in a very low 17-ketosteroid level in their urine. A series of one-hour high altitude treatments will increase the production of 17-ketosteroids and of other corticosteroids.

(iv) The blood-producing mechanisms are stimulated: in particular the total number of erythrocytes and the haemoglobin content of the blood is considerably increased, the number of eosinophils is usually decreased.

(v) Improvement of the overall thermoregulation efficiency: one of the most important effects of high altitude, after a long high altitude treatment (either natural or simulated high altitude), above 1500 m, is the rapid improvement of the thermoregulatory efficiency. It is particularly striking in subjects with poor thermoregulation such as asthmatics, rheumatics and allergic patients. With the objective water bath test (see p. 56), it is possible to measure the improved efficiency.

(vi) Increased sensitivity to drugs and toxic substances: certain drugs have a much greater effect at altitude than at sea level and may cause a greater mortality rate among animals if used in a high dosage, as compared with lowland experiences (see Section 5.3.1.2).

(vii) Explosive development of latent infectious diseases: pneumonia and latent infectious diseases, appendicitis, common cold, inflammation of dental roots etc. may develop explosively if an apparently healthy subject is brought to an altitude above 1500 m.

(viii) Water retention: Hasselbach and Lindhard[34] observed decreased urinary secretion at high altitudes. Their observations were repeated by Smith,[34] who carried out experiments on small bitches.

Effects after a long stay (of several days or weeks) at altitudes above 1500 m. Apart from the altitude effects mentioned above, after an initially increased fibrinogen level of the blood and an increased erythrocyte sedimentation rate the fibrinogen level and the blood sedimentation rate are both reduced; a reduction in acid production in the stomach is also observed and changes in the composition of the blood take place. A stay of several weeks at high altitude causes an altitude adaptation which considerably reduces the effects described above.

5.3.1.2. Effects on response to drugs

For seven years Macht and his colleagues[35] (in the Pharmacological Research Laboratory at Baltimore (USA)) studied the possible correlations between specific meteorological factors and drug action. In particular, the difference in toxicity of digitalis was studied. Macht[35] noticed that pronounced differences in toxicity occur if great changes in barometric pressure and humidity take place. For example, a tincture of digitalis became more toxic during severe storms in the USA with rapidly falling barometric pressure. To kill a cat less digitalis tincture per kilogramme of bodyweight was required during stormy weather than during quiet weather conditions. Similar results were obtained in the Blue Ridge Mountains during quiet weather which suggests a true barometric effect.

The high altitude effect observed by Macht was confirmed in the Austrian Alps by Jarisch and by Lehman and Hanzlik[36] in the USA. In all these experiments there was no question of deterioration of the drug itself, as was proved by simultaneous trials at sea level. Lehman and Hanzlik[36] studied the effect of high altitude on the emetic and fatal doses of digitalis in pigeons and cats. The same tincture was used at different altitudes: in San Francisco, at sea level (760 mm Hg) and at Tioga Pass in the Sierra Nevada mountains, California, at 3000 m (526 mm Hg). A total of 288 emetic tests with six doses of digitalis, ranging from 5 to 30 mm kg^{-1} body weight, were made in 144 different pigeons. Except for the 5 mg dose, a considerably higher percentage of vomiting occurred at Tioga Pass with the same dose of tincture.

5.3.2. Effects of normal daily barometric changes (changes in atmospheric pressure of less than 50 mm Hg)

In many studies correlations have been reported between minor changes in barometric pressure (less than 50 mm Hg), as observed during the approach of low pressure areas, and certain clinical phenomena.

The effect on water retention and restlessness has already been mentioned. Many authors, for example Czarniecki *et al.*,[37] Tromp[32] and Nedzel,[38] observed considerable changes in blood pressure during periods of rapid barometric fall or rise, accompanied by weather fronts and changing air masses. However, the fact that blood pressure in a healthy population group (as observed by Tromp in The Netherlands[10]) rises both with a falling and a rising barometer, but that the rise in blood pressure is always associated with falling temperatures (e.g. due to the influx of cold air masses) suggests that the barometric correlation is not a causal one.

The assumption that atmospheric pressure has only an indirect effect is supported by the observation, made by Alvarez[39] in Puerto Rico (Caribbean Sea), that in this area blood pressure increases with falling barometer, whereas in Germany, according to Betz and Sarre,[33] the blood pressure falls with falling atmospheric pressure. In other words, a change in atmospheric pressure is only a biometeorological indicator of a change in weather, such as changes in air mass, temperature, humidity, air movement, etc.

Although small changes in atmospheric pressure in themselves apparently do not seem to cause directly any clinical phenomena, the studies by Rentschler *et al.*,[40] in the Mayo Clinic in the USA (1929), on barometric pressure and arthritis, the observations by Tromp[41] in The Netherlands and Krypiakievicz[42] in Germany on atmospheric pressure and the restlessness of schizophrenics, and the studies by Stocks[43] in England, who observed a highly significant correlation between high atmospheric pressure and heart failure, indicate that changes in barometric pressure may nevertheless be important biometeorological indicators of changes in weather which are responsible for various clinical phenomena.*

Recently Kreithen and Keeton[44] were able to demonstrate that homing pigeons have the ability to detect air pressure changes. The threshold of detection was 10 mm H_2O or less, equivalent to a change in altitude of 10 m^2. This has not been confirmed so far in other animals or in man.

5.4. EFFECTS OF NATURAL AIR IONIZATION

Despite the many studies carried out on the possible physiological effects of air ions very few authors were able to demonstrate, in an unrefutable way, any biological effects of a surplus of positive and negative ions. Minor uncontrolled changes in temperature, humidity, traces of smoke, presence of metals etc. can change the ratio of positive and negative ions in a room fundamentally. If we add to these technical problems the complexity of recording accurately the number and type of ions near a biological subject in a cage, it is understandable that many contradictory reports have been published on this topic.

The existence of air ion effects was established scientifically particularly by Krueger *et al.* in Berkeley, California,[47] whose results were later summarized in *Medical Biometeorology*,[6] p. 351 and *Progress in Human Biometeorology*, *Part IA*,[45] p. 335. They used special tritium ion sources capable of producing high yields of unipolar ions free from gaseous radioactive contaminants. Some of their results are summarized below.

5.4.1. Effects on the respiratory system

After exposure under strictly controlled conditions, to atmospheres containing high densities of positive air ions Winsor and Beckett[46] observed dryness, burning and itching of the nose, nasal obstruction, headache, dry, scratchy throat, difficulty in swallowing, dry lips, dizziness, difficulty in breathing, and itching of the eyes. Negative ions did not produce all these symptoms, but occasionally caused mild discomfort. The influence of air ion inhalation on the maximum breathing capacity was also studied. The individuals tested were able

*Recent studies by H. Richner in Zurich, on the possible influences of rapid fluctuations in atmospheric pressure on human comfort are discussed in *Biometeorological Survey (1973–1978)*, Heyden, London, 1979.

to breathe on the average 35 l min^{-1} under normal conditions. After a preliminary exposure to positive air ions the maximum breathing capacity dropped to 25 l min^{-1}; similar pre-treatments with negative air ions produced no detectable deviation from the controls. The reduction was apparently due to swelling of the nasal mucosa induced by positive air ions. Also the response of the mammalian trachea to unipolar small air ions has been studied. The trachea is lined with a mucous film. Any inhaled particulate matter such as pollens, dust, or bacteria is ultimately trapped in the mucous film coating the epithelium and becomes a passive passenger in its flow. The flow is largely dependent on the wave-like action of the cilia beating at rates approximately 800–1000 min^{-1}. The general effects observed in excised tracheal strips from rabbits, guinea pigs, mice and rats after treatment with positive ions are the following:

(a) The membranous posterior wall of the trachea contracts; treatment with negative ions reversed the positive ions effect, the posterior wall relaxed and the peristaltic reflex was re-established.
(b) The mucosal surface becomes dry. Negative ions either had no effect on the surface or brought about the appearance of a watery fluid.
(c) Ciliary activity of an area treated with positive ions stops.

5.4.2. Effects on micro-organisms

Studies by Krueger *et al.*,[47] Kingdon,[48] Pratt,[49] and Philips *et al.*[50] suggest that positive and negative ions may have bacteriostatic, bactericidal and fungicidal properties. High concentrations of negative ions could retard the growth of various pathogens under rigorously controlled laboratory conditions.

5.4.3. Effects on endocrine glands

According to Deleanu and Frits[51] adrenalectomized rats treated with moderate concentrations of mixed negative and positive ions lived longer than controls while displaying more reactivity of the respiratory centre and improved stability of the serum proteins. Gualtierotti[52] noted that the adrenal glands of guinea pigs exposed to high densities of negative ions showed a marked increase in the width and volume of the nuclei. The histological evidence suggests stimulation of the production of glucocorticoids. In similar experiments on the thyroid gland Gualtierotti detected a substantial reduction of colloids as a result of exposure to negative ions. Histological examination of the gonads of animals treated similarly revealed stimulation of the maturation process of testicular cells and ovarian follicles.

5.4.4. Effects on biochemical changes in the body

According to Ostarova[53] cats treated with high densities of positive ions show an increase in the blood calcium level, a decrease in blood coagulation time and prothrombin time. Negative ions had the opposite effect. According to Minch[54] negatively ionized air enhances the metabolism of water-soluble vitamins. Krueger *et al.*[47] observed changes in the blood level of serotonin (5HT) if mice were exposed to high concentrations of positive or negative air ions. Negative ions induced a decrease of 5 HT while positive ions raised it.

5.4.5. Possible effects of Sharav, Chamsin and Bora winds

Studies by Sulman *et al.*[55] in Israel have shown that the hot dry desert winds, like the Sharav or Chamsin, or cold rainy winds (Bora) can produce changes in neurohormonal secretions (urinary 17-ketosteroids, 17-OH, adrenaline and noradrenaline, serotonin etc.).

Serotonin over-production seems to correlate with migraine attacks in the patients used in these studies (500 female weather-sensitive patients). According to Sulman *et al.* this serotonin irritation syndrome can be prevented or cured by negative air ionization treatment in 75% of the cases. Further studies with other subjects should be carried out to confirm these findings because it is extremely difficult to exclude the suggestion factor in such clinical experiments in hospitals. Apart from this the same hormonal changes can be produced by changes in the cooling index in the atmosphere.

5.4.6. Effects on plants

Recent studies by Elkiey *et al.*[56] in Canada suggest reduction of the number of lesions and percentage of leaf area covered by lesions of 'Net Blotch' disease (caused by *Helminthosporium teres*) of barley. Treatment with positive ions delayed the appearance of disease symptoms by three days. Negative ions had no effects.

5.4.7. Possible causes of air ion effects

Krueger *et al.*[57] analysed the possible causes of these physiological effects. A series of tests was conducted on the ciliary activity of tracheal strips, replacing the air in the ionization chamber with various gases. Negative ion effects were observed only when oxygen was present; none occurred when the chamber was filled with nitrogen or carbon dioxide. When conditions for positive ion formation were imposed, typical results were obtained with carbon dioxide and none with nitrogen or oxygen. It was concluded, therefore, that negatively charged oxygen and positively charged carbon dioxide are the mediators of physiological effects occurring in the trachea as a result of atmospheric ionization.

The tracheal changes attributed to positive air ions can be duplicated by the intravenous injection of 5-hydroxytryptamine (5-HT). The 5-HT effects can be reversed by treatment with negative air ions. On the basis of these facts Krueger assumed that positive air ions are serotonin releasers and that a local accumulation of 5-HT in the trachea is the immediate cause of positive ions effects. Negative air ions reverse the positive ion effects by speeding up the rate at which free 5-HT is oxidized.

Unfortunately the various biological effects observed by Krueger and his colleagues[57] can only be established under very strictly controlled environmental conditions with constant humidity, temperature, lack of smoke particles etc. Apart from this the effects are very small as compared with the various thermal and radiation effects described in the previous sections.

Therefore, in our opinion, the many commercial claims of producers of different types of expensive ionizers, to be recommended for therapeutic purposes, must be studied with great suspicion.

5.5. EFFECTS OF AEROSOLS

5.5.1. Structure and size of aerosol particles

The atmosphere can be considered to be an aero-colloidal system, in which very fine solid or fluid particles are dispersed between the gas molecules of the air. Concentrations of these particles, both of terrestrial and cosmic origin, in gases are known as aerosols. The extensive distribution of these very fine solid and fluid particles in the atmosphere may explain why a truly gaseous atmosphere composed of gases only, exists only under laboratory conditions.

The properties and distribution of aerosols are determined by the physicochemical laws of aero-colloids and by the various meteorological factors. The atmosphere contains a large number of small ions, i.e. electrically charged clusters of 10–30 molecules of atmospheric gases (particularly oxygen and hydrogen molecules) with great mobility and an electric charge per ion of 15.9×10^{-20} Coulomb (on the average there are 500–600 small ions of each sign per ml of air). These small ions are usually attached to the solid and fluid particles of the atmospheric aerosol and form the secondary ions, of which the largest group is known as Langevin ions. These are less mobile (0.0005 cm s^{-1}) and very frequent in areas polluted with dust. Near towns values of 50 000 and more Langevin ions ml^{-1} have been observed. These secondary ions move around as a result of electric fields in the atmosphere. According to the Stokes-Millikan law, the movements are determined by the size of the particles, the density and the electric charge (both sign and quantity). The number of small ions is closely related to the electric conductivity of the air, the large ions to the electric potential gradient (i.e. the difference per metre in electrostatic potential in the atmosphere between the ground and a point in the air) of the atmosphere.

The size of naturally or artificially dispersed aerosol particles varies usually between 2 nm and 5 μm. The latter size corresponds to a mass of 10^{-9} g, which enables the particles to float in the air. Unless special turbulent conditions prevail, particles above 50 μm are no longer stable.

The lowest boundary of aerosol particles is not very well known because of their usually short period of existence (30 s or less). In the case of fragments of quartz crystals aerosol particles of 0.5 nm have been reported.[58] The number of particles varies even more than their sizes. An aerosol can be considered to be stable if the average particle content of an atmospheric aerosol is about 10^4 particles of 0.1 nm ml^{-1}. This condition prevails in a dustfree continental air mass. However, air masses are known with more than 4×10^6 particles ml^{-1}.

5.5.2. Physical properties of aerosol particles

The relatively small mass of aerosol particles as compared with their surface explains the specific properties of aerosols, in particular the property that the physico-chemical reactions between gaseous fluid and solid phases are considerably accelerated in aerosols. Due to the second law of thermodynamics, the Law of Minimum Free Energy, the fluid or hydrated solid aerosol particles will tend to become globular in shape because the surface tension is directly proportional to the surface, which for a constant mass is smallest in a globular body.

The electric charge of the outer surface of aerosol particles is different from the central parts. For example, in pure water droplets, the negative surface charge will be present to a depth of 8×10^{-7} cm, the positive charge goes down as far as 13.3×10^{-7} cm; in other words an electrostatic double layer is formed, in thickness equal to 40 molecules. Due to air turbulence this double layer may be destroyed. The originally electro-neutral particle may then obtain a surplus charge either positive or negative. This facilitates the adhesion of new ions with opposite electric charge and causes the space charges in aerosol clouds.

Apart from the surface tension and electric double layer the vapour pressure of fluid aerosol particles also plays an important role in the physico-chemical behaviour of aerosols.

In the atmosphere, e.g. in clouds or fog, droplets of different diameter occur, the larger ones of which grow at the expense of the smaller ones. Due to the greater vapour pressure of the small droplets the smaller ones evaporate and after passing the saturation point the fluid condenses on the larger droplets, a process known as isothermal distillation. This process is assisted by the surplus charge of aerosol particles, which lowers the vapour pressure. These various properties of aerosol particles explain the increased hydration capacity of aerosols with increasing surplus charges of the particles.

5.5.3. Biological properties of aerosols

With decreasing size of particles in an aerosol cloud of constant mass, the speed of various chemical reactions is considerably accelerated due to the increase in total surface of the particles. Dirnagl *et al.*[59] pointed out that 1 ml of water, composed of spherical drops of 1.24 cm diameter, has a total surface of 4.84 cm^2. With a particle size of 0.1 μm the number of particles increases to 523×10^{12}, the total surface to 60 000 cm^2.

Depending on the size and electric charge of the aerosol particles they are able to penetrate to different levels in the lungs. However, a concentration of

10^{19} molecules ml^{-1} may considerably exceed the threshold of physiological resistance to a poisonous gas. This may explain why even very small concentrations of poisonous substances in fog or industrial aerosols, particularly during atmospheric inversion, may have devastating effects on the lungs of the inhabitants living in such an area. This could explain why certain aerosols in districts with large chemical industries, containing poisonous organic compounds in the air, may be responsible for lung cancer. Normally the protective mechanism of the nasopharynx and bronchi is sufficient to keep the lower air passages free from dust, at least as far as mineral particles larger than 10 nm are concerned. More than 75% of particles reaching the alveoli are less than 3 nm (length of the tubercle bacillus). The smaller particles (0.5–3 nm) are the most injurious. Of the larger particles only asbestos dust, as large as 200–400 nm, can slide down the bronchi, with damaging effect.

Extensive studies on the penetration depth of aerosols were carried out by the physician Bisa[58] in the Institute of Aerobiology in Germany, by the physicist Dirnagl,[59] at the Balneological Institute of Munich (Germany), by the physician Andersen and the physicist Lundqvist[61] at Århus (Denmark) and by several others.

The knowledge of the great chemical activity of aerosols has been applied in medicine for more than 100 years, particularly in Germany, France and Italy, in inhalation therapy, a name recently changed into the more scientific term of aerosol therapy.

5.6. EFFECTS OF ELECTROSTATIC AND ELECTROMAGNETIC FIELDS

5.6.1 Electrostatic fields

Studies on the biological effects of electrostatic fields on living organisms have been carried out both with constant and with slowly fluctuating electrostatic fields. Most experiments carried out so far were with animals and, in particular, with insects. Despite the technical facilities, which are available in all modern high tension laboratories, due to lack of co-operation between physicians or physiologists and electrical engineers or physicists so far no systematic studies have been made in relation to man.

It has often been assumed that, as the electrostatic potential gradient in the atmosphere is practically zero in houses, possible effects could have no practical importance. However studies, over a period of four years in Leiden,[9] of the daily effects of atmospheric cooling on a number of physiological parameters have shown that, even if a subject stays most of the day in an air conditioned environment, staying in the open air for even less than one hour is enough to create, for 12 h or more, a physiological pattern belonging to this particular atmospheric environment and not related to the conditions in the air conditioned room.

Therefore if strongly fluctuating potential gradients in the open air would seriously affect the human body these effects can be expected to be observed even if the subject stays most of the day in a building, provided that he has been in the open air for an hour or, perhaps, even less. The critical time limit is not yet known.

If one wants to discuss the effect of electrostatic fields on living organisms one should first consider the natural electric fields created by the organism itself, its so-called electric aura. In particular Burr,[62] formerly Professor of Neuro-Anatomy at Yale University, USA, demonstrated, as early as 1932, that there is a relatively steady state of voltage difference between any two points on a living organism. These electric gradients are remarkably stable and are changed only by alterations in fundamental biological processes such as growth, injury, ovulation, etc. Most organisms show also a clear electric polarity. According to Burr and his co-workers[62] the standing potentials in the living organism define the electrodynamic field, the fundamental pattern which determines the organization of the living being and reflects in significant ways, its fundamental biologic activity.

Experiments in Germany by Heydweiller,[63] Sauerbruch and Schumann[64,65] and Von Ardenne[66] have shown that the human body is also surrounded by an electric aura which despite its low energy can be recorded at a considerable distance from the body.

Oppenheim[67] and Koopman[68] demonstrated that the human body, in a dry atmosphere—below 70% relative humidity, and particularly below 30%, could develop electrostatic fields, due to friction of shoes or chairs, of up to 14 000 V. Subjects seated in an insulated chair developed 3000 V by violent movements of their arms, causing a discharge current of about 10^{-7} amp and electric energy of up to 3×10^{-4} W.* Despite these and other studies we still do not know the biological implications of all these phenomena.

Experiments by Murr[69] in University Park, Pennsylvania, have demonstrated the biological effects of electrostatic fields on plants. Numerous plant species were treated in both reversed electrostatic (soil positive) and conventional electrostatic (soil negative) fields, and in electromagnetic (alternating current) fields. The reversed electrostatic potential gradient varied between 0 and 120 kV m^{-1}. Plant and leaf damage was observed for high potential gradients in the reversed field. Spectroscopic analyses of the damaged plant sections revealed micro-chemical activity essentially identical to that observed

*NOTE: Frictional contact between wool, cotton, silk or synthetic fibres, and the human skin (not the hand) makes the body usually electrically positive, the wool or silk clothes being negative. If the person undresses and he stands insulated, considerable electric voltages can be registered (a few hundred volts). If the atmosphere is very dry, crepitating noises can be heard during undressing, particularly with clothes of synthetic fibres; in a dark room light phenomena can be seen and small neon tubes start glowing.

During strong inspiration and expiration alternating glowing phenomena can be observed. High electric potentials are obtained only if the atmosphere is dry, i.e. less than 70% humidity, particularly at about 30%.

for conventional electrostatic and electromagnetic field growth. The severity of leaf damage was less in plants grown in the reversed fields as compared with the conventional field. The critical dynamic potential in corn and bean plants proved to be 100 kV m^{-1} for 10 h treatments.

Sidaway[70] carried out experiments in Cardiff, UK, using seeds, of the lettuce *Lactuna sativa*, placed in electrostatic fields with a potential difference of 180 V. Positive fields inhibited germination of the seeds whereas negative fields promoted this process. Later studies established that under certain experimental conditions respiration of seeds increased 50% above the control levels in negative electrostatic fields.

The possible effect of electrostatic fields (potential gradients) on insects has been discussed by many authors. Fabre,[71] Uvarov[72] and Wellington[73] described the effects on behaviour, in particular during thunderstorms. Schua[74] showed that worker bees increased their number of foraging flights during periods of electrical activity in the atmosphere. Guard bees became hostile when alternating potentials were placed at either side of the hive entrance. Flying insects are repulsed when they approach a positively charged screen. Levengood and Shinkle[75] observed, in *Drosophila melanogaster Meig.*, an increase in the average progeny yields of 34.4%, as compared with the controls, if the insects were subjected to electric fields.

They believed that the effect was due to a change in the natural ion levels. Field observations by Maw[76] and others have shown that there is an apparent relationship between the location of male swarms of *Aedus trichurus Dyar* and the natural electrical gradients that exist around trees and shrubs in a forest.

Andersen and Vad[77] in Århus (Denmark) studied the influence of electric fields on bacterial growth of *Serratia Marcescens* and *Escherichia coli.* The latter had previously been studied by Wehner[78] in Texas in 1964 using fields with a strength of 555 V cm^{-1}. Changes in growth were expressed in terms of the diameter of colonies after incubation at 37°C for 46 h. In the case of *S. Marcescens* the mean diameters of the colonies were reduced significantly at constant field strengths of +940 V cm^{-1} as compared with field-free conditions. In the case of *E. Coli* the diameters were reduced at field strengths of ~3125 V cm^{-1}. Andersen[79] also studied the effect of electric fields in a room on the number of light ions in the inhaled air. If subjects walk in a low humidity room they will develop a certain electric potential relative to their surroundings. Previously, in 1956, Bach[80] had pointed out that such fields would affect the output of light ions from an ion source and therefore also the number of inhaled ions. Andersen[79] was able to confirm that the presence of electric fields in the environment reduced the uptake of light gas ions and changed the proportional uptake of positive and negative ions.

These observations support the view that fluctuations in the natural potential gradient may affect various living organisms and perhaps also man. However, considerably more research work is required before the effect of fluctuating potential gradients on man can be accepted.

5.6.2. Pulsating and alternating electromagnetic fields

Evidence for a direct or indirect biological effect of pulsating electrostatic or electromagnetic fields in the atmosphere on the living organism is supported by a number of studies.

Schua[74] demonstrated the sensitivity of golden hamsters to alternating fields of 900 V m^{-1} and a frequency of a few Hz. Animals moved their nests if electric fields were switched on above these nests. Previously Saito[81] had shown that rabbits placed between an electromagnet become restless after 8–10 min, the capillary vessels being dilated.

In 1953 Reiter[82] observed changes in the pH of subcutaneous tissue in guineapigs if the animals were placed between plates 1 m apart with alternating fields of 50–100Hz. Similar observations were made during natural thunderstorms. If the animals were placed in a Faraday cage no pH changes were observed.

Lotmar and Ranscht-Froemsdorff[83] studied changes in skin respiration of rabbits in relation to the weather and in particular to atmospheric electromagnetic waves of 3–50 kHz. During high pressure conditions respiration was accelerated, during low pressure atmospheric conditions it was decreased. The electromagnetic wave patterns under these conditions were stable and unstable respectively. A direct correlation with the electromagnetic waves was assumed because the experiments were carried out under constant temperature and humidity conditions. More studies are required to confirm these findings.

Thompson, Dunlap and others[84] described rapid contractions of the eye muscle if the head of a subject is surrounded by electric wires creating alternating electromagnetic fields of 60 Hz. Owing to these muscle contractions, the subject experiences flickering of the eye. Prolonged experiments even brought on headaches.

Studies by Haine and König[85] have shown that moulting of the aphid *Myzus persicae Sulz* was reduced when electromagnetic waves of 9 Hz and high intensities were applied. Moulting increased with medium intensities. Low electric intensities were followed by a delayed increase in moulting.

Direct effects on nerves are suggested by the experiments of Gengerelli and Holter.[86] They observed a contraction of frog's nerves if these were placed in an alternating field of 10 000 V and 60 Hz, provided that the nerve was not surrounded by a saline solution. Similar observations were made by Hermann as far back as 1887.

Recent studies by Frazer and Krupp at the USAF School of Aerospace Medicine in Texas, showed that if rhesus monkeys (*Macaca mulatta*) were exposed for six hours to radiofrequency radiation of 15–27 MHz, at power levels up to 1000 mW cm^{-2}, a modest rise in rectal temperatures was observed.

A summary of the most recent studies in the field of electromagnetic effects can be found in the abstracts of the papers submitted to a special study group, under the chairmanship of R. Reiter, during the 7th International Biometeorological Congress at College Park (USA), August 1975, published in the Proceedings of this Congress.[87]

The various reported experiments support the assumption that electromagnetic and fluctuating or pulsating electrostatic fields in the atmosphere, within certain frequency intervals, may have important biological effects on the nervous system of man and animals. However, considerably more research is required in this still under-developed area of electrobiometeorology.

5.7. EFFECTS OF COSMIC RADIATION

The discovery of cosmic rays goes back to 1908 when Hess in Switzerland, during a balloon flight, discovered an increase of particle radiation with increased altitude. Later balloon experiments showed a background radiation of more than ten times that of the earth. These early experiments led to the conclusion that radiation of very great energy and penetrating power reached the earth from a source beyond the earth's atmosphere.

Ordinary cosmic radiation is called at present 'galactic' cosmic radiation. The primary flux consists of about 85% hydrogen nuclei and 14% helium nuclei. The remainder is made up of heavier nuclei, the exact composition of the elements being still relatively unknown. It is known that cosmic rays are a complex of radiations derived largely from the incidence of extremely high energy protons along with some heavier particles with atomic numbers up to 40 (zirconium) and perhaps, beyond. Heavy cosmic particles are stripped of electrons and produce very broad paths of dense ionization as they enter matter.

The origin of cosmic radiation is not precisely established. It is believed, however, that it originates in interstellar space and that the source is that of remnants of *super novae*. Cosmic particles are believed to have been accelerated to their present energies by pulsating interstellar magnetic fields. With the exception of solar-induced variation,[5] galactic cosmic rays are remarkably constant. The energies of cosmic particles appear to average about 6×10^9 eV and probably extend as high as 10^{18} eV. Although the energies of these particles is high, the particles are low in intensity. This intensity in free space, of course, is substantially higher than that of their secondaries at sea level on earth, where man is protected by an atmospheric shield equivalent to 10 m of water, and in medium and low altitudes by the magnetic field of the earth, which deflects these particles back into space. The least energetic cosmic ray primary nuclei arrive at the earth's magnetic field with an energy of nearly 10^9 eV per nucleon.

Observations made in the Soviet Union and USA with space capsules since 1972[88] have shown (in outer space), gravity and radiation effects on both animals and plants. In general little is known about cosmic ray effects. However, a number of interesting experiments have been carried out since 1947 near the surface of the earth and these may give a clue to possible cosmic ray effects at very high altitude.

Figge[89] made some interesting experiments in 1946 on the influence of cosmic rays on cancer. 148 male mice were injected with 0.25 mg of methylcholanthrene in sesam oil. The mice were equally distributed in eight aluminium cages 11 × 11 × 4 in. Lead plates, $\frac{1}{4}$ inch thick, were placed over five out of the eight cages, about 3 in above the mice. When cosmic rays pass through thin sheets of metal they produce shower bursts of ionizing radiations, which intensify the influence of cosmic radiation. The optimum thickness for the production of small cosmic radiation showers was found to be 0.6 cm, for large showers 2 cm. Four of the cages were covered with a lead plate 0.6 cm in thickness, the other with lead plates of thickness 2 cm. The cages were placed on different floors of a ferro-concrete building. Figge found that mice exposed most to the effect of cosmic radiation developed the greatest number of tumours. After 10 weeks, of the 111 injected mice still alive and exposed to increased cosmic radiation, 84 had developed carcinoma, whereas with 67 mice, exposed to cosmic radiation in a normal way, only 22 had developed cancer.

Barnothy and Forro,[90] in Budapest studied the lethal effects of cosmic ray showers on the fertility of Angora rabbits and white mice by placing a lead cover of 16 mm thickness above the cages. Eugster and Hess[91] in Switzerland kept test animals in cages covered with 18 mm of lead at 2340 m altitude and studied also changes in fertility. However, their studies were not conclusive.

Ong[92] in Peking, in a similar way studied the influence of cosmic radiation on tuberculosis at high altitude and at sea level. Tuberculous mice exposed to cosmic radiation at sea level, under 2 and 10 cm thick sheets of lead, showed significantly greater mean survival times than those of tuberculous mice exposed directly to cosmic radiation. Also the virulence and immunizing properties changed at an altitude of 1897 m.

Planel *et al.*[93] studied the growth of cultures of *Paramecium caudatum* and *Paramecium aurelia* in the well-known underground laboratory in Moulis (Ariège) belonging to the Centre National de Recherches Scientifique in France. Those cultures subjected to strongly reduced cosmic radiation showed a delay in development and a lengthening of the cellular cycle. These studies confirmed previous findings of the same authors using 5–10 cm thick lead plates above the cultures in the laboratory above the surface of the earth.

Similar results were found by Planel *et al.*[94] in 1967, using *Drosophila melanogaster*. Planel *et al.*[95] also studied the effect of reduced cosmic and other ionizing radiations on the growth of human cancer cells of the type KB and HEP-2. Lead plates of 5 cm thickness reduced the growth rate by 8%; under 10 cm plates the reduction was 14%.

More recent studies with the US satellites have shown biological effects on *Clostridium sporogenes*, *Neurospora conidia*, *Escherichia coli* and other plant organisms. Human γ-globulin showed an increased reactivity after cosmic radiation in satellites. Soviet scientists believe that blood coagulation of man is affected by cosmic radiation, however, gravity changes and stress conditions may also be involved.

The results of the various reported biological effects of cosmic radiation must be studied more systematically, particularly in view of the probably increasing number of space flights in the future.

5.8. EFFECTS OF MAGNETIC FIELDS

The average field strength of the natural magnetic field near the surface of the earth is about 0.5 Oersted (or Gauss; 1 Gauss is the force created by a magnetic field which exerts a force of 1 dyne on a unit pole). The magnetic field strength on the surface of the earth is subject to a number of rhythmic changes: a daily variation; a thirteen day period; a lunar period of 27–28 days (causing a change in amplitude of 2 γ; 1 $\gamma = 10^{-5}$ Oersted); a yearly variation; a secular variation of 960 years; a period of about 11 years related to the sunspot period; and magnetic storms during great sun activities and solar flares.

The south pole of the earth's magnetic field is near the geographic north pole, the north pole is near the geographic south pole.

The intensity of the geomagnetic field decreases approximately by 0.04% for every km elevation and reaches, in the region called the magnetopause (between 8 to 14 earth radii on the earth's sunlit side and up to 31 earth radii on its darkened side), about 1% of the geomagnetic field intensity. In the interplanetary region around the earth the field strength is usually about 5 γ, but may change in other parts of the solar system. On the moon the field intensity is about 100 γ, on Venus presumably 500 γ and on Jupiter probably 1000 γ. Therefore man during his lunar and interplanetary journeys will be exposed to magnetic environments considerably lower than exist on earth (i.e. 50 000 γ). Moreover these planetary field intensities are not merely lower, but they may exhibit rhythmic variations with frequencies different from those on the earth.

To 'remove' an organism on the earth's surface from the ever present geomagnetic field is not simple. One has either to compensate the geomagnetic field by an artificial field of opposite direction and compensate also for all variations in the geomagnetic intensity, or one has to shield the specimen from the ambient field. The first can be accomplished with elaborate Helmholtz coil systems, the latter by using enclosures made from mu-metal (an alloy which has a very high permeability at very low field strength). Boxes made of mu-metal, nested in each other with non-magnetic spacers, may reduce the intensity inside the enclosure to less than 100 γ. Experiments by Beischer[96] with human subjects did not show noteworthy changes in most of the physiological and psychological parameters investigated if human subjects were exposed to such a 'zero field' environment for a period of 12 days, with the exception that in the absence of the geomagnetic field some decrease occurred in the peripheral critical flicker-fusion frequency of vision. On the other hand many animals show a geomagnetic tropism even with very weak magnetic fields. For example Conley[97] observed, in a field 1000 times weaker than the geomagnetic one, a reduction of the response of young male mice to induced foreign biopolymers, measured through the effect of enzyme activity.

Experiments by Halpern and Van Dyke[98] seem to suggest that mice reared in mu-metal cages exhibited signs of premature aging, early death, behavioral changes and development of *alopecia* (loss of hair). Orientation effects of the geomagnetic field on the motion of planarians were reported by Badman,[99] Brown,[100,101] Barnwell[102] and others. Snails seem to be capable of distinguishing geographical directions and orientations of an artificial magnetic field. The curve of emergences of snails from a narrow opening in an artificial field were mirror images of emergences of the same snails in the geomagnetic field. Also the orientation of the bee dance on a vertical honeycomb seems to be affected by experimentally induced small alterations of the ambient magnetic field.[103]

The orientation of birds during their period of migratory unrest has been shown to be altered by experimental manipulations of the magnetic field directions of their environments.[104] Geomagnetic tropism has also been observed[105-108] in plants with a tendency of the roots to align in the north pole direction of the magnetic field. Summaries of the most important reported biomagnetic effects have been published by Barnothy,[109] Caldwell, Persinger, Russo,[110] and Tromp.[111]

Under strictly controlled laboratory conditions some of these biomagnetic effects in very strong or very weak magnetic fields are biologically interesting phenomena. However, under normal conditions at the surface of the earth, biomagnetic effects on man are very small and are negligible as compared with other physical environmental stimuli.

5.9. EFFECTS OF GRAVITATIONAL FIELDS

The force of gravity (g) is one of the most characteristic parameters of our environment. Since all living organisms on earth are continuously exposed to the force of gravity they are responsive to changes in the magnitude of acceleration forces. All terrestrial organisms are required to exert some effort against gravity in supporting and maintaining their body weight, the principal component of which is water. Therefore it is reasonable to assume that gravity influences the optimum balance of body fluids because of the mobility of water with respect to gravity. Despite this gravitational dependence still very little is known of its effect on human physiology.

A special type of gravity influence is known as weightlessness, a condition that is believed to occur in a gravity-free space. However, there is no gravity-free space because the force of gravity extends to infinite distances and never ceases to exist entirely. Therefore weightlessness must be defined as a condition in an environment of greatly reduced gravity, the degree of reduction being dependent on the orbital distance from the earth. About 320 km above the surface of the earth the gravity force is only 10^{-5} g. Therefore during space flights weightlessness is the principal environmental variable.

Duplication of weightlessness on earth has been possible for only brief periods (less than 60 s) in aircraft flying a parabolic flight. These short exposures to weightlessness did not indicate any changes of physiologic significance, however, no conclusions could be drawn relative to the consequences of exposure to the weightless state for long periods. Some analogy to weightlessness has been achieved through studies involving immobilization such as bed rest and water immersion, the effects of which will be discussed later in this section.

Exposure of different organisms to different accelerative forces in large centrifuges considerably increased our knowledge of gravity effects. For example cells of *Escherichia coli*, exposed to 110 000 g, were larger than control cells due to an enlargement of the entire cell as well as of its intracellular organelles. Cell growth was dissociated from cell division, the latter being suppressed more than the former. Experiments with *Drosophila melanogaster* larvae showed that both the decrease in growth rate and the decrease in size become more marked as the g force increased from 1630 to 4000 g.

White mice, *Mus musculus*, continuously exposed to centrifugation (4 g) for four weeks, showed a decrease in body-mass and definite changes in skeletal growth.

Hamsters exposed to 4–5 g for four weeks, showed evidence of curtailed growth. However, the heart, diaphragm, gastroenemius muscle, lungs, kidney and head showed a relative increase in mass, while the femurs showed a relative increase in length.

Several experiments were carried out with man. In 1960 Clark[112] exposed human subjects for a few hours to a 2 g acceleration. The changes observed included a decrease in urinary output and an increase in the white blood count. Some tingling sensation remained in the little fingers up to months after the experiment.

Laboratory studies carried out on the effects of immobilization, achieved through bed rest and water immersion of the body, have shown that the lack of physical stress and strain on the musculo-skeletal system in the weightless state results in significant losses of calcium, nitrogen and related elements.

Immobilization in bed leads to an excess of *osteoblastic* activity (an osteoblast is an immature bone-producing cell). There is an imbalance of bone removal and its replacement which gives rise to low-grade hypercalcaemia (excess of calcium in urine), hypercalciuria (excess of calcium in blood) and an increased incidence of renal malfunction and renal calculi.

After three weeks in bed, healthy persons exhibit an increase of more than 20% in heart rate, a reduction of 10% in total blood volume (due primarily to a loss of plasma volume) and a decrease in heart size. A 9% reduction in basal metabolic rate occurs. In a series of seven 14-day bed-rest studies, it was shown that, regardless of the amount of dietary calcium (300 to 2000 mg) which is consumed per day during bed rest, there is some loss of bone mass in every case. Because of the immobilization experiments in beds it was believed that significant losses of calcium, negative balances of nitrogen and other substances might occur during space flight for several weeks, leading to serious weakness of the musculo-skeletal system.

Water immersion experiments in darkened, heated tanks have produced a variety of disorienting and hallucinatory experiences with considerable subjective variation. The precise physiologic mechanisms responsible for loss of orthostatic (upright position) tolerance are unknown, but may be related to diminished blood volume, decreased muscle tone in the extremities or to functional alterations in the sympathetic nervous system. A group of Navy divers were studied under total water immersion. They remained under water continuously for 18 h to simulate weightlessness and were tested thereafter on a human centrifuge for blackout threshold. All subjects exhibited a reduced tolerance for acceleration and reflected some decrease in cardiovascular function.

These various experiments, utilizing centrifuges, water immersion and other immobilization experiments, suggest that the principal difficulties in the space environment would be manifest in the central nervous system, the cardiovascular system and the gravity receptor organs. The chief consequences are believed to be disorientation and the problem of combined stresses of altered gravity. On the other hand there is a strong conviction that biochemical and physiological adjustments are able to compensate for the change in accelerative force and to permit physiological adaptation to hypogravity.

For further detals see articles by Saunders.[113]

5.10. EFFECTS OF RADIOACTIVE FALL-OUT

An important factor in our modern environment, with dangers of nuclear wars and an increasing number of nuclear plants, is the so-called radioactive fall-out, a concept which originated in 1945, after the first atomic bomb, and refers to the radioactive debris produced in a nuclear detonation which falls or settles back onto the earth. The amount of radioactive material and its characteristics are dependent upon the size of the weapon and the height of the burst. The geographical distribution of the fall-out is determined by weapon size and meteorological conditions.

Casarett and Thompson described the events in more detail.[114] The principal types of radioactive fall out were described by these authors as follows:

'If a burst occurs at ground level or at very low altitudes, large quantities of soil or water are sucked into the updrafts of the hot fireball; radionuclides which have been produced in the fission process will deposit on the particles and water droplets. The larger particles soon fall out on the ground in the vicinity of the explosion site, usually within 24 h after the detonation. This is "local" fall-out. Small particles and condensed nuclei rise with the cloud. They may remain in the lower layer of the atmosphere (troposphere) or, if the detonation is large, they may be injected into the stratosphere. The tropospheric material is carried around the earth, primarily at the latitude of the detonation, and is deposited within four to six weeks in snow, rainfall or as dry deposition. This is "intermediate" fall-out. Stratospheric debris may remain in the atmosphere for months or years before it returns to earth, usually in the hemisphere of the detonation, as "delayed" or "worldwide" fall-out.'

Of the more than 200 radioactive nuclides produced in fission processes, only a few are known to represent a real hazard to man through their appearance in fall-out. The hazard from a radionuclide depends on several factors:

(a) *Amount produced*: the more there is of a given radionuclide, the greater the potential hazard from it;

(b) *Radioactive half-life*: the half-life of a specific radionuclide is the time required for one-half of it to undergo radioactive decay. Radionuclides with half-lives of days or a few weeks (such as iodine-131, with an eight day half-life) are important only in local and tropospheric fall-out. Radionuclides with intermediate half-lives such as strontium-89 (half-life of 51 days) can be important when weapon testing is in polar latitudes, because of the rapid removal of the debris from the stratosphere in these latitudes;

(c) *Transfer to man*: most radionuclides from fall-out are transferred to man through various pathways in terrestrial and aquatic food chains. Some are deposited directly on the surfaces of plants which are eaten by man; this is a simple plant-to-man food chain. Others pass through a long food chain with potential loss or removal at each step. Aquatic food chains contribute less than terrestrial chains to human intake of fall-out products, since the water environment provides significant dilution. However, radionuclides such as tritium, which are produced in some peaceful applications, may become increasingly important in aquatic food chains;

(d) *Metabolism in the body*: radiation exposure (and biological effect) from internally deposited radionuclides depends on their retention time and also on the site of localization. For example strontium-90, which is deposited in bone and removed slowly, is potentially more hazardous than caesium-137, which is deposited in soft tissue and is removed relatively rapidly.

Based on these considerations, the radioactive materials in fall-out which are most likely to represent significant hazard to man are iodine-131, strontium-90, strontium-89, caesium-137 and carbon-14.

Iodine-131 enters the body through the consumption of milk and edible plants or plant products, as a result of the contamination of plant surfaces. It concentrates in the thyroid gland and provides the greatest radiation exposure to that tissue. Since iodine-131 has a half-life of only about eight days, however, it is of serious importance only during the first two weeks after release.

Strontium-90 is considered of greatest importance because of its long half-life and the presence of a short-lived high energy daughter (yttrium-90). Strontium is introduced into the food chain by direct deposition on plant surfaces or by uptake from the soil. It follows the metabolic pathways of calcium; in the body, strontium is incorporated in bone.

Caesium-137 enters the food chain if it is deposited on plant surfaces. If it enters the soil, however, it tends to be retained there rather than entering the food chain through plant roots. In the body caesium follows the same metabolic routes as potassium, and localizes in soft tissue. It can represent a genetic hazard as a result of exposure of the gonads but, because it is removed relatively rapidly from the body (several months), caesium is not considered to represent as great a hazard as strontium-90 in terms of life-time exposure.

Carbon-14 is produced by neutron activation in the atmosphere when a weapon is detonated. It is considered a hazard because of its long half-life (about 5800 years) and because it is readily incorporated into plants and animals.

For further details we must refer to the Studies by Casarett and Thompson[115] and to the many references mentioned in their papers.

5.11. EFFECTS OF AIR POLLUTION

The most important aspects of air pollution, caused by organic and inorganic particles and chemical pollutants in the atmosphere, have been summarized already.[116,117] It is surprising that direct evidence for important biological effects of air pollution is very scarce. It is well known that the smell and the itching effect on the eyes create considerable discomfort. However, important biological effects are difficult to prove.

Biersteker[118] reviewed the various reported biological effects. He distinguishes four different types of pollution in relation to their biological effects: reducing pollution, oxidizing pollution, industrial pollution and natural pollution.

5.11.1 Reducing pollution

This can be defined as the pollution which has occurred for many ages in coal and oil burning communities mainly due to the production of sulfur oxides and smoke as a side effect of the use of these fossil fuels. Communities which burn large quantities of soft coal in open fires are the worst off because of the high smoke concentrations (up to a few mg m^{-3}) which are developed in the ambient air during windless, foggy weather. The concentrations of benzpyrene and trace elements reach levels which are five to ten fold higher than normal during such conditions. Carbon monoxide and dioxide concentrations follow the same trend. According to Lawther *et al.*[119] definite relations between this type of carbon monoxide and dioxide pollution, and peak breath flow rates or bronchoconstriction (as observed by Nadel *et al.*[119] with sulfur dioxide concentrations of 4–6 ppm) could not be established. Since measures were taken in London against the use of soft coal the relationship between daily mortality fluctuations, hospital admissions for respiratory diseases and well-being of bronchitics with fluctuations in air pollution level disappeared. Also reduced sputum production by chronic bronchitics was observed in the London area. Studies of school children showed that air pollution may affect the prevalence of bronchitis and the incidence of other respiratory diseases. This was found not only in England but also in Japan.

5.11.2. Oxidizing pollution

This can be defined as the pollution which manifests itself in communities with relatively little smoke and sulfur dioxide pollution, but with a high emission of nitrogen oxides and hydrocarbons (especially olefins) from cars under strong sunlight and low wind speeds. The result is the production of so called secondary pollutants: peroxides like peroxyacetylnitrate (PAN) and related compounds (peroxyproprionyl nitrate (PPN) and peroxybutyryl nitrate (PBN)) and ozone. This oxidizing pollution or photo-chemical smog was first observed in Los Angeles in the early 1940s, where Haagen-Smit, Middleton, Leighton and others[120,121] showed that it depends on the photo-chemical breakdown of hydrocarbons mainly emitted by automobiles. This type of pollution was reported in the 1960s from many other communities in the USA, and has also been reported from Rotterdam and Moscow. It is not clear yet whether the registered rise in oxidants, which in Los Angeles reaches peak levels of approximately 0.5 ppm in the middle of the day, represents the same biologically active ingredients in different communities. The plant damage can be of the PAN type or of the ozone type. The haze formation possibly depends partly on the presence of sulfur dioxide in the ambient air at the same time that oxidants are being formed. Because motor vehicle exhausts play such an important role in the formation of this type of smog, correlations with carbon monoxide and lead levels of pollution can be expected. Ozone levels usually reach a peak later in the day than the other pollutants. It has not yet been possible to single out the specific intermediate products of the photo-chemical processes which produce biological effects. Experimental data show that subjection to concentrations of oxidants as little as 0.15 ppm of ozone for a few hours lowers the resistance of mice against streptococcal aerosols. It has also been shown that older guinea pigs in Los Angeles develop increased airway resistance during smog and that mice developed more lung adenomata.

Some human volunteers exposed to 2 ppm of ozone for two hours experienced very unpleasant effects which lasted for several days. Some subjects developed bronchoconstriction on exposure to 0.1 ppm of ozone. Biersteker[118] pointed out that 'it is difficult to draw conclusions about the exact mechanism through which oxidizing pollution affects man's health. The intermittent exposures, the differences which the ambient temperatures make to the dose-effect relationship, the cross-tolerance for acute but not for chronic exposures, the high vulnerability under physical exercise, and the effects of diet and age all have helped to illustrate the complexity of the mechanism which operates in real life'. Epidemiological observations in the Los Angeles area have shown that eye irritation occurs in about 25% of the population when oxidants reach concentrations of 0.1 ppm; the exact lachrymators in the smog have not been found. The possibility exists that oxidizing smog without lachrymators occurs.

5.11.3. Industrial pollution

This can be described as air pollution which occurs down-wind of factories and also as the air pollution which occurs down-wind of certain agrarian activities, for example the keeping of hundreds or thousands of animals in bio-industries and the spraying of pesticides. It may contribute very significantly to the levels of sulfur dioxide, particulates, and hydrocarbons which occur in a residential area, but the effects are based on the same considerations which have been discussed above. Apart from that, industries can temporarily or chronically emit pollutants with biological effects, which are relatively well known due to the experience of industrial hygienists with these pollutants.

In Table 5.2 a list of some important air pollutants is given together with their toxicities according to the American Conference of Governmental Industrial Hygienists.

TABLE 5.2
List of air pollutants prepared by the American Conference of Governmental Industrial Hygienists

Pollutants	Maximum permissible concentration	Duration	Criteria
Iron oxide	10 mg m^{-3}	8 h	Haziness of atmosphere
Beryllium	0.0002 mg m^{-3}	8 h	Berylliosis risk
Ozone	0.05 ppm	short	Odour threshold
Aldehydes	0.01 ppm formaldehyde	short	Irritation of eyes, nose and
	0.01 ppm acroleine	short	throat
	0.2 ppm total aldehydes	short	
Ethylene	0.5 ppm	1 h	Plant damage
	0.1 ppm	8 h	
Lead	0.2 mg m^{-3}	30 d	Human health
Fluorides	0.001 ppm	30 d	Plant damage
	0.0035 ppm	24 h	
Carbon monoxide	20 ppm	8 h	Human health
	70 ppm	1 h	
Phenol and cresol	100 mg m^{-3}	short	Odour threshold
Particulate matter			
(suspended)	5 mg m^{-3}	30 d	Haziness of atmosphere
(settled)	1 mg cm^{2}	30 d	
Sulfur	80 mg m^{-3}	Annual mean	Medical and others
	260 mg m^{-3}	24 h maximum	
	800 mg m^{-3}	1 h maximum	

5.11.4. Natural pollution

Natural pollution of the air with ozone, dusts, pollen, sulfur dioxide, hydrogen sulfide, carbon monoxide, hydrocarbons etc. occurs in many areas. The concentrations usually remain low and they have therefore attracted little attention, except for the inconvenient effects they may produce.

Although the number of people suffering from natural allergens (particularly spores of fungi, dust, pollen etc.) is very large, the cause is considered natural and therefore less threatening than the anthropogenic pollutants mentioned above.

For further details see the publications mentioned above and the references given in these papers.

5.12. EFFECTS OF ACOUSTIC WAVES (NOISE)

Noise is a unwanted sound. Its intensity is usually measured in *decibels* (dB) on a relative scale. An increase of 10 dB corresponds to a doubling of loudness. 100 dB is very loud, 120 dB can cause actual pain or damage. The extreme limit of noise tolerance is 140 dB. Some examples of sound levels are: a whisper (20 dB), conversation (30 dB), television (73 dB), crying child (80 dB), alarm clock (85 dB), underground train (90 dB), heavy lorry (93 dB), circular saw (105 dB), unsilenced motorcycle (120 dB), pile-driver (130 dB).

Many governments have passed legislation setting maximum permissible decibel levels for road vehicles, usually around 75 dB for cars and 80 dB for lorries.

A member of the French National Academy of Medicine showed[122] that the reduction of noise in an office by some 20 dB resulted in a 9% increase in working capacity and a 29% decrease in spelling mistakes. US business firms are estimated to lose over four million dollars a day through loss of efficiency as a result of office noise.

The influence of sound waves (i.e. noise) on the human body can be due to direct effects upon the auditory system, on non-auditory phsyiological processes and on purely psychological mechanisms.

The direct influence on the auditory system can be a 'temporary' threshold shift, i.e. short term effect which may follow exposure to noise. If the limit of noise exposure compatible with recovery is exceeded, recovery will effectively cease after a particular period of time after the exposure, and the residue is called 'permanent' threshold shift.

The term 'persistent' threshold shift is used to indicate the threshold shift remaining after at least 40 h. The word 'permanent' is used for conditions from which no recovery seems to be possible.

Finally, acoustic trauma is a condition of sudden aural damage resulting from an intense, short term exposure.

The mechanism involved in ear damage is the mechanical bending of the ear structures, repeated thousands of times per second and with degrees of displacement and acceleration in excess of the normal, which will damage the cochlear structures. One can observe effects varying from minor changes to obvious damage and even complete breakdown and disruption of the organ of Corti. People who have been exposed to very intense noise, explosions for example, may sustain ruptured eardrums and damaged ossicles, such middle ear damage may also be associated with the damage of the cochlear structures.

The following effects have been observed in non-auditory physiological systems. Sound waves of 90 dB can elicit vegetative reactions such as changes in the peripheral circulatory system and the pupillary function. Constant dilatations of the pupil can be observed during exposure to noise stimuli. The stroke volume of the heart decreases together with the noise-induced peripheral vasoconstrictions. Vasoconstriction in the skin due to noise stimuli can be easily demonstrated with a plethysmograph. In tested subjects cumulative vegetative reactions can be observed due to repeated noise stimuli. The vasoconstrictive effects as well as pupil dilatation depend upon the intensity and bandwidth of the noise spectrum. Beginning at 70 dB, both effects increase with increasing sound intensity. With a constant intensity level the greater the bandwidth of the noise spectrum, the greater the vasoconstrictive effects. These reactions are not dependent on the psychological emotions of the subjects. It is not yet certain whether these vegetative reactions to sound waves of 70 dB could affect human health seriously.

Several studies[123-127] indicate that noise stresses of more than 100 dB produced increased activity of the adrenal cortex in animals if the exposure lasted several days to several months. Hypertrophy of the adrenal cortex was observed, together with increased weight and lipid content of the fascicular zone. Apart from increased ACTH and corticosteroid release, increased adrenaline and noradrenaline excretion has also been observed. Studies by Singh and Rao[124,125] suggest that prolonged stimulation stimulates gonadotropin output, increased ovarian and uterine weights etc. Numerous studies (e.g. by Woodbury and Vernadakis[124,125]) indicate marked alterations in brain excitability of animals due to increased adrenocorticoids, particularly of the cortisone type, as a result of strong noise stresses.*

In this connection a recent study by Yodlowski *et al.*[128] should be mentioned. It is known that acoustic waves of frequencies below 10 Hz are common in the atmosphere. They are generated by various sources, such as wind, thunderstorms, weather fronts, magnetic storms, aurorae, ocean waves, earthquakes etc. Many of these atmospheric oscillations are of high amplitudes well above 100 dB at frequencies below 5 Hz, and above 120 dB below 1 Hz.

Yodlowski *et al.*[128] demonstrated that homing pigeons (*Columba livia*) can detect sound energy below 10 Hz at amplitudes well within the range of naturally occurring sounds at these frequencies. They could detect frequencies

* Recent research by I. Singh and I. S. Chohan on the effects of noise on blood coagulation is reported in *Biometeorological Survey (1973–1978)*, Heyden, London, 1979.

below 1 Hz. Further experiments suggested that the receptors for the infrasound are in the inner ear, but it is not clear which organ is responsible.* Atmospheric infrasounds can travel very large distances, often thousands of kilometres, without much attenuation. Such infrasounds enable birds to sense approaching weather fronts by detection of infrasounds transmitted to the earth's surface or by detection of the infrasound components of thunder associated with storms that are still hundreds of miles away. The localization of several remote infrasound sources may serve as a distant landmark system for triangulation of the home site and of the location of a release site for homing pigeons. Yodlowski *et al.*[128] assumed that infrasound information could thus assist in almost every aspect of avian navigation, in both homing and migration.

A third important effect of noise may be due to psychological stresses created by a continuous, high noise level. These psychological effects may be classified in the following categories: interference with sleep or rest; interference with working performance; interference with any form of communication based on sound. Together they represent the influence of sound waves on human behaviour. It is possible that sound waves, causing chronic psychological reactions (annoyance, disturbance of sleep and concentration, irritation, etc.) act harmfully on human health through the well known physiological stress mechanisms. Studies by Sommer and Harris[124,125] suggest that there are effects of noise and vibration on mental performance as a function of the time of day. They observed high efficiency of mental performance around 3.00 pm, low efficiency around 6.00 am.

The subject's susceptibility to noise increased during the low period of efficiency. The sensitivity to a 5 Hz vibration combined with 85–110 dB noise had a differential effect on mental arithmetic performance depending on the time of the day of the test.

Finally it has been suggested by research workers in the USA that noises could effect the behaviour and possibly also the physiological functions of a foetus in pregnant women. These and other important social implications of noise stresses urgently require additional physiological studies.

During the last 20 years the physiological effects of complete silence (so-called 'sensory isolation') have been studied, particularly in the USA, Canada, Japan, USSR and Germany. Complete sensory deprivation was used by ancient Greeks in their oracles, by Indian tribes and by Hindu or Buddhist priests or yogis, to improve meditative states. It enabled them to relax completely, to reduce the activity of their autonomic nervous systems and to slow down their metabolic rhythms. Heart and respiration rates decreased, and brain waves showed a predominance of alpha patterns.

The subjects seldom entered a state of deep sleep; they remained wakeful. However, some subjects become panic stricken under these conditions and had to leave the room where they were exposed to complete sensory deprivation. Sensory deprivation is reputed to be a regular 'brain washing' technique.

* See also studies by H. Richner mentioned on p. 112.

5.13. FACTORS AFFECTING THE RESULT OF A BIOMETEOROLOGICAL STIMULUS

Seven important factors may change the outcome of a meteorological stimulus. They are: the former history of the subject (i.e. the antecedent conditions experienced by the subject); the phase of the biological rhythms during the time the meteorological stimulus affects the body; the degree of acclimatization; the sex, age and typology of the subject; type of clothing used; the nutrition consumed and psychological stresses.

5.13.1. Previous history of the subject

In 1931 Wilder[129] introduced a physiological principle, the 'law of the initial value'. This states that the result of a stimulus depends on antecedent conditions: the higher the initial value of a physiological process, the less the tendency for the initiated process to increase. Thus the higher the excitability of the autonomic nervous system resulting from weather stress, the smaller the increase after renewed excitation. This principle, related to the phenomenon of 'accommodation' during electrical excitation of nerves, explains why a certain biometeorological effect caused for example by a cold front or by the influx of cold air masses, may differ if the cold front is preceded by a warm front or if a number of cold fronts pass an area at short intervals. In the latter case quite contrary reactions may be observed.

5.13.2. Biological rhythm

Apart from previous biological events the biological rhythm of an organism also plays an important part. For instance, various studies suggest that during the night parasympathetic stimulation dominates. Therefore, the same weather stress can have an entirely different result, depending on the hour of the day that a certain meteorological phenomenon occurs. In other words the effect of a stimulus depends on the phase of the biological cycle at the time at which the stimulus affects the body. See also Section 7.4.3.(b).

5.13.3. Acclimatization

It was pointed out, in Section 3.5, that acclimatization is not a reflex action, but is the result of slow physiological adjustments. Previous exposure to thermal stress assists adaptation at a later period. The time interval differs with different individuals and depends on the degree of thermal stress; it is usually a matter of weeks. This explains why higher sweat rates can be elicited in summer than in winter and why an influx of cold polar air after a period of heat adaptation will have a greater effect than after a previous cold period. Repetition of stimuli, as experienced during acclimatization, may be accompanied by gradual changes in responses to those stimuli (habituations).

5.13.4. Sex, age and typology

Other factors which affect the outcome of a meteorological stress are the sex and age of a subject. Young children and people in old age usually lack an efficient thermoregulatory mechanism. It follows that the result of a thermal stress in these groups will differ from the effect in normal healthy adult subjects. Studies by Karvonen[139] in Finland indicate that the aerobic power of women, i.e. the maximum amount of oxygen to be absorbed by the lungs which determines the maximum working capacity, is less than in men. In Sweden, in men of 20–29 years, it is 3.01 l h^{-1} of oxygen; in women 2.23. This difference is partly due to a difference in height, and partly it is the result of the smaller number of erythrocytes per ml of blood in females. As a result the oxygen absorption is less. Studies by Hardy,[131] Kawahata,[132] Sargent,[133] Wyndham[134] and Kuno[135,136] indicate that at a given temperature women perspire on the whole less than men per unit of body surface. Differences are small on the cheek, breast, forearm and calf. Usually perspiration starts in women at higher skin temperatures than in men. Shortly before and during menstruation the thermoregulatory efficiency of females is considerably reduced. In view of these thermoregulatory problems in relation to sex and age it is evident that the outcome of a meteorological stress could be entirely different under apparently similar laboratory conditions. Apart from sex and age, the differences in body built and personality pattern could affect the outcome of a meteorological stress.

5.13.5. Type of clothing

Three clothing factors determine the outcome of meteorological stimuli: the material, the shape and the colour of clothing.

(a) *The material of clothing* determines the degree of heat insulation (i.e. poor heat conduction), of heat absorption and the magnitude of electrostatic charges which are caused by the friction between the skin and the clothing.

The heat insulation is not only determined by the thermal conductive properties of the textile fibres but particularly by the air in the pores of the material. Wool consist of fine curled scales which keep in the air and prevent air movement. Cellulose and protein fibres have a great sweat absorption capacity. This explains why cotton and woollen clothes can absorb an amount of water equal to 20–40% of their weight without feeling humid. Nylon, terylene, dacron, orlon etc. have little water absorption capacity and prevent heat loss during thermal stress. They also create high electrostatic charges during dry conditions, particularly if the subject wears leather shoes. Charges of more than 1000 V have been recorded due to friction between shoes and carpets. Friction of wool, cotton and silk create positive electrostatic charges on the skin, negative charges on the clothes. The biological effect of these high electrostatic charges on the peripheral nerves is still unknown.

(b) *The shape of clothes* is very important as it determines the number and size of air passages between open air and the body, which in turn affects the heat and humidity exchange between the body and the meteorological environment.

(c) *The colour of clothes* is of little importance for the reflection of long wavelength infrared radiation. However, the short wavelength infrared radiation is reflected by light coloured clothes. Coloured fine metal scales on the inside of winter coats and blankets, reflects the heat radiation of the body to the skin.

5.13.6. Nutrition

Nutrition determines the degree of metabolism and therefore the efficiency of the thermoregulatory system.

5.13.7. Psychological stress

It was pointed out in Section 3.3 that psychological stress could affect the thermoregulation rewarming curve and therefore the thermoregulation efficiency of a subject and its sensitivity to thermal changes in its environment.

REFERENCES

1. J. E. Greenleaf, *Int. J. Biometeorol.* **10**, 71 (1966).
2. F. Becker, W. Catel, E. Klemm, G. Straube and A. Kalkbrenner, *Arztl. Forsch.* **3**, 436 (1949).
3. H. Brezowsky, private communication, 1960.
4. S. W. Tromp, *Int. J. Biometeorol.* **11**, 105 (1967).
5. *idem*, *J. Interdiscipl. Cycle Res.* **4**, 207 (1973).
6. *idem*, *Medical Biometeorology*, Elsevier, Amsterdam, 1963.
7. *idem*, *Rep. Biometeorol. Res. Centre*, Leiden, No. 6, 1967.
8. *idem ibid.* No. 8, 1967.
9. *idem*, *Int. J. Biometeorol.* **7**, 59 (1963).
10. *idem*, *Rep. Biometeorol. Res. Centre* Leiden, No. 13, 1969.
11. M. J. Halhuber, K. Inama and H. Jungmann, *Z. Angew. Bäder-u. Klimaheilk.* **15**, 3 (1968).
12. J. Benoit, *Colloq. Int. Centre Nat. Res. Sci.* **172**, 121 (1970).
13. I. Assenmacher, *C. R. Soc. Biol.* **245**, 2388 (1957).
14. R. Miline, *C. R. Ass. Anat.* **87**, 431 (1955).
15. *idem*, in *Pathophysiologia Diencephalica* Springer Verlag, Vienna, Austria, 1958.
16. *idem* and M. Ciglar, *C. R. Ass. Anat.* **91**, 1035 (1956).
17. F. Hollwich, *Ber. Tagg. Dtsch. Ges. Inn. Med.* **26**, 278 (1965).
18. *idem*, *Ann. Inst. Barraquer* **9**, 134 (1969).
19. W. F. Ganong, M. D. Shepherd, J. R. Wall *et al.*, *Endocrinology* **72**, 962 (1963).
20. M. Radnot and E. Wallner, *Acta Opthal. (Kbh.)* **43**, 164 (1965).
21. F. Hollwich and B. Dieckhues, *Klin. Mbl. Augenheilk.* **149**, 840 (1966).
22. S. W. Tromp and J. J. Bouma, *Proc. 5th Int. Biometeorol. Congr.*, Montreux, Switzerland, 1969, p. 88.
23. F. P. Ellinger, in Ref. 6, p. 336.
24. E. W. Dempsey and E. B. Astwood, *Endocrinology* **32**, 509 (1943).

25. G. M. Higgins, *Anat. Rec.* **58**, 205 (1934).
26. S. W. Tromp (Ed) *Progress in Human Biometeorology III*, Swets & Zeitlinger, Amsterdam, 1974.
27. D. Raab, *Z. Biol.* **39**, 524 (1898).
28. H. Von Tappeiner, *Dtsch. Med. Wschr.* **16**, (1904).
29. H. F. Blum, *Photodynamic Action and Diseases Caused by Light*, Reinhold, New York, USA, 1941.
30. H. Haxthausen and W. Hausmann, *Strahlentherapie*, Sonderband XI, 1929.
31. S. Epstein, *J. Invest. Dermatol.* **2**, 43 (1939).
32. S. W. Tromp and J. J. Bouma, *Monograph Series Vol. 13*, Biometeorol. Res. Centre, Leiden, 1974.
33. E. Betz, *Rep. Physiol. Inst.*, Univ. Marburg, Germany, 1960.
34. in ref. 6, p. 298.
35. D. J. Macht, *Amer. J. Pharm.* **106**, 135 (1934).
36. A. J. Lehman and P. J. Hanzlik, *Proc. Soc. Expl. Biol.* **30**, 140 (1932).
37. W. Czarniecki *et al. Biometeorology 5* Part 1, 44 (1972).
38. A. J. Nedzel, *J. Lab. Clin. Med.* **22**, 1125 (1937).
39. A. M. Alvarez, *Puerto Rico J. Publ. Hlth. Trop. Med.* **10**, 374 (1935).
40. E. B. Rentschler, F. R. Van Zant and L. G. Rowntree, *J. Amer. Med. Ass.* **92**, 1995 (1929).
41. S. W. Tromp, *Monograph Series Vol. 5*, Biometeorol. Res. Centre, Leiden, 1959.
42. J. Krypiakievicz, *Jb. Psychiat. Neurol.* **11**, 315 (1892).
43. P. Stocks, *Rep. Cent. Bur. Statist.*, London, UK, 1925.
44. M. L. Kreithen and W. T. Keeton, *J. Comp. Physiol.* **89**, 73 (1974).
45. S. W. Tromp (Ed), *Progress in Human Biometeorology I*, Swets & Zeitlinger, Amsterdam, 1972.
46. T. Winsor and J. C. Beckett, *Amer. J. Phys. Med.* **37**, 83 (1958).
47. A. P. Krueger *et al. J. Gen. Physiol.* **41**, 359 (1957).
48. K. H. Kingdon, *Phys. Med. Biol.* **5**, 1 (1960).
49. R. Pratt and R. W. Barnard, *Amer. Pharm. Ass. Sci. Ed.* **49**, 634 (1960).
50. G. B. Philips, G. J. Harris and M. W. Jones, *Int. J. Biometeorol.* **8**, 27 (1964).
51. M. C. Deleanu and T. Frits, *Arch. Phys. Therap (Lpz.)* **19**, 109 (1965).
52. R. Gualtierotti in *Aerionotherapy*, Carlo Erba Found. Milan, Italy, 1968, p. 56.
53. Z. D. Ostarova, *Nerv. Sist. (Leningrad)* **5**, 144 (1964).
54. A. A. Minch, *Vestnik. Akad. Med. Nauk.* **18**, 33 (1963).
55. F. G. Sulman, S. Levy, Y. Pfeiffer, E. Superstine and E. Tal., *Int. J. Biometeorol.* **19**, 202 (1975).
56. Private communication, 1976.
57. A. P. Krueger, S. Kotaka and P. C. Andriese, *Special Monograph Series, Vol. 1*, Biometeorol. Res. Centre, Leiden, 1966.
58. K. Bisa, *Z. Aerosol. Forsch. Therap.* **4**, 288 (1956).
59. K. Dirnagl, K. Bisa and R. Escher, *Siemens Z.* **28**, 341 (1954).
60. A. P. Wehner, *Amer. J. Phys. Med.* **48**, 119 (1969).
61. I. B. Anderson, in *Biometeorology 5, Part II*, S. W. Tromp, W. H. Weike and J. J. Bouma (Eds), Swets & Zeitlinger, Amsterdam, 1972, p. 229.
62. H. S. Burr and F. S. C. Northrop, *Proc. Nat. Acad. Sci. (Wash.)* **25**, 294 (1939).
63. Heydweiller, *Ann. Physik* **4**, 227 (1902).
64. F. Sauerbruch and W. O. Schumann, *Z. Techn. Physik.* **9**, 96 (1928).
65. W. O. Schumann, *ibid.* **9**, 315 (1928).
66. M. Von Ardenne, *ibid.* **9**, 288 (1928).
67. G. Oppenheim, *Westdtsch. Artztez.* **24**, 15 (1933).
68. L. J. Koopman, *Tijdschr. Parapsychol.* **8**, 97 (1936).

69. L. E. Murr, *Int. J. Biometeorol.* **10**, 135 (1966).
70. G. H. Sidaway and G. F. Asprey, *ibid.* **12**, 321 (1968).
71. J. H. Fabre, *The Sacred Beetle and Others*, Dodd Mead, New York, USA, 1918.
72. B. P. Uvarov, *Trans. Entomol. Soc. London* **79**, 1 (1931).
73. Wellington, *Amer. Rev. Entomol.* **2**, 143 (1957).
74. L. Schua, *Z. Vergl. Physiol.* **34**, 258 (1952).
75. W. C. Levengood and W. P. Shinkle, *Science* **132**, 34 (1960) and **133**, 115 (1960).
76. M. G. Maw, *Proc. Entomol. Soc. (Manitoba)* **18**, 1 (1962).
77. I. Andersen and E. Vad, *Int. J. Biometeorol.* **9**, 211 (1965).
78. A. P. Wehner in *Progress in Human Biometeorology II*, S. W. Tromp (Ed), Swets & Zeitlinger, Amsterdam, 1977, p. 233.
79. I. Andersen, *Int. J. Biometeorol.* **9**, 149 (1965).
80. Ch. Bach, in *Proc. 3rd Biometeorol. Congr., Pau, France,* 1964, p. 1.
81. Saito, *Jap. J. Obstet. Gynec.* **19**, 381 (1936).
82. R. Reiter, *Meteorobiologie und Elektrizität der Atmosphëre*, Akad. Verl. Gesellsch. Geest & Portig, Leipzig, East Germany, 1960.
83. R. Lotmar and W. R. Ranscht-Froemsdorff, *Z. Angew. Bäder-u. Klimaheilk.* **15**, 3 (1968).
84. in *Proc. Roy. Soc. B* **82**, 396 (1910).
85. E. Haine and H. König, *Z. Angew, Entomol.* **47**, 459 (1960).
86. J. A. Gengerelli and N. J. Holter in *Proc. Soc. Expl. Biol. Med. (New York)* **46**, 523 (1941).
87. *Proceedings 7th International Biometeorological Congress, College Park (USA) Aug. 1975*, Swets & Zeitlinger, Amsterdam, 1976.
88. in ref. 45, p. 381.
89. F. H. J. Figge, *Science* **105**, 323 (1947).
90. J. M. Barnothy and M. Forro, *Experientia (Basle)* **4**, 1 (1948).
91. J. Eugster, *Weltraum Strahlung (Hans Huber Verlag, Bern)* **60**, 109 (1955).
92. S. G. Ong, *Sci. Sinica, Biophys. Sect.* **2**, 263 (1964).
93. H. Planel *et al.*, *C. R. Soc. Biol. (Paris)* **164**, 654 (1970).
94. *idem ibid.* **161**, 467 (1967).
95. *idem ibid.* **165**, 2189 (1971).
96. D. E. Beischer, *Astronautics* **7**, 24 (1967).
97. C. C. Conley, in *Biological Effects of Magnetic Fields,* M. F. Barnothy (Ed), Plenum Press, London, 1969, p. 29.
98. M. H. Halpern and J. H. Van Dyke, *Aerospace Med.* **37**, 222 (1966).
99. D. G. Badman, *J. Psychol. Forsch.* **29**, 360 (1966).
100. F. A. Brown in *Proc. 3rd Biomagn. Symp., Univ. Illinois* 1966, p. 6.
101. *idem*, *Nature (London)* **209**, 533 (1966).
102. E. H. Barnwell and F. A. Brown, in ref. 97, p. 263.
103. M. Lindauer and H. Martin, *Z. Vergl. Physiol.* **60**, 219 (1968).
104. H. L. Yeagley, *J. Appl. Physiol.* **18**, 1035 (1947).
105. Pittman, *Biomed. Sci. Instrum.* **1**, 117 (1963).
106. *idem*, *Canad. J. Plant Sci.* **43**, 513 (1963).
107. *idem ibid.* **44**, 283 (1964).
108. *idem ibid.* **45**, 549 (1965).
109. in Ref. 97, p. 392.
110. in *Progress in Animal Biometeorology*, S. W. Tromp (Ed), Swets & Zeitlinger, Amsterdam, 1976.
111. in *Psychical Physics*, Elsevier, Amsterdam, 1949, p. 73 and p. 206.
112. in T. M. Fraser, *Human Response to Sustained Acceleration*, NASA SP-103 Washington, DC, 1960.

113. in ref. 45, p. 112.
114. *ibid*., p. 356.
115. A. P. Casarett and J. C. Thompson, in ref. 45, p. 356.
116. in ref. 6, p. 106.
117. in ref. 45, p. 61.
118. in ref. 45, p. 89.
119. P. J. Lawther, *et al.*, *Lancet*, 745 (1955).
120. A. J. Haagen-Smit, *Ind. Eng. Chem.* **44**, 1342 (1952).
121. *idem ibid.* **48**, 65A (1956).
122. E. Hammelburg, verbal communication, in ref. 45., p. 409.
123. G. Jansen, *Int. Z. Angew, Physiol.* **20**, 233 (1964).
124. in G. Lehmann and J. Tamm, *Forsch. Ber. Wirtsch. u. Verkehrsmin.*, NRW 517, Cologne, West Germany, 1958.
125. in G. Lehmann, *Arbeitsgemeinschaft Forsch.*, Landes, NRW, Cologne, West Germany, 1961, p. 94.
126. S. Mathias and G. Jansen, *Int. Z. Angew. Physiol.* **19**, 201 (1962).
127. G. Jansen, *Amer. Sp. & Hear. Ass. (Wash.)*, 89 (1970).
128. M. L. Yodlowski, M. L. Kreithen and W. F. Keeton, *Nature (London)* **265**, 725 (1977).
129. J. Wilder, *Z. Ges. Neurol. Psychiat.* **137**, 317 (1931).
130. Karvonen, *Rep. W.H.O.*, Geneva, Switzerland, 1970.
131. J. D. Hardy and C. Muschenheim, *J. Clin. Invest.* **13**, 817 (1934).
132. A. Kawahata, *J. Mie Med. College, Japan* **1**, 25 (1950).
133. F. Sargent, *Int. J. Biometeorol.* **9**, 229 (1965).
134. C. H. Wyndham, *J. Appl. Physiol.* **6**, 681 (1954).
135. Y. Kuno, *Human Perspiration*, Thomas, Springfield, Illinois, USA, 1956.
136. *idem*, *Proc. Int. Union Physiol. Sci.* **4**, 3 (1965).

CHAPTER 6

INFLUENCE OF METEOROLOGICAL STIMULI ON THE PRINCIPAL HUMAN DISEASES

A great majority of human diseases is either triggered, stimulated or even caused by meteorological stimuli. Most of these meteorotropic phenomena have been fully described in *Progress in Human Biometeorology, Part II.*[1] The present discussion will be limited to a small number of the most important human diseases and to some other phenomena related to human health.

6.1. SEASONAL DISEASES

The effects of meteorological stimuli on different basic physiological processes in man were reviewed in Chapter 5 and summarized in Table 5.1, pp. 94–96. For most of these parameters in addition to direct meteorological effects, seasonal or pseudo-seasonal changes in the magnitude of the effect were also given. Table 5.1 clearly shows that practically all basic physiological processes in man change during the different months of the year. As the infectious agents and the meteorological factors causing or triggering diseases also fluctuate in number and intensity during the different months of the year, it is evident that most diseases show a seasonal or fluctuating occurrence. These seasonal and pseudo-seasonal phenomena have been summarized for a number of diseases, in Table 6.1.

6.2. ALLERGY

Allergy is a phenomenon which can be defined as the altered reactivity of an individual to a specific substance, usually resulting from prior exposure to the same, or a chemically related, substance. The term was first used by Von Pirquet and Schick[2] in 1905.

A state of altered reactivity which is characterized by increased susceptibility to a protein or toxin, following ingestion or injection of the protein, or to

TABLE 6.1
Seasonal and other long-periodical relationships observed in various meteorotropic diseases

Diseases	Observed biometeorological relationships[a]
Infectious diseases (see p. 164)	
Angina tonsillaris (Quinsy, Ludwig's Angina)	Increasing after sudden influx of cold air.
Cerebrospinal meningitis	In western Europe beginning in autumn, high in December–April; in the USA maximum in February–March; in tropical west Africa maximum in dry season in dry areas.
Cholera	Rare in Europe but maximum in August. Common in India and surrounding countries in the hot humid tropics except during January–February in those areas where the temperature drops considerably in these months.
Common cold	Minimum in September, maximum in February–March (in northern Hemisphere). In The Netherlands and England, increase in incidence with falling temperature and rise in humidity.
Diphtheria and croup	In western Europe, beginning in August, maximum in November–beginning of December, but still high till February.
Dysentry	Bacillary dysentry in Germany; maximum in summer; in Great Britain February–April; in the tropics in the rainy season.
Influenza	In colder parts of northern hemisphere maximum in December–February; in southern hemisphere June–August; in 1918 in France and England epidemic related to dry, cold periods. In 1956 in The Netherlands: maximum in September–October.
Lobar pneumonia	In northern hemisphere maximum December–February; in tropical Africa maximum during dry season. High mortality if temperature during preceding week has been lower.
Poliomyelitis	In northern hemisphere minimum in March and beginning of April; increasing in April with maximum in August–September. In southern hemisphere: minimum September–October; near the equator, no seasonal fluctuations.
Scarlet fever (Scarlatina)	In western Europe beginning in August, maximum end of October–beginning of November; ending in December. In New York and Chicago: maximum in February–March. In northern USA incidence three times that of southern part; in the tropics rare. Scarlatina and other streptococcal infections highest during years with low humidity.
Small-pox (Variola)	In northern hemisphere beginning in August, maximum in March. In India and Ghana, maximum in dry periods.
Tuberculosis	Increased sensitivity to tuberculin test in March–April, low sensitivity during autumn; mortality increases in spring.
Typhoid (or Enteric) fever	In western Europe, increasing from May till September, followed by rapid decrease from middle of September. In Java maximum in wet season.
Typhus	If transmitted by ticks, peak of infection May–July (in northern Europe) coinciding with peak of breeding-season among ticks.
Whooping cough (Pertussis)	No particular period for all countries in northern hemisphere; different for different years, but often in spring and autumn.

[a] For references, see Ref. 4. For causes and symptoms of diseases, see Ref. 5.

TABLE 6.1—contd

Diseases	Observed biometeorological relationships
Non-infectious diseases	
Appendicitis	In 1933 Rappert studied 1000 cases. During 20 days, of which 16 coincided with weather-front passages, the daily number of cases doubled. Maurer, using 1398 cases in Münich (period 1933–1935), observed in 78.5% of the cases change in air-mass (normal chance in this period was 42.3%). Jaki using 2185 cases in Debrecen (period 1930–1939) did not observe a meteorotropism in that area. During heat-waves in USA Mills (1934) observed an incidence twice the winter values.
Arteriosclerotic heart diseases (coronary thrombosis, myocardial infarction, angina pectoris) and apoplexy	In western Europe and northern part of USA: maximum in January–February, minimum in July–August. In southern part of USA highest during hot summer, low during winter (see Section 6.5).
Asthma (non-infectious)	Maximum August–November in The Netherlands; highly significant correlation with atmospheric cooling; no consistent relationship with allergens (see Section 6.3)
Bronchitis	Maximum in winter, particularly in air-polluted areas; low in spring and summer (see Section 6.3.2).
Colics	
Hepatic colic (due to gallbladder stones: cholelithiasis)	According to Mauer, 83% of the colics occur in Munich (Germany) during changes in air masses (normal chance 42.3%). According to Hentschel and Von Knorre (1959) particularly common after influx of cold humid air masses.
Renal colic (due to kidney stones: nephrolithiasis	According to Maurer, 85% of the colics occur in Münich during changes in air mass (normal chance, 42.3%).
Urinary colics (due to urinary bladder stones: urolithiasis	Hauck found a highly significant correlation between fronts, changes in air mass, and renal colics in Leipzig (Germany). Very common in a 'stone belt' in northern and north east Thailand, about 40/100,000 against 0–10/100,000 in south Thailand. Ratio of urinary stones in males and females 8:1, for renal stones 3:1 (Ketusinh, Unakul). During March–May very hot and dry, urine 30% more concentrated, high colic incidence[6].
Dental caries	Meteorotropic relationships are uncertain. Anderson, Erpf, Lathrop, McBeath and Zucker noticed seasonal variations in dental caries in the USA, maximum incidence in late winter and early spring; minimum in summer (see Section 6.13.1).
Diabetes	Conditions often deteriorating in late autumn and winter; greater need for insulin; greater tendency to acidosis and coma (see Section 6.13.2).
Eclampsia	Meteorotropic relationships with cold fronts suggested by a number of studies but not statistically significant. Studies of Jacobs using 666

TABLE 6.1—contd

Diseases	Observed biometeorological relationships
	cases in Germany, Austria and Switzerland rather suggest cold-front relationship. The only statistically significant study was carried out by Berg in Köln (Germany).
Eye diseases	
Glaucoma	High in winter (maximum in November); low in summer (see Section 6.6, Table 6.2).
Retinal detachments	Maximum in June (The Netherlands) or March–May (Switzerland); minimum in winter (see Section 6.6, Table 6.2).
Föhn disease	Very common in Switzerland, Austrian Tyrol and southern Bavaria due to föhn weather, but also observed in western Canada, western USA, Australia etc. (see Section 6.13.3).
Goitre	In simple goitre meteorotropism is related to fluctuations of the iodine content of the thyroid and serum iodine; maximum in winter. In Graves' disease or exophthalmic goitre studies by Breitner, Hutter, Jacobowitz *et al.* indicate a minimum in summer, increasing in winter till spring (maximum in March).
Hay fever	Different seasons depending on the flora of the area; close relationships with atmospheric turbulence (see Section 6.2).
Hernia (inguinal, lumbar and umbilical)	In 1935, in Kyoto (Japan), Mori observed in 860 three-months-old babies with umbilical hernia that the maximum number occurred in those born in January; the incidence was minimal in April. Meteorotropism not so far established in cases of inguinal hernia; according to Mori, the condition is most frequent in babies born in winter, and less frequent in those born in April; lumbar hernia has probably same meteorotropism as rheumatic diseases, but no accurate studies are available.
Malformations	More anencephalics born in October–March (maximum December) Patentductus arteriosus cases show a marked seasonal fluctuation amongst girls only: maximum May–December, minimum January–April (in Birmingham), or maximum October–January, minimum February–August (in Massachusetts) (see Section 8.2).
Mental diseases	Maximum mental stress (unrest) in November–January, often second maximum in spring. Birth frequency of patients with schizophrenia maximum in January–March (see Section 6.8.2).
Multiple sclerosis	Studies by Allison suggest higher incidence in cold temperate climates and low incidence in sub-tropics (Section 6.13.5).
Rheumatic diseases	Complaints usually more frequent during the cold, humid, windy seasons, particularly in the case of arthritis (see Section 6.9).
Rickets	Observed mainly between latitudes 40° and 60°; absent in the tropics except in rich families keeping their young children in the darkness; common in heavily polluted industrial areas; incidence decreases with altitude; in high mountain valleys observed only on the shady side; according to Schmorl and others most frequent in winter, decreasing from March, minimum in July–August.
Skin diseases	A high incidence of eczema in spring. High incidence in summer of dyshidrosis, miliaria rubra, dermatomycosis etc. From July–August (often increasing till November–December) higher incidence of

TABLE 6.1—contd

Diseases	Observed biometeorological relationships
	furuncles, carbuncles, erysipelas etc. (see p. 178).
Tetanus	In recent years with new forms of prophylaxis a very rare condition. Formerly pronounced incidence peak January–March, very low in summer, autumn and early winter in central Europe and in the USA. According to Moro, tetanus common after drastic weather changes (e.g. föhn conditions after cold weather). Also György observed tetanus particularly on warm sunny days in early spring.
Thrombosis and embolism	The start of thrombosis is usually not sharply defined, unlike embolism. Therefore, accurate meteorotropic studies are only possible with embolism. Despite many studies on post-operative lung embolism and weather, very few statistically significant studies have been reported. Tivadar, using 156 cases in Budapest, found a significant correlation with fronts. Raettig and Nehls studied 489 cases in Pommern, Germany. A statistically significant correlation with both cold and warm fronts and lightning storms was found. Reimann and Hunziker could not confirm a significant front relationship in Basel, Switzerland. On the other hand, Maurer, Kayser, Sandritter and Becker confirmed a statistically highly significant correlation between lung embolism and the passage of weather fronts in different parts of Germany, especially with sudden changes of air mass. They affect the physico-chemical state of the blood and the blood-clotting mechanism (see Section 6.5). Verbal reports from surgeons suggest that post-operative thromboses are missing or very rare in tropical areas.[7]
Ulcer (peptic)	No clear seasonal relationships. Maximum incidence during months with strong atmospheric cooling, particularly after long period of heat adaptation. According to Hutter, in Germany serious complaints most common in December–February. According to Apperly in Australia (period 1924–1926), incidence decreasing from Tasmania (40° latitude) to Queensland (25° latitude) and western Australia (with warm tropical climates). According to Einhorn, in USA rise in recurrences in the fourth week of February and first week of March, with maximum in May; from the fourth week in May a decline until second week of June; gradual rise from the third week in August, until second week of September (maximum). Scheidter observed in Bavaria (period 1928–1932) a significant relationship between passages of warm fronts and perforations of peptic ulcers. Gebhardt and Richter (period 1927–1933) showed for peptic ulcer a minimum in April, maximum in October. Sallström in Sweden (period 1937–1942) found a minimum in the number of cases in June, maximum in September. De Franchis in Italy (period 1935–1945) observed the highest incidence during autumn, followed by winter, spring and summer. Boles analysed all the cases of haemorrhage from gastric and duodenal ulcer admitted at the Philadelphia General Hospital (period 1949–1953). The incidence was highest in January, February, April, May and June. Schedel in Munich (period 1948–1951) studied the relationship between perforated ulcer and the weather types of Ungeheuer (see Ref. 4, pp. 24–25). Most perforations occurred during phase 3, 4 and 5. A maximum occurred during phase 4; the minimum during phase 6. The correlations were statistically significant. Bodhe and Mokashi studied 986 cases of peptic

TABLE 6.1—contd

Diseases	Observed biometeorological relationships
	ulcer perforations in Poona (India) during the period 1958–1972. Incidence increased from May (hottest month in the year) onwards with maximum during rainy season in August and September. In October sudden drop until April.[8]
Ulcer (duodenal)	According to Hutter, in Germany, maximum in May (particularly in men below 40), secondary maximum in November (only in males). According to Einhorn, similar to gastric ulcer. In Leipzig Gerhardt and Richter observed a minimum in April, maximum in January, with secondary maximum in June–July. Sallström in Stockholm observed a minimum in June, maximum in August. Boles in Philadelphia observed an increase in the incidence of haemorrhage from duodenal ulcer during January, February and March, and during October, November and December.[8]

infection by a toxin-forming disease, is known as anaphylaxis, described for the first time by Richet[3] in 1902. The altered reactivity or allergic reaction is explained as an antigen-antibody reaction. An antigen is any substance (usually a protein) normally not present in the body, which, when introduced parenterally (i.e. otherwise than through the alimentary canal) into the blood-stream or animal tissues, stimulates the production in the body of a substance, described as an antibody, and which, when mixed with that antibody, reacts specifically with it in some observable way. The antibody-antigen reaction induces cell degranulation and liberation of pharmacologically active substances, such as histamine, serotonin and bradykinin.

Of those compounds the effect of histamine is best known and it is mainly responsible for what is observed in the so-called skin test. Its importance in causing eye irritation and rhinitis in hay-fever is shown by the relief afforded by antihistamine drugs. It is found that most likely antibodies are modified serum globulins, in particular those resembling γ-globulins. The globulins, like all proteins, are characterized by complex amino-acid structures. Small structural changes in these amino acids may affect the chemical properties considerably. If the new molecules of the antibody enter the blood-stream and meet the antigen, the globulins may be sufficiently modified to react with the antigen.

As both the liver and the thyroid affect the globulin level of the serum, and as both organs are considerably influenced by weather and climate, it is logical that allergy has several meteorotropic aspects. It has been shown (p. 98) that daily, monthly and yearly fluctuations in the serum γ-globulin content of a constant group of healthy blood donors were due to such meteorotropic processes. It was also found that all allergy patients had a poor thermoregulation efficiency (pp. 152, 250). Normalization of this mechanism through high altitude therapy (see p. 250) may cure or reduce the allergy complaints of patients and make them resistant to meteorologic effects.

Allergens in the environment affect an allergic subject by combining with an antibody bound to the surface of cells, especially the mast cells in the moist mucosa of the skin and the nasal mucosa.

A considerable amount of literature has been published on a special allergen known as house-dust of which the allergenic properties increase with increasing humidity of the atmosphere. This has been attributed to the increased action of micro-organisms on certain constituents of dust during humid weather. It is believed that humidity favours fungi development. The fungi break down the house-dust and decompose wool, feathers, paper and so forth. Other micro-organisms may join in this process.

In recent years several allergists are inclined to believe that the allergic reactions are due to mites. Mites living in house-dust depend in their development on the temperature and humidity of the environment. Optimum conditions seem to be 25 °C and 85% relative humidity.

Many allergists believe that asthma is caused by house-dust. However, it can be easily demonstrated in long range studies[9,10] that these apparent correlations are not true relationships.

Several allergic diseases are known, for example allergic rhinitis (inflammation of the nasal mucous membrane accompanied by considerable fluid excretion resembling a common cold, however, it can appear and disappear in a very short time due to thermal changes in the environment), urticaria (a condition marked by the appearance of red-white elevations of the skin, accompanied by intense itching, persisting for hours or days), allergic dermatitis and allied dermatoses and hay-fever.

One of the most extensively studied allergic diseases is hay-fever (pollinosis or seasonal vasomotor rhinitis). It is a seasonal form of allergic rhinitis caused by pollen or by mould spores of plants. Characteristic symptoms are: excessive sneezing, stuffiness and intermittent, profuse, clear, watery nasal discharge; temperature elevation is very rare, secretion of tears and itching of the eyes are common; also itching of the nose, roof of the mouth and posterior pharynx are reported. Dull frontal headaches and a feeling of fullness over the paranasal sinuses are common.

The term 'hay-fever' is actually incorrect, because the clinical symptoms are never caused by hay but by seasonal allergens produced by plants, usually flowering during the hay season. Outdoors, the atmosphere contains all the major groups of aero-allergens, consisting of microbes and suspended particles of fungal or other origin. According to Davies,[11-13] in Britain and in similar areas of temperate climate, tree pollens such as those of *Alnus*, *Betula* and *Corylus* become airborne in the latter part of the winter when allergy to *Betula* occurs. Numerically, the most important airborne tree pollen encountered in London is that of the plane, *Platanus acerifolia (Ait) Willd*, which is a cause of pollinosis in late spring. In north western Europe the *Gramineae* (grasses) constitute the most important cause of summer rhinitis; in England not less than 0.5% of the population are affected.[11-13] Autumnal ragweed pollenosis caused by *Ambrosia*, which affects 5% of the population in the USA, does not occur in north west Europe, the causal species of *Ambrosia* are not indigenous in this area. In London the spores of *Ustilago*, a pathogen of cereals, attain atmospheric concentrations of thousands per cubic metre in July.[11-13] Between June and September the spores of *Cladosporium* are abundant, concentrations of many thousands per cubic metre occurring between July and late August. Ascospores are recorded throughout the summer and autumn, and Basidiospores from toadstools are also common. Davies[11-13] pointed out that, as temperature is a limiting factor for plant growth, it determines when the pollen season occurs, its length and, over a period of years, the severity of a particular season for patients with allergic sensitivity to fungal spores and pollens.

In general the availability of water is less important than temperature (and sunlight) in determining the atmospheric pollen season. A high rainfall in the early part of the growing season whilst leading to prolific leaf development and heavy flowering may not be productive of a season with high pollen concentrations. Rain washes the air free from particulate matter. Very heavy rain washes pollens, and even the anthers which bear them, to the ground. Prolonged rain in the grass pollen season not only brings relief from pollinosis, but may so reduce the pollen concentration of the air that grass fertilization is reduced and seed production affected.

Atmospheric grass pollen concentrations exhibit diurnal periodicity. In London airborne grass pollen concentrations rise from a few at noon to a peak at about 5.00 p.m., and then drop until by the early hours of the morning the air is free from pollen.[11-13] Although a heavy shower of rain in the morning will delay the take-off and development of the diurnal peak, it has little effect on the mean daily concentration. However, a shower of rain in the late afternoon when the pollen clouds are developed washes the air and significantly reduces the mean daily concentration.

Rain and atmospheric moisture have an even greater effect on the fungal spore content of the air. Highest pollen and spore concentrations are obtained on clear sunny days with light convection and atmospheric instability. High concentrations may also be produced by the development of a temperature inversion which inhibits the dilution of spore clouds (for further details see Ref. 13).

A phenomenon similar to allergic rhinitis and hay-fever, but with a different non-allergic origin, is known as vasomotor rhinitis. It is probably due to an instability of the autonomic nervous system regulating the blood flow in the nasal mucosa. Slight cooling, emotional trauma, fatigue, etc. could create symptoms similar to those of hay-fever. As pollen, mould-spores, house-dust and most other allergens in our environment are light substances, which require only slight air turbulence to become floating particles in the air, it is evident that meteorological factors must to a considerable extent control the distribution and quantity of pollen and spores in the atmosphere and must therefore affect the incidence of hay-fever attacks. In view of the interrelationship between the internal micro-climate of rooms and buildings and external weather conditions, hay-fever patients in closed buildings and rooms will also experience these various factors.

We discussed these various forms of allergy rather extensively because thousands of people suffer from allergic diseases, particularly during the hay season. Through biometeorological studies of the occurrence in different localities, the changes in incidence can be explained and a natural therapeutic method has been developed for reducing the complaints by strengthening the thermoregulatory mechanism.

6.3. ASTHMA AND BRONCHITIS

Amongst the various meteorotropic diseases in northern Europe and the USA the respiratory diseases are the most common ones, in particular asthma and bronchitis.

It is estimated that at least one or two out of every 100 inhabitants in western Europe is suffering from these diseases. Studies in different parts of the world and particularly long-range studies in The Netherlands since 1955, using large groups of children and adults, have shown a number of significant relationships between weather and climate and different forms of asthma.[14]

6.3.1. Asthma

Asthma is a very complicated disease. At least four major theories have been put forward to explain the cause of the periods of breathlessness: the infectious theory, which assumes early bronchial infections of a child; the allergen theory, the most common theory, particularly in the USA, which attributes all the complaints to allergic reactions; the endocrinological theory which attributes the asthmatic complaints to endocrinological disturbances, which may be partly true in the case of children during puberty and of women around the age of 50; and the psycho-somatic theory explaining all complaints in terms of psycho-somatic disturbances. A careful examination[15,16] of the life history of hundreds of asthmatics shows clearly that psychic factors do not cause asthma (asthmatics fundamentally differ physiologically from healthy subjects, see Section 6.3.4), but once a subject becomes an asthmatic, psychological factors may become very important due to his continuous respiratory problems.

An internationally accepted classification of the various asthmatic complaints does not exist but, from a practical point of view, one can distinguish between bronchial asthma (a non-infectious form of asthma characterized by periods of wheeziness and breathlessness, often beginning within the first year of life and often accompanied by a special eczema which usually disappears at a later age), and bronchitic asthma resulting from a neglected bronchitis. This brings us to a second serious, meteorotropic, lung disease, bronchitis.

6.3.2. Bronchitis

Bronchitis is a disease caused by inflammation of the bronchi as a result of infectious agents (e.g. influenza virus, pneumococci, etc.) or air pollutants. It is more common in males than in females and seems to run in certain families. The clinical manifestations are largely due to bronchial obstruction caused by tenacious sputum which cannot be removed sufficiently by the cilia of the trachea.

The death rate varies considerably in different countries. For the same age group it is, for example, ten times higher in England than in Denmark.

Acute bronchitis is usually characterized by fever, sweating, coughing, substernal pain, choked-up feeling, paroxysms of dyspnoea following spells of coughing. Chronic bronchitis is a chronic affection of the bronchi and bronchioles causing a prolonged cough, which is usually exacerbated by a change in the weather. A special form of bronchitis is known as asthmatic bronchitis. It is due to the rapid accumulation of tenacious sputum and the associated swelling of the mucosa, causing a severe dyspnoea, a wheezy chest and an unproductive cough. Fever is due to secondary infection.

6.3.3. Asthma–weather relationships

In large constant groups of asthmatics the observed peaks in the attacks of the group, both the daily number and degree of the attacks, is sharply correlated with rapidly changing specific meteorological conditions. However, similar rapid changes (often in a matter of hours) in the number and type of allergens are usually not observed, which is one of the many arguments suggesting the minor importance of allergens in the causation of asthma (for further evidence see Reference 9).

Every month periods of high and low percentages of asthma occur, the periods varying in length from a few hours or days to a week or even longer. In The Netherlands every year the same pseudo-seasonal rhythm in asthma frequency is usually observed in both children and adults.

In the western part of The Netherlands the fluctuations in the daily amplitudes of asthma frequency suddenly increase at the end of June. The maximum is usually reached in September or October. The shift in the month of maximum values has proved[15] to be related to a shift in the onset of winter in The Netherlands, with its autumn storms and strong atmospheric cooling.

Contrary to general belief, fog has no asthma-producing effect in patients suffering from non-infectious bronchial asthma. The contrary is rather the case. Periods of fog (characterized in western Europe by periods of little atmospheric turbulence and usually high barometric pressure) show very low asthma frequency. However, this is only true for patients suffering from asthma without bronchitic components. In bronchitic patients and bronchitic asthmatics fog and air pollution increase the number of attacks considerably.

A highly significant statistical relationship was found[16] between asthma frequency and atmospheric cooling, i.e. the joint effect of temperature and wind speed. Cooling can be calculated for example with the formula of Bedford *et al.* (Formula 1.1, Section 1.5.3.(a)).

The asthma frequency increases rapidly after a sudden increase in general turbulence of the atmosphere if combined with the influx of cold air masses. The increase is most striking after a preceding quiet period and, in western Europe, is often characterized by high atmospheric pressure. The change in turbulence is usually characterized by a sudden rise or fall in atmospheric pressure (Figures 6.1 and 6.2).

In particular, periods of rapid barometric fall, due to rapidly approaching low pressure areas, have a marked asthma-increasing effect in western Europe if these weather changes are accompanied by a sudden influx of cold polar or continental air masses followed by one or more active cold fronts. The latter are usually accompanied by great speed of movement, a sudden change in atmospheric pressure and temperature after the passage of the front, strong precipitation (rain, snow or hail), considerable disturbances in the electric field of the atmosphere (fluctuations of the electric potential gradient between high positive and negative values), sudden changes in the direction and velocity of the wind, and so forth. Owing to these specific weather conditions, a sharp increase in the cooling effect of the atmosphere is observed. However, not only cooling but also excessive heat stress, particularly if combined with a high humidity of the air, may create severe asthma attacks. This is rarely observed in western Europe except during heat waves or shortly before thunderstorms.

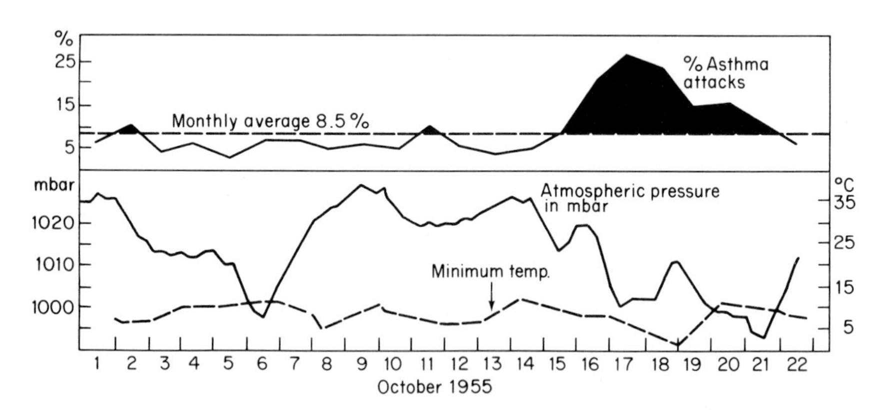

Fig. 6.1 Daily changes in the percentage of asthma attacks in a group of 90 children at the asthma clinic 'Heideheuvel' in Hilversum, The Netherlands, October 1955. During this period an area with high atmospheric pressure and quiet atmospheric conditions prevailed over western Europe till 14 October 1955 (see Fig. 6.2). The rise in asthma attacks started around 15 October 1955 after the first group of a series of cold fronts with rain and hail, reached the coast of western Europe (after: S. W. Tromp, *Biometeorological Analysis of the Frequency and Degree of Asthma Attacks in the Western Part of The Netherlands (period 1953—1965)*. Monogr. Series 6, 1966, Biometeorological Research Centre, Leiden, The Netherlands

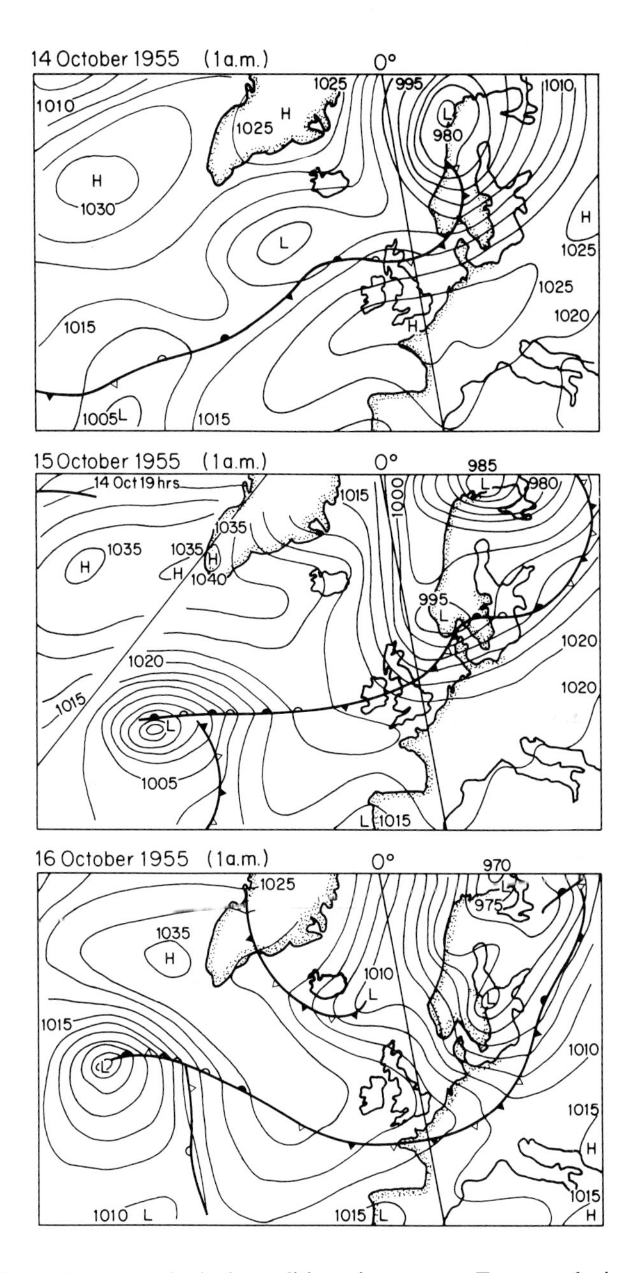

Fig. 6.2 General meteorological conditions in western Europe, during the period 14–16 October 1955, based on weather charts of the Royal Netherlands Meteorological Institute. The dark black line is a cold front

The principal reason for the great weather sensitivity of asthmatics is a fundamental difference in at least three major physiological functions as compared with normal subjects.

6.3.4. Physiological characteristics of asthmatics

(a) Asthmatics have a poorly functioning hypothalamic thermoregulation mechanism. In asthmatics, without exception, the thermoregulation rewarming curve (see Fig. 3.3, p. 61) shows a clear deficiency in rewarming capability. This explains why asthmatics react strongly to abrupt changes in the thermal environment, both to cold and to heat stress. In humid tropical regions like India the high humidity combined with high temperature before the beginning of the rainy monsoon also creates a high incidence of asthmatic attacks due to thermoregulatory disturbances.

(b) Asthmatics have an insufficient functioning of the adrenal gland. For example the average 17-ketosteroid excretion in urine in 24-h urinary samples, collected daily during 7 days in asthmatics not using drugs, is often 1/3 to 1/5 below the normal values for males and females of the same age group.

(c) Most asthmatics have a relatively low blood pressure for their age. The apparently dominating influence of the parasympathetic nervous system often causes parasympathetic stomach and intestinal complaints in asthmatics.

(d) The threshold for the contraction of the bronchi, after atmospheric cooling, is much lower in asthmatics than in normal subjects.

Of these four fundamental differences in physiological pattern between asthmatics and normal subjects the first two can be improved or even cured after a long series of high altitude treatments in low pressure climatic chambers,[17] with simulated altitudes above 1500 m or, preferably, above 2000 m (see pp. 109, 249).

It is outside the scope of the present publication to explain in greater detail the deeper physiological mechanisms involved in the asthma-weather relationship. The interested reader is referred to Reference 18.

Finally it is worth mentioning that Schook and Schüller[19] in the asthma clinic in Hilversum (The Netherlands) studied the possible relationship between month of birth and asthma. They analysed the birth dates of 2544 children suffering from chronic non-specific lung diseases such as bronchial asthma and asthmatic bronchitis. The following results were obtained: most children, both boys and girls, were born in July, August, September and October. About 6.5% of the children born in January–April were asthmatic, against a yearly monthly average of 8.4%. From July till October the proportion increased to 10.3% a month; this relationship was observed during 13 of the 15-birth years studied; the July–October maximum was particularly high (11.3%) in children who suffered from infantile chronic eczema during the first year of their life. They assumed that contact with mites in house-dust during the first weeks of a child's life would make the child sensitive to the allergen.

6.3.5. Bronchitis–weather relationship

This has been studied extensively, particularly in Great Britain, where Pemberton and Goldberg[20] observed an increase in the incidence of bronchitis in the cities as compared with rural areas. Death rates from bronchitis, of males of the age group 45–64, per 100 000 of the male population of England and Wales in 1953 were 61 in rural districts against 114 in the more industrialized cities of 100 000 inhabitants and over; for females the same figures were 17 and 32. Experiments by Amdur *et al.*[21] and by Cralley[22] suggest that the increased sulfur dioxide content of the air of most industrial cities may be at least partly responsible for a decrease in the ciliary activity of the human trachea. Goodman *et al.*,[23] have shown that the areas of high bronchitis mortality in England lie to the north of a line drawn from Bristol to the Wash, areas representing high industrial activity with considerable air pollution.

In Boyd's studies[24] a close association was found in the London area between the weekly deaths from respiratory diseases, of people of 45 and older, and the temperature and humidity in the preceding week; mortality increased with falling temperature, particularly below 0 °C during thick fog. Such periods are usually characterized by an increase in the sulfur dioxide concentration of the air. Sulfur dioxide in very humid air, will change into droplets of sulfuric acid, which may be responsible for the inflammation of the bronchi. More recent studies by Martin and Bradley[25] confirmed this observation.

Lawther and his colleagues[26] (at St Bartholomew's Hospital in London) for many years studied the influence of environmental factors on bronchitis. Studies with 180 patients living in Greater London showed that practically all days with sharp peaks in smoke and sulfur dioxide concentrations coincided with marked peaks in the degree of illness of the patients. Many patients are affected before wet fog is formed. Fog may even be absent during such a period of deterioration of health if there is still a high concentration of smoke. Lawther pointed out that the close relationship between the smoke concentration and the degree of illness of bronchitis patients is only true for the winter months. It seems to disappear in spring and summer. This suggests that other factors must be involved apart from air pollution, a conclusion supported by the geographical distribution of this disease in The Netherlands. The highest mortality occurs in the eastern, least industrialized parts of The Netherlands. A N.E.–S.W. climatic boundary separates both areas, the eastern part having more foggy days. Considerable research work on weather and bronchitis was carried out by Reid and Fairbairn,[27] at the London School of Hygiene and Tropical Medicine. They studied the medical records of 565 postmen retired prematurely because of chronic bronchitis during the period 1950–1954, and matched these against the records of similar age groups of postmen still in the Service. The following results were obtained:

(a) postmen who were ultimately disabled or who died prematurely from chronic bronchitis had more and longer absences than their controls even in early adult life;
(b) after the age of 45 the 'bronchitics' had 8–10 times more attacks of pneumonia, pleurisy and asthma than their 'controls';
(c) bronchitics have more disability from peptic ulcer and more deaths from coronary disease and lung cancer, which may be due to the fact that, apart from smoking, various stress conditions and over-stimulation of the autonomic nervous system play an important part in the development of bronchitis;
(d) bronchitics above the age of 45 suffered most from bronchitis, pleurisy and pneumonia in the areas of the UK with the highest fog frequency and degree of air pollution, this correlation was more consistent than that with temperature;
(e) the bronchitis ratio between postmen and clerks (working indoors) also became larger in the areas with high fog frequency and a high degree of air pollution;
(f) in the London area the invalidity rates from chronic bronchitis for postmen (working under the same material conditions and practically the same environmental temperature and humidity) rise steadily from south-west to north-east.

Despite the many studies carried out on bronchitis and weather it is still not possible to explain all the observed relationships. One of the following factors may be involved:

(i) in industrialized areas with high carbon dioxide and sulfur dioxide contents of the atmosphere, foggy conditions apparently reduce the ciliary activity of the trachea, either through the formation of sulfuric acid droplets or perhaps on account of a surplus of positive air ions. Depending on the size and electric charges of the fog droplets, the environmental agents can penetrate the lungs to different depths;
(ii) the susceptibility of the lungs of bronchitics to these external agents seems to be different in males and females, the female being more susceptible, in early adult life, than the male to respiratory infection; the susceptibility of both sexes also decreases in spring and summer, probably because of the rise in environmental temperature (which reduces foggy conditions) and because of increased resistance of the body to infectious diseases;
(iii) in non-industrialized areas, the cooling effect of fog may increase the capillary resistance of the membranes of the respiratory system and reduce the blood flow to these parts of the body, thus favouring the growth of infectious agents; in this connection the studies of De Vries[28] in The Netherlands and of Hsiah *et al.*[29] are of interest. De Vries tested chronic bronchitic subjects grouped according to the threshold concentration of

inhaled histamine required to elicit bronchial obstruction. In one series of tests dry cold air was administered at −55 °C, in another artificially produced 'clean' fog (condensed water vapour) at −10 °C was used. At the end of the three minute exposure the vital capacity, and the one second forced expiratory volume, were determined spirographically. Using cold fog larger decreases in vital capacity were observed than with cold dry air. Forced expiratory volumes were also reduced to a greater extent by cold fog than by cold dry air.

Hsiah *et al.*[29] in China compared subjects with chronic bronchitis who did and did not complain about increased breathlessness. After cold air inhalation those who had complained showed increased respiratory obstruction; those who had not, and the normal controls, experienced no such increase. This result shows that chronic bronchitic subjects are far from uniform in their response to environmental stimuli.

(iv) it may be that people with a hereditary or subsequently acquired unbalanced autonomic nervous system are more susceptible to these various weather effects, in view of the greater proneness of bronchitics to peptic ulcer and to coronary disease.

These and other still unknown factors may be responsible for the meteorotropic behaviour of bronchitis and asthmatic bronchitis.

6.4. CANCER

Various studies by Tromp and others[4,30] suggest that cancer in general may be partly a meteorotropic disease, but before discussing this aspect the most common, and typically meteorotropic, form of cancer, skin cancer, will be described. This has the highest incidence per head of the population in Australia. Excellent reviews of the skin cancer weather relationship have been prepared by B. S. Mackie and Leila E. Mackie,[1,4] according to which skin cancer provides more than half of all diagnosed cancers.

The first outstanding feature of skin cancer is that it occurs almost solely on those areas of the body exposed to meteorological stimuli. This fact of the cancer's extreme predilection for exposed skin is highly significant, as it is strongly indicative of an external causative factor, in particular of solar radiation.

The second important point is the variability in resistance of different subjects. Skin cancer develops readily in fair skin, but less readily in the case of a high degree of skin pigmentation. The immunity caused by melanin, the pigment of the skin, depends upon heredity and race. Humans can be divided into three main types, according to their degree of pigmentation. Type 1 has little or none, Type 2 has a well-developed basal layer pigment and Type 3 has ample pigment in all epidermal layers. Cancer is seen mainly in Type 1, while races with browner complexions are seldom affected. The predominant occurrence on exposed parts, in inverse proportion to the degree of pigmentation, suggests that solar radiation is causally involved.

According to the Mackies[1,4] a number of other observations confirm this assumption. Experiments with animals have shown that sunlight readily produces skin cancer in albinos and that the most active wavelengths lie in the ultraviolet range between 290 and 320 nm. A similar ultraviolet irradiation effect is probably active in humans as well, since the incidence of skin cancer is much greater in regions closer to the equator when similar population groups are considered. This latitude effect has also been observed by several workers in the USA and Australia, and is presumed to depend on the increased proportion of ultraviolet radiation which penetrates the earth's atmosphere when the sun approaches the vertical.

Weather may affect the distribution of skin cancer indirectly because pleasant weather conditions encourage the discarding of clothes and the disuse of weather protectors. On the other hand, excessively high temperatures and humidity discourage open air activities as also do cold, rain and wind. High standards of living with ample outdoor space and diverting geographical features such as lakes and beaches all increase exposure and resultant cancers. Rural and seaside occupations act in the same way. Thus people working or living outside are more commonly affected than those who spend less time in the sun. Altitude would be expected to aid cancer production because the intensity of the ultraviolet radiation increases with elevation and these wavelengths are well reflected by snow. However, there are no statistics to support this view. Better skin protection due to lower temperatures at altitude may explain this discrepancy.

Strong ultraviolet sun radiation over several years may initially cause three pre-cancerous changes known as solar keratosis, keratoacanthoma and lentigo. Solar keratosis is a very slowly growing, horny excrescence above the surrounding skin; keratoacanthoma resembles squamous cell carcinoma (see below); lentigo is a brownish or black, irregularly shaped, flat spot showing an increase of the pigment-forming cells in the epidermis.

True skin cancer only develops after repeated exposure of the skin to the sun over a period of years. It consists of an uncontrolled multiplication and penetration into deeper tissues of cells normally found only within the epidermis. The three forms of cancer which may be caused by ultraviolet radiation are squamous cell carcinoma, malignant melanoblastoma and basal cell carcinoma. Squamous cell carcinoma usually develops from a pre-existing solar keratosis as a hard, pink to white, rounded, elevated lump, either surmounted by adherent keratosis or showing an ulcer in the centre. Malignant melanoblastoma is an uncommon but dangerous type of cancer, in the cause of which ultraviolet radiation appears to act as a contributing carcinogen; it arises as a black nodule in an area of lentigo, mostly in elderly people on the cheeks, and solar exposure appears to be the major causative factor. Basal cell carcinoma is the most common form of skin cancer. It is seen more on the face and neck than on other exposed areas and usually appears first as a small pearly nodule which progresses slowly into a lump or an ulcer. The deeper mechanisms of these solar induced skin cancers are fully described by the Mackies.[1,4]

In the case of other cancer sites meteorological stimuli (particularly the thermal stimuli), may also play an important role. This assumption is supported by the observation of Tromp and Bouma[31] that cancer patients have a very poor thermoregulatory efficiency (based on the waterbath test,[32] Section 3.3). Empirical studies also support the assumption that thermal stresses may affect cancer development. Lea[33] studied a possible association between the death-rate, from cancer of the breast in females, and the mean annual temperature in Norway, Sweden and Great Britain. He found that a highly significant correlation exists between breast cancer and mean annual temperature.

McKay[34] found a similar correlation in 32 countries between the death-rate for malignant neoplasms of the intestinal tract and the mean annual temperature. He assumed that these correlations may be related, at least in part, to the prevalence and endemicity of a carcinogenous viral agent.

Lee[35] pointed out that there is a pronounced seasonal incidence in the case of mortality from malignant disease, which may be partly due to a general deterioration in health during the winter, particularly in older people.

Krasnow[36] studied 50 counties in the USA, 25 with the highest and 25 with the lowest death-rates of cancer of the digestive system, breast and respiratory system. It was found that malignancies of the large intestine and rectum, the digestive system, the stomach and the breast showed a highly significant negative correlation with temperature. On the other hand the respiratory system (lung, trachea and bronchitis) malignancies showed a direct relation with temperature. In non-urban areas environmental cooling (the combined effects of temperature and wind speed) is an important factor in malignancy death-rates. However, in metropolitan areas low temperature alone was practically as significant as low temperature plus wind velocity.

Young[37] demonstrated the effect of temperature on induced mammary tumours in rats. These are less frequent at 5 °C than in an environment at 32 °C. Implanted Sarcoma 180 tumours in mice are smaller in mice kept at 8 °C than in those kept at 22 or 33 °C.[37]

Pomp *et al.*[38] observed that hyperthermia had a retarding effect on the growth of malignant tumours. The observed retarded growth is probably due to the raised tumour cell metabolism, and to changes in enzyme activity and the antibody system as a result of hyperthermia.

In 1956 De Sauvage Nolting[39-41] observed seasonal fluctuations in birth dates of cancer prone patients after correction for the average monthly birth frequency of the total population in The Netherlands. Most cancer patients were born in the winter, in particular in the period December–March, fewest were born in June–July. The 1242 cancer cases analysed by Keogh in Australia showed the same phenomenon, however with a 6 months shift as compared with the northern hemisphere. In other words most cancer prone children in

Australia were born in June–July. Identical seasonal variations were observed in 1961 in The Netherlands. The seasonal difference was highly significant. This seasonal variation in frequency of cancer birth months was also reported in the USA by Stur[42] and others.

In addition to temperature high altitude stress also seems to affect cancer development. Warburg[43] exposed cancerous rats to a gas mixture containing 5% oxygen for 40 h, simulating an altitude of about 12 000 m. The tumours became markedly, but not completely, necrotic. The experiment was repeated by Campbell and Kramer[44] at a simulated altitude of 5100 m (about 410 mm Hg). Rats were kept alive for two weeks. The tumours did not regress but the growth was markedly inhibited. Sandstroem and Michaels[45] observed regressions in Walker tumours in all experimental species kept at 360 mm Hg atmospheric pressure (about 6000 m altitude). The average regression rate, as compared with controls, was 39%. At pressures below 300 mm Hg it was 45%. The effect is explained by the authors as being due to stimulation of the adrenal function by the reduced oxygen pressure at high altitude, thus changing the level of corticosteroids in the blood serum.

Meteorological stimuli seem also to affect the growth of transplanted tumours. Brezowski *et al.*[46] studied the causes of fluctuations in the number of rats in pure strains in which no cancer development occurred after tumour transplantation, 737 rats of the Walker-carcinoma strain being studied during the period 1961–1963. They found that the number of resistant animals was greatest during quiet high pressure atmospheric conditions. The number decreased during the approach of cold fronts and increased again after the front had passed. The observed differences were statistically significant.

Finally it is possible that thermal stresses affect carcinogenic viruses as observed for example in animal leukaemia. More than 400 papers have been published since 1974 on the possible influence of thermal stresses on oncogenic (i.e. tumour causing) and enteric (intestinal) viruses. Further details on the possible effects of weather and climate on cancer can be found in Reference 47.

6.5. CARDIOVASCULAR DISEASES

Most heart diseases are caused by arteriosclerosis and have an important meteorotropic aspect. They can be classified as coronary thrombosis (formation of a blood clot within the heart or in one of the coronary arteries of the heart), myocardial infarction (development of an acute local deficiency of blood supply in the cardiac muscles, usually due to a blood clot, followed by injury of the cells and usually involving the left ventricle of the heart), angina pectoris (paroxysmal thoracic pain, usually radiating to neck and left arm and due to spasms of the heart muscles and a temporarily insufficient blood supply), and death through heart failure. A considerable percentage of the total mortality in the population represents death from heart diseases. Another important vascular disease is apoplexy or stroke (an extravasation of blood in the brain) usually caused by lack of elasticity of the brain capillaries as a result of arteriosclerosis (see also p. 167).

Former studies by Bartels,[48] Beljajen,[49] Kisch,[50] Kolisko[51] and many others strongly suggest a meteorotropic relationship between the passage of weather fronts and the increased occurrence of these diseases. The studies of Amelung,[52] Fladung[53] and Ströder *et al.*[54] in Germany, in particular, have given considerable statistical support to this observation. For example, Ströder[54] showed that 87% cases of myocardial infarction followed by sudden heart failure, of which the exact diagnosis was established electrocardiographically and afterwards by autopsy, occurred on days of considerable meteorological disturbance characterized by the passage of active weather fronts, either cold or warm.

Studies by other cardiologists have clearly shown a seasonal incidence of both morbidity and mortality from arteriosclerotic heart diseases. For example Wood and Headley[55] studied 133 electrocardiographically established cases of coronary thrombosis (not immediately followed by death) at the University of Pennsylvania (USA). The highest incidence was found in autumn and winter, the lowest in spring (March–May) and summer, with a clear minimum during June. The Bureau of the Census in Washington DC studied 27 390 cases of angina pectoris and coronary heart diseases and found a similar seasonal relationship. The same seasonal incidence was established in the USA by Master[56,57] in New York, by Bean[58,59] in Boston, by Mullins[60] in Pittsburgh, by Bean[58,59] in Cincinnati, by Mintz and Katz[61] in Chicago, by Miller[62] in Rochester, by Hoxie[63] in Los Angeles and by Billings[64] in Nashville.

A similar study by Koller,[65] at the Statistical Department of the W. G. Kerckhoff-Herzforschungs-Institut of Bad Nauheim (Germany), using 1 600 000 sets of mortality data from the UK (period 1921–1933), showed a clear maximum occurrence of arteriosclerotic heart diseases in January–February with a minimum around July–August. No difference in the seasonal occurrence between males and females was observed, despite the great difference in total mortality from heart diseases between the sexes.

All these studies were carried out in northern countries with temperate climates. However, studies by Heyer *et al.*[66] (using 1386 cases of which the time of onset was known and the diagnosis was established electrocardiographically) in Texas where the summers are very hot, showed a different picture. It was found that acute myocardial infarction in Texas is more frequent during the hottest season of the year, i.e. during July and August, with maximum temperatures often exceeding 37 °C and with moderate humidity (60–70%). During the rather mild winters in Dallas the incidence of myocardial infarction is low.

Further meteorological studies by Teng and Heyer,[67] using the same material, showed a statistically significant increase in the incidence of acute myocardial infarction during, or immediately after, the sudden influx of polar or tropical air masses, causing fluctuations in temperature of from 2 °C to 17 °C in cases of cold air masses. The vast majority of the patients suffered the onset of their illness while asleep or at rest, from which it may be inferred that physical activity was unlikely to have been an important factor in these cases.

Studies by Tromp[68] in The Netherlands have shown that both phenomena observed in cold and hot countries may occur in the same area even at northern latitudes. A statistical study in The Netherlands of mortality from angina pectoris, coronary thrombosis, myocardial infarction and apoplexy for each year in the period of 1935 to 1958 points to the following:

(a) mortality is highest in January–February (usually the coldest months of the year in The Netherlands) and lowest in July–August. Winters with abnormally low temperatures are characterized by very high mortality, warm winters by relatively low mortality but still considerably higher than in summer;

(b) there is an almost perfect inverse relationship for all the months of the year between the curves representing the average monthly temperature and the mortality from arteriosclerotic heart diseases;

(c) the generally low mortality figures in summer are highest during the warmest summers, which are often accompanied by short spells of heat waves;

(d) during years with a double temperature minimum, e.g. the winter of 1943–1944 with a temperature minimum both in December 1943 and February 1944, a double peak in the heart mortality curve is observed, both in December and February;

(e) there is usually little difference between the monthly mortality curves of the male and female population; only the total percentage of females who die from apoplexy is higher than that of males, whereas the reverse is true for angina pectoris, coronary thrombosis and related heart diseases.

These observations suggest that both hot and cold stress increase the mortality from coronary thrombosis and related heart diseases. These findings have been confirmed in climatic chamber studies, in particular by Burch *et al.*[69-73] in New Orleans.

The various meteorotropic correlations described above may be summarized as follows:

(a) the mortality rate of coronary heart disease in middle and high latitude countries, is considerably higher in males than in females, and is invariably highest in the months of January and February and lowest in July and August.

(b) the mortality rate for apoplexy, which is higher in females than in males, shows the same seasonal pattern as does coronary heart disease.

(c) a comparison between average monthly temperatures and mortality rates for both coronary heart disease and apoplexy reveals the following significant correlations:

(i) exceptionally cold winters are characterized by very high mortality; warm winters have a relatively low mortality.

(ii) although mortality is at a minimum in summer, there are variations which appear to be related to peak temperatures. The higher the temperature the greater the mortality from apoplexy and coronary heart disease.

(d) in those regions which experience very warm conditions, e.g. the southern part of the USA, the highest incidence of mortality is in the summer; the lowest is in the winter. Extreme cold and heat stress increase mortality rates. The correlations between average cooling and mortality are so significant that approximate mortality predictions can be made, solely on the basis of meteorological data.

Considering the great sensitivity of coronary heart patients to environmental thermal stresses it is logical to assume that these patients have a poor hypothalamic thermoregulation mechanism. A study by Tromp and Bouma[74] in 1965 confirmed this assumption. It was also found that in elderly heart patients treatment by anticoagulants improved their thermoregulatory functions.

A detailed description of the different physiological mechanisms involved in the relationships between weather and heart disease would be outside the scope of the present book. However, the following mechanisms due to meteorological stimuli should be mentioned:

(a) the elasticity and peripheral resistance of the blood vessels is affected causing a change in blood pressure;
(b) the activity of the sympathetic nervous system will change;
(c) the physico-chemical state of the blood will change, in particular the blood viscosity, blood clotting time and fibrinogen content etc.;
(d) the capillary fragility is affected (during winter it is usually high, partly due to vitamin P deficiency; this may explain the higher incidence of cerebral haemorrhages in newborn infants during winter).

One or more of the above mentioned mechanisms may be involved during a meteorotropic heart attack (for further details see also pp. 88–98 and Ref. 1).

6.6. EYE DISEASES

Several eye diseases show a meteorotropic behaviour; acute-glaucoma retinal detachments, conjunctivitis, hordeolum, keratitis and rheumatic iritis. In Table 6.2 the principal observed relationships are compiled.

Despite the relatively large number of observations the basic mechanisms of the meteorotropic behaviour of eye diseases are still insufficiently known. At least three factors may be involved:

(a) Increased blood pressure and interoccular pressure during sudden cold periods and after excessive heat stress (e.g. in the case of glaucoma and haemorrhages).
(b) Increase in capillary permeability after the influx of warm air masses or after the passage of fronts, which may release a small thrombus after a cataract operation.

TABLE 6.2
Meteorotropic relationships of eye diseases

Disease	Daily relationship	Seasonal relationship
Acute glaucoma[a]	In Germany highest incidence during coldest months of the year (November–February) and on very hot days in summer. Normally low incidence in summer. High incidence after cold fronts or influx of cold air masses. Same in Italy and Switzerland. In India maximum in July during wet monsoon. In China 2 maxima: in winter and first part of May (often cold storms). According to Fisher[76] and Huerkampf[77] lowest interocular pressure between 3 and 6 h, highest between 8 and 17 h.	High in winter, low in summer.
Retinal detachments[b]		In The Netherlands (Utrecht) maximum in June, minimum in winter.[1,71] In Switzerland (Geneva) maximum March–May, minimum November–January.
Acute conjunctivitis[c]	In The Netherlands[1,71] 74% of all cases during pleasant sunny weather (only in May–September).	
Phlyctenular conjunctivitis[d]	In 78% reported cases on cloudy days without sunshine.	In south Germany most frequent between January–April (maximum).
Trachoma[e]	According to Cooper[78] four main factors: 1. Arabian winds entering India, laden with sand dust; 2. Dust storms in north west India; 3. High incidence of house flies; 4. Locust swarms before the monsoon.	No clear seasonal incidence in middle East but most common in India March–May and August–October.
Hordeolum[f]	In The Netherlands,[1,71] in 73% of reported cases, common during rainy weather.	
Blepharitis[g]	Same as Hordeolum.	
Keratitis dendritica[h]	In 82% of reported cases, a few days after drastic changes in weather, e.g. after an approaching	

TABLE 6.2—contd

Disease	Daily relationship	Seasonal relationship
	depression. In K. marginales and K. scleritis no weather relationship.	
Herpes cornea[i]	Frequent during drastic weather changes with frontal passages and steeply falling barometric pressure.	Maximum in February and March.
Rheumatic iritis[j]		In Switzerland (Geneva) most frequent April–August, minimum in February and October.
Haemorrhage[k] in the anterior chamber of the eye (after cataract operation)	Most frequent in early morning and in females (70%). In Germany most frequent with falling barometric pressure and warm slightly humid weather. In Hungary most frequent after passage of tropical air masses from north Africa; lowest incidence after maritime arctic air masses.	In Germany most frequent in period January–May; in Hungary maximum number of cases in June–July; minimum in April.

Glossary

(a) Glaucoma: excessive intraoccular pressure causing hardness of the eye and even blindness. More common among females, most frequent in right eye, particularly above the age of 50.
(b) Retinal detachments: common in the age group 60–70; in The Netherlands 60% occurs in myopic eyes.
(c) Conjunctivitis: inflammation of the membrane lining the eyelids and covering the eyeball.
(d) Phlyctenular conjunctivitis: severe inflammation of the eye marked by small vesicles surrounded by reddened zone.
(e) Trachoma or granular conjunctivitis: a chronic contageous viral conjunctivitis affecting the entire cornea and reducing vision to zero. In India 3 500 000 blinds, i.e. 1/5 of the blind population of the world, due to trachoma. Dust particles cause micro-fissures of conjunctival epithelium which become a host for bacteria and viruses transmitted by the house fly and the Indian locust.
(f) Hordeolum: infection of a gland of the eyelid margin.
(g) Blepharitis: inflammation of the ciliae and glands of the eyelid margin.
(h) Keratitis dendritica: tree-shaped inflammation of the cornea.
(i) Herpes cornea: kind of keratitis.
(j) Rheumatic iritis: rheumatic inflammation of the iris.
(k) Haemorrhage: bleeding of the blood vessels of the eye.

(c) Changes in the intensity and the wavelength distribution of light in combination with differences in light sensitivity of the eye in different months of the year, the sensitivity being highest in the spring and early autumn (see p. 104). This so-called photo-sensitivity is related to a certain extent to the colour of the hair and of the eyes of the subject. The sensitivity of eyelids, conjunctiva and cornea to ultraviolet and infrared radiation is known as photo-ophthalmia. Persons with blue or brown eyes have a different sensitivity of the cornea (the transparent anterior part of the eye); those with blue irises have the most sensitive corneas. Females are 10% less sensitive than males (Millodot[75]). Also diurnal differences in corneal sensitivity have been observed.

6.7. INFECTIOUS DISEASES

The influence of weather and climate on infectious diseases is a very complex problem. Infectious diseases can be affected by several meteorological factors which can be classified into two main groups. The first lowers the resistance of the human body to infection, the other affects the ease of spread of an infection. Factors in the first group affect the local resistance, i.e. the site of entrance of micro-organisms or viruses into the body and the place of settlement and further development. Others affect the general resistance of the body, determined by the specific physico-chemical state of the body at a given time. To the second a large group of factors (the ease of spread of infections) belong:

(a) the influence of weather and climate on the social habits of man, such as crowding in rooms, shutting of windows and doors and changes in clothing and diet during certain seasons;
(b) the influence on the development of micro-organisms and viruses; and
(c) the effect of meteorological factors on the spread of these infectious agents through the atmosphere.

According to Stallybras and Winkler,[79] Hemmes[80] and others, the meteorotropic infectious diseases can be classified into three groups:

(i) *Summer diseases*: cholera, typhoid fever, poliomyelitis and bacillary dysentry, with increase in incidence in Europe from May to September. In the tropics, the maximum incidence is during the rainy seasons and in the humid areas;
(ii) *Autumn diseases*: streptococcal infections, scarlatina, and diphtheria in western Europe, with increase in incidence from August to November;
(iii) *Winter diseases*: the common cold, influenza, lobar pneumonia, epidemic cerebrospinal meningitis and smallpox increase from September to March. In the tropics, the maximum incidence is in the dry season and in dry areas.

These seasonal fluctuations are related to differences in humidity and temperature, as suggested by survival experiments of Hemmes[80] and others. The summer diseases occur during periods when environmental temperature and humidity are high; the winter diseases occur when the outside temperature is low and the humidity high, the humidity indoors, however, being low. According to experiments performed by Hemmes in The Netherlands, the influenza virus dies more rapidly at high relative humidity. The virus is, therefore, more readily transferred by infected particles in dry rooms in winter. However, no simple relationships are found for the autumn diseases. This supports the assumption that the meteorologically induced seasonal fluctuations, in the general resistance of the population, may play a more important part, in the epidemiology of diseases, than is generally assumed.

6.7.1. Meteorological effects on the local resistance of the body to infections

The local resistance of the body can be affected in different ways:

(a) The permeability and capillary resistance of membranes are affected by thermal stresses. Regli and Stämpfli[81] have demonstrated a decrease in capillary resistance (increase in permeability) following the influx of warm air masses and an increase following cold stress. These observations were confirmed by Lotmar and Häfelin[82] and by Arimatsu.[83]

(b) The acid coating on the skin is affected by alkaline aerosols in the atmosphere. Marchionini *et al.*[84] demonstrated that the acid coating produced by the eccrine sweat glands protects the skin against infections. The pH of the skin is usually from 4–6; after evaporation of sweat it is from 3–5. Spots with low acidity usually develop in skinfolds due to the lack of evaporation of sweat. All meteorological factors affecting the rate of sweating and evaporation and thus the pH of the skin may influence the growth of micro-organisms and their penetration into the skin.

(c) The dryness of membranes (especially the nasal mucosa) is affected by the temperature, humidity and ionization of the air. Low atmospheric humidity is believed to cause a withdrawal of water vapour from the body cells which in turn may lead to microfissures in the nasal mucosa, particularly on cold frosty days and often in centrally heated rooms. Excessively prolonged cold induces constriction of peripheral blood capillaries and reduced blood flow, also causing drying and cracking of the nasal mucosa. Humidity affects the secretion of the nasal mucosa. Armstrong[85] was able to show that a greater secretion of the mucosa parallels increased concentration of antibodies in the nose. Cold dry air stimulates secretion by the nasal mucosa. The gradual drying of the membranes tends to concentrate these antibody secretions. According to Howitt and Bell[86] the content of antibodies in nasal secretions parallels that in the serum.

(d) The humidity of the air also affects the survival of bacteria and viruses. Gram-negative micro-organisms die more quickly at low humidities; gram-positive bacteria and influenza viruses die more rapidly with high humidities and increasing air movement. Thus winter months with very low humidities and little air-movement in centrally heated buildings are favourable for the transmission of infectious respiratory diseases.

(e) Ozone, if present in high concentrations, appears to lower the resistance of mice to respiratory infections.

6.7.2. Meteorological effects on the general resistance of the body to infections

The body's general resistance to infections can be affected indirectly by the meteorological environment and by seasonal changes in diet, or directly because:

(a) There may be seasonal or short-term variations in the physico-chemical state of the blood. Cold stress increases the γ-globulin level of the blood. Changes in the γ-globulin level affect the immunity of the body to infectious diseases.

(b) The effect of the same chemical substance differs if administered under different meteorological conditions. Rabbits kept at temperatures from 26 to 29 °C escaped infection after inoculation with a dose of myxoma virus which was fatal when applied at winter temperatures (−3 to +15 °C).

(c) There is, according to Zimmermann,[87] a correlation between the 17-ketosteroid level of the blood and resistance to infectious disease. Weather stress such as cold periods, causing a rise in 17-ketosteroid production by the adrenal cortex, increases the resistance whereas warm weather, causing a low 17-ketosteroid level, lowers the resistance against infections.

(d) Weather stress may cause dysfunction of the thermoregulatory system and appears to facilitate the development of such infectious diseases as the common cold.

(e) Strong ultraviolet light appears to have both direct and indirect effects on the resistance to infections. It affects endocrine function and the functioning of the autonomic nervous system. It is lethal to bacteria, particularly when the relative humidity of the atmosphere is less than 60%.

In Table 6.1 (p. 141) a summary is given of the principal seasonal and pseudo-seasonal infectious diseases. Further details can be found in References 5 and 88.

6.8. MENTAL PROCESSES AND MENTAL DISEASES

The various studies discussed in Sections 5.1–5.4, clearly indicate the great direct influence of meteorological stimuli on basic physiological processes in man and the possibly indirect effect on their mental processes. However, it has proved to be extremely difficult to demonstrate beyond doubt any direct effects.

A considerable number of studies have been devoted to the possible direct influence of weather and climate on mental processes. This problem is a very old one but, in spite of the many publications which have appeared on the subject, most of the studies are inconclusive either because the clinical material on which the conclusions were based was too limited or untrustworthy or because the meteorological data used were not sufficiently accurate.

Still a number of trustworthy observations were made on the direct effects of meteorological stimuli on various mental processes, not only pathological aberrations of normal physiological processes in the brain, the mental diseases (schizophrenia, oligophrenia, epilepsy, apoplexy and migraine) but also on a number of social and psychological phenomena closely related to our mental processes, such as suicide and attempted suicide, restlessness of school children during certain weather conditions, reaction time, performance capability, road and industrial accidents, intelligence of school children, human behaviour and the reproduction of man.

6.8.1. Definitions

It may first be appropriate to define each of the five above mentioned mental diseases according to the generally used Dutch classification, the more so as no international and often even no national agreement amongst psychiatrists exists concerning the definition of a mental disease. As the most extensive studies to date were made by Tromp and Bouma[89-91] in The Netherlands (assisted by 21 psychiatrists) we shall follow the definitions used by this group of psychiatrists which are more or less in agreement with the classification used by the W.H.O.

(a) *Schizophrenia* or *dementia praecox* comprises a large inhomogeneous group of psychoses characterized by disorientation, loss of contact with reality, splitting of the personality etc. Several kinds of schizophrenia are distinguished: katatonia, hebephrenia and dementia paranoides.

(b) *Oligophrenia* (weak-mindedness) comprises three larger groups of mental diseases: debilitas mentis (persons suffering from mental deficiency, having difficulty in learning and in controlling themselves or managing their own affairs), imbecillitas (more serious form of mental debility, with 25–50 I.Q.) and idiocy (a severe mental deficiency, the person being incapable of guarding against common physical dangers, I.Q. less than 25).

(c) *Epilepsy* is a mental disease marked by seizures with convulsions and loss of consciousness.

(d) *Apoplexy* or stroke is characterized by sudden paralysis and coma from cerebral effusion (escape of fluid) or extravasation of blood into the brain tissues. This disease was already briefly mentioned under the arteriosclerotic heart diseases.

(e) *Migraine* or hemicrania is a one-sided headache, usually accompanied by nausea, vomiting, dizziness, yawning and increased sensitivity of the eyes, etc. One of the first indications of a coming attack is a disturbance of vision, often in the central part of the visual field. Throughout the attack the urine is strongly acid. Prior to an attack the morning urine is more alkaline than usual. This may occur 48 h in advance of the attack.

6.8.2. Meteorological effects on mental diseases

(a) *Apoplexy*. As pointed out on p. 160, there is a similar meteorotropic relationship with apoplexy to that described for arteriosclerotic heart diseases.

(b) *Migraine*. This very common disease may have various pathological causes, including purely psychological ones. It is particularly common after a previous concussion of the brain, especially if the patient has not taken sufficient rest. Migraine attacks occur very often during drastic changes in weather, just before heavy snowfall, thunderstorms, the passage of fronts, sudden changes in light intensity after a period of cloudiness, etc. Very little is known, however, about the physiological mechanisms involved.

According to Fanchamps[92,93] two physiological processes must occur at the same time to produce a migraine attack:

(i) vasodilatation of the extracranial arteries and contraction of the capillaries in the brain;

(ii) an increased sensitivity of the pain receptors (i.e. a lowered pain threshold) in the arterial walls due to plasmakinin.

Two substances play an important role in these processes: serotonin (5-Hydroxytryptamine, 5-HT) and histamine.

According to Lance[94] serotonin in the serum increases before attacks of migraine. The serotonin causes a constriction of large arteries and a dilatation of small blood vessels. It plays an important role in the hypothalamic thermoregulation (see Section 3.3). Once the migraine attack has started the total plasma serotonin level drops sharply. It is followed by a liberation of serotonin by the blood platelets during a migraine attack which increases the membrane permeability and lowers the pain threshold in the arterial walls.

Histamine* is normally present in all organs and tissues in small quantities and probably derives from the eosinophil leucocytes. It is one of the most powerful dilators of capillaries and arterioles, but causes a constriction of arteries. It increases capillary permeability. Injury of the skin, infections, emotional and meteorological stresses may increase the liberation of histamine. Thermal stresses affect the liberation of serotonin.

Apart from the already mentioned occurrence of migraine attacks during drastic weather changes, snowfall, thunderstorm and the passage of fronts, several other more specific meteorotropic relationships have been observed.

(i) According to Pearce[95] in 30% of migraine attacks excess light stimulation plays an important role;

(ii) Atmospheric chilling after the passage of fronts etc. may cause vascular and metabolic disturbances in the brain in poorly thermoregulated subjects which may cause an attack in those predisposed to migraine;

(iii) Gomersall and Stuart[96] in a long study in Scotland observed that with rising barometric pressure and low relative humidity the attacks increased.

*Histamine is formed from the amino acid L-histidine by the action of the enzyme histidine decarboxylase. It has important actions on the circulation, the smooth muscles, on secretions and in producing itch or pain.

Despite the many drugs which have been developed most migraine attacks are difficult to suppress. Normalization of the thermoregulation efficiency through high altitude treatment (Section 5.3.1) might be a possible solution in many instances.

(c) *Schizophrenia*. A four-year study was carried out by Tromp and Bouma[89-91] in The Netherlands on the possible influence of atmospheric and other unknown physical environmental factors on the restlessness and ill-temper of about 200 mental patients, in particular schizophrenics. The study covered seven psychiatric institutes with the co-operation of 21 psychiatrists attached to these institutes and of the matrons of the wards. The only patients studied were those with a clear-cut diagnosis, i.e. all psychiatrists of the same institute agreed on the probability of the diagnosis. Daily records (observed during four 6-h periods of the day) were collected, particularly of the degree of motor unrest in the patients, for which purpose a number of objective criteria were used. The unrest was recorded as the degree above or below the average daily level. Several criteria were used to determine this degree of unrest. A patient was rated restless if he was cursing, insulting other patients or nurses, raging etc. and as a result other patients became irritated. He was considered very restless if he was throwing things, slapping people, and the nursing staff was being forced to lock the patient up or to start special drug treatment.

Of the various groups of mental patients the schizophrenics, particularly females, gave the clearest results. A large number of the observed patients did not receive any medication during our period of study.

The daily percentage of restlessness of this large group of schizophrenic patients fluctuated as a rule between 7% and 43%, the average monthly percentages between 2% and 16%. In a large constant, rather homogeneous group of schizophrenics (homogeneous from the point of view of sex, age and social level) the following short- and long-term periodic or pseudo-periodic changes in restlessness were observed:

(i) Each year, during the four years of observation, the highest degree of unrest was observed in November, December and January. In some psychiatric institutes a secondary peak of unrest was observed in spring. The high degree of unrest in winter showed up both in the high average monthly percentage and in the greater number of days with high daily percentages of unrest.

(ii) Statistically significant daily fluctuations are correlated with the influx of warm continental, tropical maritime or warm maritime air masses, causing a gradual rise in temperature of the atmosphere and an increase in unrest. Short-term influxes of cold air masses have a reverse effect. Unpleasant weather conditions such as heavy rainfall and snowstorms, affecting seriously the mood of health people, do not seem to affect the restlessness of the schizophrenic patient.

The following studies carried out in Leiden since 1960 suggest that these short- and long-term, mainly meteorologically triggered, fluctuations in behaviour seem to be related to a poorly functioning hypothalamic thermoregulation mechanism in schizophrenics. With the water bath test[32] it can be shown that in a healthy, well-thermoregulated subject the temperature of the hand palm (after 2 min cooling in water of 10 °C) reaches the initial value in about 6 min, in schizophrenics it may take 20 min or more. The initial temperatures are often very low, the difference between initial temperature and the rewarming temperature after 6 min is often 4 °C or more.

In the Henschel test[97] the feet of schizophrenics were put in water of 45 °C. The time required before the skin temperature of the fingers rose was much longer for schizophrenics than for normals.

In normal well-thermoregulated subjects the urinary output increases during atmospheric cooling and decreases when the atmosphere is warming up. In schizophrenics one observes the opposite effect.[4]

Another interesting observation concerning meteorological effects on schizophrenics is the relationship found between month of birth and schizophrenia.[4]

In recent years in different countries evidence has been collected that schizophrenia and the incidence of mental deficiency in children are related to the month of birth, but this needs confirmation.

In 1957 Pasamanick (Director of the Research Division of the Columbus Psychiatric Institute at Columbus, Ohio) and Knobloch[98] traced the month of birth of mentally deficient children born between 1913 and 1948 and institutionalized in the Columbus State School. It was found that schizophrenics are born significantly more often in January, February and March as compared with the standard monthly birth distribution in the USA. Since for this group the third month after conception, the period when the cerebral cortex of the unborn child is becoming organized, would be June, July and August, any environmental factor causing a foetal injury in these very warm months might affect intellectual functioning. According to these authors, very high temperatures, causing metabolic disturbances and decreased protein intake during these months, might be one cause of such a trauma. They also found that the number of mental defectives born in January or February after a very hot summer is greater than after a cool summer. The same authors were able to show that the number of abnormalities in pregnancy are significantly higher in January, February and March than in the summer months, July to September.

De Sauvage Nolting,[39–41] former psychiatrist at the Psychiatric Institute Den Dolder (The Netherlands), listed dates of birth of 8000 schizophrenics in The Netherlands and of 7000 in Great Britain. He found, after the normal seasonal birth rate correction, that a mathematically significant higher birth rate occurred during February and March, a minimum in June and July. No connection with the month of birth was found for epilepsy and other mental diseases.

Studies by the same author, using material from Australia, showed a similar relationship but with a six-month shift due to its being in the southern hemisphere. Other authors have confirmed De Sauvage Nolting's observation in other northern countries.

(d) *Oligophrenia*. The following observations were made by Tromp and Bouma[91] in The Netherlands in connection with oligophrenic patients.

(i) Daily and monthly percentages of restlessness are of the same order of magnitude as in schizophrenics.

(ii) Pseudo-seasonal fluctuations, though less pronounced than in schizophrenics when averages for each month are compared, show similar early winter peaks when the average for three-month periods is taken.

(iii) Daily fluctuations in the restlessness curves sometimes coincide with those of the schizophrenics, particularly during the winter season; often they are considerably different.

Whereas schizophrenics are hardly affected in their restlessness by unpleasant weather conditions, oligophrenics are considerably affected. This purely psychological effect may also give rise to considerable restlessness when there is an influx of cold air.

(e) *Epilepsy*. Two different weather relationships were studied: its influence on the restlessness of epileptics and its effect on epileptic seizures.

(i) Influence on restlessness

(a) The pseudo-seasonal fluctuations suggest a maximum in November–December and a minimum in summer; there may be an occasional secondary maximum in May.

(b) There was no convincing relationship between daily restlessness and weather.

(ii) Influence on epileptic seizures. In certain instances the seasonal fluctuations in seizures were very pronounced. The lack of consistent seasonal phenomena was not however surprising since most of the patients observed belonged to a group which had lived for many years in a psychiatric institute and were under continuous treatment with drugs.

6.8.3. Meteorological effects on suicide and suicide attempts*

These are difficult to establish because they do not correlate with such meteorological factors as temperature, atmospheric pressure, hours of sunshine or precipitation. However more sophisticated meteorological analyses have revealed certain relationships. De Rudder[99] and Tholuck[100] studied 200 cases of suicide which had been recorded at the Institute of Forensic Medicine

*Attempted suicides are very difficult to establish with certainty. Reasonably accurate data can be obtained only by studying carefully the clinical reports of the physicians who visited the subject and by examining the police reports.

at the University of Frankfurt in Germany during 1939. No relationships were established with atmospheric pressure, temperature, hours of sunshine or precipitation but there was a highly significant correlation with both cold and warm weather fronts. Studies by Rohden[101] in Switzerland have indicated a significant increase in the number of suicide cases and crimes when the hot, dry föhn wind prevails. Reiter[102] was able to demonstrate a statistically significant correlation between periods of high frequency of suicide and days of strong disturbances of long wave electromagnetic radiation in the atmosphere.

Extensive studies were carried out in the western part of The Netherlands by Tromp and Bouma[89–91] during the period 1954–1969 using about 10 000 well established cases of suicide and attempted suicide. The ratio between suicide and attempted suicide ranges from 1:6 to 1:10. Because the number of cases per day was very small the method of 'cluster days' analysis was adopted. In other words, only those days with an incidence more than three times the daily average for the year were analysed in relation to the meteorological conditions of the days in question. No simple correlation was observed between suicide or attempted suicide and any one particular meteorological factor. However, a highly significant correlation was observed between cluster days and days with strong atmospheric turbulence associated with the approach or passage of depressions and/or weather fronts. Sudden rises or falls of temperature appreciably above or below the average monthly temperature, to which the subject is acclimatized, and changes in wind speed (i.e. the cooling index), have this triggering effect. It seems likely that a serious disturbance of the hypothalamic-pituitary system, caused by drastic changes in weather, contributes to the 'triggering off' of suicide and attempted suicide in those subjects predisposed to a serious mental disturbance following certain psychological, sociological and other forms of stress. A possible relationship between month of birth and suicide attempts will be discussed in Section 8.1.2.

6.8.4. Meteorological effects on other mental processes

As mentioned before the restlessness of school children during certain weather conditions, changes in the reaction time and performance capability of healthy subjects due to certain meteorological stimuli, the occurrence of certain road and industrial accidents as a function of weather, the intelligence of school children as a function of their month of birth, the effects of weather and climate on human behaviour and on reproduction, also indicate the effect of weather and climate on different mental processes in humans. These effects will be briefly discussed in Chapter 8.

6.9. RHEUMATIC DISEASES

As rheumatic diseases do not represent a sharply defined group of diseases, physicians do not adhere to the same classification and definition either nationally or internationally. Rheumatism is often defined as a disease affecting the connective tissues, marked by pain or stiffness in joints, muscles and related structures, usually recurrent and often due to exposure to certain environmental conditions which may be acute or chronic. The disease seems to be more common among females than males. The incidence of this group of diseases varies considerably not only in different parts of the world but also in the same country, as will be discussed later.

Studies in Sweden and The Netherlands, where the population does not migrate to any large extent, suggest that about 20% of the morbidity in these countries is due to one or more varieties of rheumatic diseases, particularly in the age group of 50 and older.

Around 1960 13 million people in the USA, of whom 300 000 were temporarily unemployable were suffering from one form or another of rheumatic diseases. In Great Britain £40 000 000 were lost in 1950 due to serious illness as a result of rheumatic diseases.

At least 1% of the population in Sweden is permanently invalided as a result of rheumatic diseases. Notwithstanding this very high incidence and the general belief, of both rheumatics and physicians, that weather and season affect the severity of the complaints, relatively little is yet known of the exact influence of meteorological conditions on the origin and development of rheumatic diseases, probably because rheumatic diseases create only a sensible discomfort and are usually not fatal like cancer or heart diseases. This also explains the relatively small funds which are available for research into rheumatic diseases.

The various rheumatic diseases have been classified by the W.H.O. into two groups, truly rheumatic diseases and para-rheumatic diseases.

(a) *The truly rheumatic diseases* are divided into two groups:

(i) Those localized in the joints; (a) due to inflammation: acute rheumatism (rheumatic fever), rheumatoid arthritis (polyarthritis), ankylopoetic spondylosis (Morbos Becheterew), etc.; (b) due to degeneration: arthrosis deformans (osteoarthritis);

(ii) Those not localized in the joints; due to inflammation: fibrositis, bursitis, myositis, tendosynovitis, and periarthritis.

(b) *Para-rheumatic diseases*: various collagen diseases (collagen is the main supportive protein of connective tissue) and other diseases with rheumatic characteristics, such as lumbago, podagra (gout), myogelosis, etc.

6.9.1. Definitions of the different rheumatic diseases

(a) *Arthritis*: an inflammation of the joints due to bacterial infections or injuries (either caused by a simple severe mechanical injury or a repetition of minor unnoticed injuries, mechanical or physical).

(b) *Acute rheumatism* (or rheumatic fever): an acute arthritis, common in children, particularly in their larger joints, often accompanied by rheumatic inflammation of the lining or muscles of the heart and by high fever.

(c) *Rheumatoid arthritis* (or polyarthritis): a polyarticular arthritis, chronic or recurrent, often symmetrical on hands and feet, common in women between 20 and 40.
(d) *Ankylopoetic spondylosis* (or Morbus Becheterew): an inflammation leading to immobility and consolidation of the vertebral column and hip-joints.
(e) *Arthrosis deformans* (or osteoarthritis): a non-inflammatory degenerative joint disease, particularly in knee and hip-joints, of unknown cause, particularly common at the age of 60 and older and leading to immobility.
(f) *Fibrositis*: an inflammatory hyperplasia of the white fibrous tissue of the muscle sheaths.
(g) *Bursitis*: an inflammation of the sac that helps to reduce the friction around a joint.
(h) *Myositis*: an inflammation of muscle tissues.
(i) *Tendosynovitis*: an inflammation of a tendon (the fibrous cord by which a muscle is attached) and its sheath.
(j) *Periarthritis*: an inflammation of the soft tissues surrounding a joint, accompanying arthrosis deformans.
(k) *Lumbago*: acute rheumatic pains in the lumbar region, usually due to one-sided muscular rheumatism.
(l) *Podagra* (gout): painful disease of the joints due to the body's inability to dispose of its purines (the basic compound of uric acid). The excretion of uric acid increases during an attack.
(m) *Myogelosis*: a hardening of the muscle substance.

6.9.2. The effects of the weather and climate

Rheumatic diseases in humans have been extensively reviewed by Burt,[103] Mills,[104] De Rudder,[105] Tromp,[4] Lawrence *et al.*[106] and others.[107,108] All studies appear to suggest considerable differences in the kind and severity of rheumatic diseases and particularly arthritis within different countries, although there seems to be little variation in the actual global incidence. However, the degree of subjective complaints and the progress of the disease is much affected by weather conditions and in particular by strong atmospheric cooling (temperature-wind effects). These factors are particularly prevalent in countries of middle and high latitudes. For example high incidences have been reported from several western European countries, whereas the incidence seems to be very low among Eskimos in polar regions.

Mills[104] observed regional differences in acute rheumatic fever and rheumatic heart diseases in the USA. Acute rheumatic fever is most frequent in the cooler, more stormy parts of the USA. In the tropical and subtropical parts with little air turbulence (therefore not in Florida) it is markedly less frequent (10 to 20 times more often in the north than in calm tropical areas). In the north it strikes most frequently in winter and early spring, seldom during summer, whereas in Cincinnati, rheumatic fever attacks in March are six times more frequent than in August and September.

According to Nichol[107] and Paul[108] the same relationships are found for children of the same race with rheumatic heart diseases, the most favourable areas being New Mexico, Arizona and California. Coburn observed a rapid recovery of children from New York suffering from rheumatic fever after they had been sent to Puerto Rico.

Cobb and Lawrence,[109] compared the cities of Pittsburgh and Leigh in the USA. They observed striking differences in the prevalence rates for rheumatoid arthritis. Above the age of 55, these are probably 12 per 100 in Leigh against 4 per 100 in Pittsburgh. The difference for rheumatoid arthritis, observable with X-rays, was even 5 against 0.3 per 100 persons. Both cities have almost the same annual mean temperature and precipitation but Leigh has a more frequent gentle precipitation against torrential rain storms (with strong cooling) in Pittsburgh. However, also racial differences between the cities may also be partly responsible for the observed differences.

McKinley[110] described the very low incidence of acute polyarthritis in equatorial regions, such as Kenya, Uganda, Nigeria and south west Africa. It is more common in Egypt and the Union of South Africa, with periods of strong cooling. In certain parts of India rheumatic fever is very common, although these areas are located in the tropics. However, cooling conditions may be very severe in tropical countries (for further details and references see Reference 4).

6.9.3 Possible meteorological factors triggering or increasing rheumatic complaints

Several factors have been studied, in particular humidity, temperature and wind.

(a) *Effects of humidity*. Humidity as such has little effect on our skin temperature. Humidity changes in the environment may, however, affect human comfort indirectly owing to:

(i) changes in thermal conductivity of the skin near the thermal receptors,

(ii) differences in evaporation of sweat on our skin and of the moisture of our clothes affecting the degree of physiological cooling and its thermoregulation consequences,

(iii) changes in the infrared radiation of the walls and floors of rooms in buildings due to humidity changes.

Several authors, e.g. Van Breemen[111] in The Netherlands, Miller[112] and Fox[113] in England, Teissier and Roque[111] in France and Natvig[111] in Norway, noted a very high incidence of rheumatic diseases among thousands of patients living in damp houses. People living in damp basements are particularly prone to chronic rheumatism. Soldiers who spent a considerable time in wet trenches during the war often returned as chronic rheumatics.

On the other hand, physical labour in a very humid environment seldom has harmful effects due to the great heat production during labour.

Studies around 1960 in Leiden, The Netherlands, indicated a statistically highly significant correlation between the incidence of rheumatoid arthritis and damp old houses where there was considerable wood-rot in the floors. This is actually due to a growing mycelium of wood-decaying fungi (particularly *Basidiomycetes*) which requires a certain, rather constant, degree of moisture of the wood, with little evaporation. Temporary high moisture content of wood followed by evaporation and dryness does not necessarily cause wood-rot. The wood-rot fungi require a special humidity of the wood (fluctuating between 20 and 70%), certain temperatures (minimum 4 °C, maximum 27–35 °C, optimum 22–27 °C), sufficiently humid air around the wood, little air movement and darkness.

It was found in The Netherlands[114,115] that people suffering from rheumatic diseases are often living in old damp houses built in poorly drained soil and low floors which were at ground-water level in winter. Observations by Tromp in The Netherlands, using infrared (heat) wave detectors, indicated that, although a wooden floor in a room may be dry everywhere and there may even be an air space of 2 or 3 ft between the floor and a damp concrete basement, the most humid parts of the basement can be detected above the dry wooden floor because of a local decrease in infrared radiation. A rheumatic person is very sensitive to such spots.

The indirect effect of humidity does not show up only through humidity of the air and humidity of the floor (wood-rot) and cellar of a house but also through the soil on which the house is built. During the Public Health Congress held in Eastbourne, England, in 1933 it was reported[116] that the greatest incidence of juvenile rheumatism occurred in England on soils along the course of rivers or near former water-courses.

A study in St Thomas' Hospital, London, indicated that a large portion of the patients suffering from rheumatic fever lived within half a mile of the Thames or other waterways. In the State of New York the smallest number of cases occurred in districts furthest away from the river.[117]

The following observations point to a possible relationship between soil, microclimate and rheumatism. Electric soil surveys by Tromp[118] in The Netherlands indicate that usually every area is crossed by one or more zones of high electric soil conductivity, or low soil resistivity. These zones are caused by various geological conditions, e.g. the presence of underground fractures or faults; old river valleys in a sandy soil filled up with clay; an abrupt local rise of a peat layer under a horizontal stratum of clay, etc. In all these cases, owing to a locally higher moisture content (and often also a higher content of humic acid), the soil resistance may suddenly drop considerably, e.g. from 70 000–5000 ohms m^{-1}.

Houses built above a moisture zone of this kind may show patches of dampness causing locally reduced infrared radiation as mentioned before.

Physiologically pre-disposed subjects spending many hours in parts of a house where such local cooling conditions prevail seem to be liable to become rheumatic.

(b) *Effects of temperature and wind.* The various observations mentioned above suggest that rheumatic diseases are rare in an environment of relatively high temperature, provided there is no considerable cooling due to great humidity and air turbulence. This may explain why people living in tropical regions, who fail to take the necessary precautions after perspiring profusely and then cooling down, present symptoms of rheumatism.

6.9.4. Principal physiological mechanisms of meteorotropic effects observed in rheumatic patients

(a) Thermoregulatory efficiency tests in The Netherlands in large groups in arthritic patients showed that 79% of the subjects had an inefficient thermoregulatory mechanism and 21% an extremely poor thermoregulatory efficiency, both in summer and in winter. Therefore atmospheric cooling affects a rheumatic patient much more than a healthy subject. In this connection the studies of Copeman[119] are very important. If normal deep tissues are cooled rapidly from 30–15 °C, deep pain is experienced. As cooling proceeds, analgesia (absence of sensibility to pain) develops, which becomes complete at a tissue temperature of 10 °C. The same cooling, taking place slowly, hardly causes any pain in a normal person whereas such slow cooling may cause pain in rheumatic subjects, because in them the adjustment of heat loss, by the control of the blood flow and capillary contraction, are poorly regulated. It is known that nervous subjects and patients suffering from rheumatoid arthritis exhibit an abnormal degree of peripheral vasoconstriction. This also explains why rapid cooling, especially during the passage of active cold fronts accompanied by strong winds, causes severe pain in rheumatic subjects or in tissues, abnormally sensitized by trauma. The same is true for cutaneous pain in the skin. In normal skin, pain results from heating above 43 °C or cooling below 10 °C, but in some stages of neuritis (i.e. inflammation of the nerves), pain may be felt even at temperatures just above 30° and below 15 °C because arteries are partly blocked and unable to allow the normal increase of blood flow in response to warming.

(b) A 4-year study by Tromp and Bouma[120] in The Netherlands has shown that the excretion of hexosamines in urine decreases with increased cooling and increases with decreased cooling. The hexosamine excretion in the urine of rheumatic patients is about half that observed in healthy subjects. Hexosamines, derivatives of mucoproteins, occur in the joints and cartilages. They are the chief agents producing the viscosity of mucin, a glucoprotein secreted by the mucous membranes. Hunter *et al.*[121,122] demonstrated that the viscosity of the synovial fluids responsible for the lubrication of human joints depends to a considerable extent on the mucin content. Accumulation of hexosamine, as occurs during cooling, raises the viscosity of the synovial fluids and increases the severity of rheumatic complaints, the more so as cooling decreases the cellular permeability and therefore the removal of metabolic waste products is retarded. The mucin content depends also on the pH of the synovial fluid. More mucin is dissolved in alkaline fluids. Atmospheric cooling causes an alkalinization of body fluids. Also the favourable effects of thermal baths in Spas is partly due to an increased hexosamine excretion after the baths.

(c) The adrenal function and the secretion of corticosteroids is usually considerably reduced. Since cooling conditions have a considerable affect on steroid levels in the body, this factor may also be involved in the meteorotropic effects of weather and climate on rheumatic diseases.

(d) Cooling effects the muscular enzymes. During muscular contraction and relaxation a series of chemical processes occurs which is responsible for part of the heat production in the body. During these exothermic (heat producing) reactions, one of the important constituents of the skeletal muscles, adenosine triphosphate (ATP), is broken down by an enzyme, adenosine triphosphatase, into adenosine diphosphate and phosphate. Studies by Racker *et al.*[123,124] in the USA have shown that strong cooling has an important effect on these enzymes. The activity of ATP-ase rapidly decreases at +4°C.

In Section 9.1, on Climatotherapy, some important practical applications of biometeorology for rheumatics will be briefly summarized.

6.10. SKIN DISEASES*

Many physicians, particularly dermatologists, have reported on both the causation and the therapeutic effects of solar radiation and high altitude on skin diseases. The interested reader is referred to a number of excellent monographs and textbooks on meteorotropic aspects of skin diseases including those written by Hausmann and Haxthausen,[125] Jausion,[126] Blum,[127,128] Rocha and Silva,[129] Kesten and Slutkin,[130] Urbach[131] and Brenn and Gehrken (in Reference 1). It is outside the scope of the present book to give even a short review of all the reported meteorotropic relationships.

A number of seasonal phenomena reported for skin diseases are summarized in Table 6.1. In spring a high incidence of eczema is observed. In summer a high incidence of dyshidrosis, miliaria rubra, dermatomycosis etc. occurs. From July–August (often increasing till November–December) a higher incidence of furuncles and carbuncles, erysipelas and erysipeloid has been reported. As pointed out in Section 6.4, excessive sunlight causes skin cancer, light dermatoses, photodynamic skin diseases, etc. Herpes zoster, which is common in the northern hemisphere in July and rare in winter reaches in Java a maximum in April and May and a minimum in January–March (wet monsoon).

*For a Glossary of the skin diseases see Section 6.10.3.

The meteorotropic effects on the human skin can be divided into two main groups, those caused by solar radiation (either direct or indirect effects) and those resulting from high altitude (both thermal and ultraviolet radiation effects and the results of physiological stimuli caused by the reduced oxygen pressure, see p. 251).

6.10.1. Solar radiation effects

The various physiological processes in the skin, resulting from infrared and ultraviolet solar radiation, which may cause certain skin diseases have been described in Section 5.2. Ellinger (in Reference 4) summarized the mechanisms of solar-induced skin diseases as follows: both direct and indirect effects of solar radiation must be distinguished.

(a) *Direct influences.* In Section 6.4 the effects of ultraviolet light on the induction of skin cancer have been discussed in detail. Möller[132] was the first to indicate the role of visible light in the causation of certain skin diseases, known as light-dermatoses.

A skin disease of unknown aetiology but affected by sunlight is known as vitiligo (characterized by oval or irregular patches of depigmented skin). It is found both in humans and in cattle. An interesting group of, partly photodynamic, light dermatoses is known as erythema exudativum multiforme. It is more common among females than males, particularly among blond females; its greatest incidence is in spring and autumn. An epidemic of a skin disease in The Netherlands and Germany, resembling erythema exudativum multiforme, known as Planta disease, is probably due to a sudden introduction in mass-consumed food of an emulsifier which proved to be photodynamic combined with an increased photo-sensibility of the general population during the period of the year that this new food product was introduced. In Section 6.11 a number of skin diseases will be discussed which are due to extreme thermal stresses.

(b) *Indirect influences*

(i) Increase in sweating due to thermal stress: The heat radiation of the sun increases sweating. A considerable increase in sweating over long periods followed by sweat-gland fatigue may cause dyshidrosis (an acute recurrent, non-inflammatory vesicular eruption limited to the palms and the soles), which may develop into dyshidrotic eczema (red itching skin with small vesicles) and *miliaria rubra* or prickly heat (acute inflammation of the sweat-glands with small red papules and vesicles and intense itching and burning of the skin). Under these conditions a skin disease may develop, due to parasitic fungi, known as dermatomycosis.

(ii) Influence of ultraviolet light on the growth and development of micro-organisms: The lethal effect of ultraviolet radiation on various micro-organisms on the skin seems to be greatest in dry air with a humidity of less than 60%. Whistler[133] and Wells and Brown[134] demonstrated this for bacteria in aerosols. Lidwell and Lowbury[135] observed the same effect with dried bacteria on glass. On the other hand, the same authors noticed that strepto- and staphylococci in dust in a dry atmosphere are more resistant. Work by Buchbiner[136] showed that ordinary daylight, even after passing through a windowpane, also had a lethal effect on streptococci in dust with a 36% relative humidity of the air. According to Edward[137] light also has a lethal effect on the influenza-A virus.

(iii) Decrease in the local resistance of the skin to infectious agents (see Section 6.7.1): Marchionini *et al.*[138] demonstrated that the acid coating of the skin acts as a protective layer against micro-organisms. This is very important because various meteorological factors, such as temperature, humidity and air movement, which affect the sweating rate and evaporation, may influence the acidity of the skin and the growth of micro-organisms on and their penetration into the skin. This explains the formation during summer of furuncles (local inflammation of the skin and subcutaneous tissue, due to the penetration of staphylococci into sweat glands (commonly in the neck or axilla), accompanied by local swelling, often also fever, and carbuncles (deeply situated, laterally spreading infection, creating multiple adjacent furuncles).

According to Ernst[139] erysipelas (acute infectious skin disease accompanied by inflammation and redness of the skin, mucous membranes etc. accompanied by high fever, caused by *Streptococcus erysipelatis*) is also a summer disease. The same applies to erysipeloid (a skin rash on fingers and hand resembling erysipelas, but no fever; often due to regular contact with infected animal products; infection due to *Erysipelothrix rhusiopathiae*). High radiation intensity of the sun combined with cooling (great turbulence and high humidity) as observed along the sea coast has a favourable effect, according to Häberlin[140] and Kalkoff,[141] on skin tuberculosis (*Lupus vulgaris*).

(iv) Influence of sea bathing: the combined effects of solar radiation on the beach and seawater bathing (thalassotherapy) will be discussed in Section 9.1.1, pp. 236–240.

6.10.2. High altitude effects

The biological effects of altitudes above 1500 m were fully described in Section 5.3. The main factors which influence skin diseases at these altitudes are the intense solar radiation (particularly ultraviolet), the small number of allergens, the reduced atmospheric humidity and the usually lower temperature.

According to Brenn and Gehrkew[142] patients suffering from eczemas, which are very resistant to different drugs, are often completely cured after a short high altitude treatment (p. 25). They pointed out that the objective changes in cutaneous allergic reactivity displayed by patients who have stayed at high-altitude resorts is impressive. According to Popchristov and Balewska[143] the results of allergy tests, which had been positive prior to high-altitude treatment, were frequently found to be negative at the completion of therapy, which suggests that a high-altitude climate has a normalizing effect on the specific allergic sensitivity of the organism in general and of the skin in particular (see Section 6.2). In Davos (Switzerland, 1554 m), the results obtained in 10 000 dermatographic tests of patients suffering from constitutional neurodermatitis, showed a marked tendency to revert to normal.

In a group of in-patients suffering from neurodermatitis, Michailov[144] carried out studies to compare the effect of intracutaneous doses of histamine, serotonin, acetylcholine and bradykinin administered before and after high-altitude treatment at Davos. He found that, after the treatment had been completed, the size of the weals was much reduced, as was also the reactive erythema, i.e. a clear-cut change had taken place in the patients' response to the mediators of allergic reactions.

Chlebarov and Borelli[145] described the changes in autonomic nervous reactivity which occurred in in-patients being treated at Davos. In 96% of the patients with constitutional neurodermatitis, sympathetic nervous reactivity was elevated at the start of treatment, whereas at the end it had returned to normal. These observations indicate that a high-altitude climate has a desensitizing effect on both the general and the allergic reactivity of the skin, which explains the many favourable effects of high altitude therapy on skin diseases. Apart from the diseases mentioned above *psoriasis vulgaris*; *parapsoriasis*; chronic relapsing urticaria; *lichen planus*; and forms of acne and *mycosis fungoides* (in its initial stages) have also been successfully treated at high altitude. The chief contra-indications for high-altitude therapy are all types of photodermatosis and *lupus erythematosus discoides*.

6.10.3. Glossary of the above mentioned skin diseases

Acne: affection of the skin with eruption of papules.
Carbuncles: deeply situated, laterally spreading infection, creating multiple adjacent furuncles.
Dermatomycosis: skin diseases due to parasitic fungi.
Dyshidrosis: acute recurrent, non-inflammatory vesicular eruption limited to the palms and the soles.
Dyshidrotic eczema: red itching skin with small vesicles.
Eczema: a skin disease with itching redness, and infiltration.
Erysipelas: acute infectious skin disease accompanied by inflammation and redness of the swollen skin, mucous membranes etc. accompanied by high fever, caused by Group A haemolytic streptococci.
Erysipeloid: a skin rash on fingers and hand resembling erysipelas, but no fever, often due to regular contact with infected animal products.

Erythema exudativum multiforme: an acute variety of a congestive or exudative redness of the skin with variously formed papules, tubercles etc.

Furuncles: local inflammation of the skin and subcutaneous tissue, due to the penetration of staphylococci into sweat glands, accompanied by local swelling, often also fever, common in neck or axilla.

Herpes zoster: an acute skin disease marked by clusters of small vesicles, caused by filterable viruses closely related to that of varicella, characterized by inflammation of dorsal root ganglia and neuralgic pain.

Lichen planus: an inflammatory skin disease with wide flat papules often in circumscribed patches caused by a combination of an alga and a fungus.

Light-dermatoses: skin diseases caused by sensitivity to light stimuli.

Lupus erythematosus: inflammation of the skin with patches having raised reddish edges covered with scales.

Lupus vulgaris: tuberculosis of the skin marked by the formation of brownish nodules on the corium.

Miliaria rubra or *prickly heat*: acute inflammation of sweat-glands with small red papules and intense itching and burning of the skin.

Mycosis fungoides: a fatal skin disease, with fungus tumours and much pain.

Neurodermatitis: a chronic itching lichenoid eruption of the skin.

Photodermatosis: skin disease due to photochemical reactions.

Photodynamic skin disease: skin disease due to photochemical reactions resulting from consumption of certain foodstuffs or from touching certain plants or substances.

Psoriasis: a skin disease characterized by scaly red patches.

Urticaria: a skin condition marked by erythematous wheals, persisting for hours or days.

Vitiligo: a skin disease characterized by oval or irregular patches of depigmented skin.

6.11. DISEASES DUE TO EXTREME HEAT AND COLD

Subjects living in hot, humid countries experience a number of physiological changes due to extreme thermal stresses (partly described already in Section 5.1) which may lead up to a number of meteorotropic thermal diseases.

6.11.1. Effects of extreme heat stress

(a) *Oedema* of the feet and lower legs is often found in unacclimatized subjects standing or sitting for long periods at shade temperatures above 27 °C. The oedema is felt most in the hands and fingers and is due to the general water retention. Raising of hands or feet above heart level reduces temporarily the complaints. Also diuretic drugs have been recommended for subjects working under these conditions.

(b) *Syncope*, due to pooling of blood in the veins of the legs, is common in people standing for long periods, particularly in hot places above 25 °C.

Fainting is due to the excessive venous relaxation and insufficient venous blood return to the heart.

(c) *Sweat gland disorders* are common in unacclimatized subjects under very warm humid environmental conditions (see Section 3.3). They may cause a number of pathological conditions.

(i) As a result of an inefficient thermoregulatory mechanism of the subject or due to sweat gland fatigue as a result of long lasting sweating (causing insufficient sweating, dyshidrosis, and therefore insufficient evaporative cooling) the deep body (core) temperature may rise above the critical level of 42°C. This form of hyperthermia is often encountered in people above the age of 50 and in infants below the age of one, particularly in an infant having an inefficient thermoregulation mechanism. An excellent summary of these hyperthermia syndromes amongst unacclimatized sailors in the Persian Gulf was published by Leithead and Lind.[146] Apart from these infrared (heat) radiation phenomena, Thompson[147] reported dyshidrosis as a result of reduction in sweating, due to a blockage of sweat glands by vesicles after ultraviolet radiation in erythemal doses. After a week of sunburn, sweating was reduced 30–40%. Not only the sweat glands are blocked but also a reduction in sweat formation seems to occur, causing a reduced cooling.

(ii) A second clinical phenomenon, caused by a sweat gland disorder, observed in hot climates is heat stroke. Pearson[148] reported on heat stroke in older, and not well acclimatised, residents of Perth in Australia after a period of 10 days during which air temperatures of 45°C and 60% relative humidity were recorded. Similar conditions can be induced by surgical interference with the anterior hypothalamic region (Section 3.3 and Fig. 3.1). With rectal temperatures reflecting the deep body temperature of 42–43°C patients will not survive unless sweating can be restored. Sprinkling the skin with water and facilitating evaporation with a fan could improve the evaporative cooling. This should be continued until the rectal temperature falls below 39°C. Application of ice-chips or ice-cold water is often inadvisable, because one may induce vasoconstriction which will prevent the blood from circulating through the cooled skin.

(iii) A third consequence of sweat gland disorders in hot climates is due to salt deficiency or the inability of unacclimatised subjects to conserve salt[149,150] as a result of heavy sweating. This reduction in sodium status leads to poor physical performance, lassitude and heat cramps. The heat cramps are characterized[151] by 'the bunching up into hard knots of the abdominal muscles and those of the arms, legs and thighs. The knots stand out on the muscle due to violent tonic local contraction. The condition is very painful'.

Studies by Ladell,[152] in the Iraq desert and under laboratory conditions in London, suggest that the cramps start when plasma chloride is reduced to 80 meq l^{-1}. Injection of 23 g of sodium chloride removed the pains. According to Macfarlane,[153] the syndrome occurs particularly in susceptible unacclimatized subjects, when over 20 g of salt is lost by sweating or when intracellular hydration occurs with dilution of the extracellular fluid by considerable water ingestion. A daily supply of 10 g of extra salt during the first months of excessive heat exposure[154] or drinking of 0.25% sodium chloride[151] improves the resistance of cramp-prone subjects.

In this connection the studies by Furman and Beer[155] and Hancock *et al.*[156] should be mentioned. These suggest that a progressive rise in the sweat sodium concentration accompanied by a slight progressive fall in potassium concentration takes place during prolonged sweating as a result of two hours heat stress. After acclimatization for two weeks this increase in the sodium/potassium ratio, after two hours heat stress, was reduced from 16.2 to 8.5. After oral administration of aldactone (the aldosterone antagonist) in doses of 50 mg at six-hourly intervals for 24 h, a renewed marked increase of the sodium/potassium ratio was observed from 8.5 to 23.8 after the 2 h test.[155]

(iv) A fourth phenomenon in unacclimatized subjects in hot humid climates is prickly heat (*miliaria rubra*) (see Section 6.10.3). The skin of the trunk and to some degree of the limbs, develops red papules which prickle, burn or itch. It is due to blockage of the eccrine sweat gland orifices by squamous (i.e. scaly or plate-like) cells. The glands produce vesicles in the epidermis and create clear papules.[157] Only air-conditioning or cool quarters are usually able to give relief although the anhidrosis may persist for several weeks. A variety of this disease is tropical anhidrotic asthenia (*miliaria profunda*) which is accompanied by insufficient evaporative cooling and a rise in body temperature to 39°C. Anxiety weakness, poor exercise tolerance and heat exhaustion have also been reported.[157,158]

(v) Affective disorders are other consequences of sweat gland and general thermoregulation disorders in hot humid areas.[153] When heat is gained from the environment, there is less drive to engender more heat by activity. This lassitude is commonly found in hot regions and is probably physiological, but it is exaggerated amongst those who have emotional difficulties. Critchley[159] has discussed psychotic reactions which seem to be commoner in hot regions, e.g. manic outbursts such as amok. Quantitatively, however, the neurotic syndromes are more important and may be grouped as tropical fatigue.[160] Macfarlane[153] pointed out that probably many factors lie behind such states. Displacement from normal surroundings and activity, discomfort, reduced food intake and increased alcohol consumption, parasites, skin rashes and lack of diversion may all contribute. It is also likely that some people are unable to make the physiological adjustments needed to live efficiently in the tropics. Full adaptation to hot climates seems to require a new generation.

A part of the neurotic, neurasthenic or tropical fatigue syndromes may be due to inadequate water, salt and metabolic adjustment as well as to the emotional disturbances.[161]

(vi) Skin diseases are very common in hot humid regions.[160,162,163] In hot environments there is a high incidence of skin infection by pyogenic bacteria and fungi. In India during the dry season a skin disease rate of 6.2/10 000/day for fungal and pyogenic lesions has been observed compared with 15 during the cooler but very humid monsoon period. Repair was slow during the wet months. A similar high incidence of skin infection has been reported in New Guinea[160] and it is clear that it presents one of the major causes of disability in the tropics. Men in warships in the tropics suffer more skin disease and accidents than those in cooler regions.[162] Statistical data from the British Navy[163] confirmed this, and showed that the sickness rate trebled as the temperature rose from 15–32 °C. The main increase in morbidity was from skin lesions, the incidence of which rose steeply as soon as sweating became common, at temperatures above 27 °C.

(d) *Other physiological effects*: apart from the different sweat gland disorders described, heat stress will have a number of other physiological consequences:

(i) the blood volume increases due to the decreased diuresis, causing a lowered haemoglobin percentage of the blood;

(ii) during mild heat stress the blood pressure first decreases and causes fatigue in subjects who usually have a low blood pressure. With increasing heat stress over a long period (e.g. during heat waves) the blood pressure then rises rapidly;

(iii) physical properties of blood (such as BSR, blood clotting time, fibrinogen content) change drastically and may cause sudden heart attacks (Section 6.5);

(iv) people not suffering from constipation during normal life in northern countries may experience constipation during periods of great heat stress and dehydration.

6.11.2. Influence of excessive cold stress

In the northern Pacific, the north Atlantic ocean, the northern parts of Canada, Europe and Siberia both sailors and others working in these areas may experience another group of meteorological diseases due to the extremely low temperatures which are often accompanied by strong winds. Particularly dysfunction of the extremities is common as they present the largest surfaces exposed to cold and thus have the greatest heat loss.[164,165]

(a) Chilblains occur due to superficial tissue damage as a result of poor skin circulation at low environmental temperature.
(b) Frostbite is the result of prolonged and severe vasoconstriction of large blood vessels due to dry cold exposure at temperatures near 0 °C. Clotting of the blood in damaged blood vessels may prevent the return of circulation.[166]
(c) Wet cold syndromes occur during exposure to temperatures below 12 °C for several days.[167] Immersion in sea water or sea water falling on deck during freezing conditions may cause such syndromes if the sailors are not quickly rescued.
(d) Affective disorders are also common in cold regions if cold acclimatization takes place rather slowly due to an inefficient thermoregulation mechanism. Such depressive psychoses have been reported by Critchley[159] and others.

In Section 6.13.4 acute hypothermia, a serious disease of elderly people, will be discussed.

6.12. HIGH ALTITUDE DISEASES

The physiological effects of altitudes above 1500 m (below 635 mm Hg) were discussed on pp. 109–111. The favourable effects of high altitude will be discussed later on pp. 249–252 in connection with low pressure climatic chamber therapy. Unfavourable effects are known as chronic mountain sickness or Monge's disease[168] and sickle-cell anaemia. These appear to be directly related to the low oxygen tension (hypoxia) resulting from exposure to low atmospheric pressure (pp. 75–76).

6.12.1. Effects of hypoxia

Chronic hypoxia may lead to pathological alterations or influence the severity of the symptoms of any illness acquired at high altitudes. These effects have been studied particularly by Hurtado[169,170] and Monge[168] and their co-workers in Peru. This field of research has not yet been properly investigated but some interesting observations have been made already. Hurtado reported a number of pathological differences between the native Indians of Peru inhabiting the plains and those living at 4540 m (at Morococha):

(a) at high altitude a certain degree of hypertension in the pulmonary artery is observed in the Indian natives, which affects the right heart;[169,170]
(b) according to recent studies[171] high altitude increases the diameter of the heart (particularly the right side) and causes a hypertrophy of the muscle of the right ventricle of the heart; it increases the pressure in the pulmonary artery. Permanent adaptation of the native population to high altitude, which is accompanied by increased pulmonary ventilation and broadening of the thorax, causes a change in the position of the longitudinal axis of the heart which creates a characteristic anomaly in the electrocardiogram;

(c) it appears that hypoxia has a constricting effect on the pulmonary vascular bed and increases the blood volume;
(d) coronary thrombosis and myocardial infarction are very rare in Peru;
(e) a high incidence of peptic ulcer is observed although a lack of hydrochloric acid is common at high altitudes;*
(f) cholecystitis (inflammation of the gall-bladder) is common at high altitudes;
(g) some types of cancer seem to be rather uncommon, leukaemia has a very low incidence;
(h) high keloid formation (i.e. new growth of skin on scars) after superficial interventions is common.

6.12.2. Chronic mountain sickness

The most interesting pathological aspect of life at high altitudes is the loss of acclimatization to the low oxygen tension. The syndrome Chronic Mountain Sickness or Monge's disease, was first described by Monge[168] in 1928. It is characterized by an abnormal degree of hypoxia and by an exaggerated level of polycythaemia. The circulating blood volume reaches high values, due entirely to a pronounced elevation of the red cell volume. The plasma volume is decreased. The etiology of chronic mountain sickness remained obscure for a long time, but it has been observed in recent years that at least some of the cases lack the hyperventilation adaptive processes present in the acclimatized subjects. This leads to a lowering of the partial pressure of oxygen in the alveolar air and a consequent decrease in the arterial oxygen saturation. The abnormal accentuation of polycythaemia appears then to be the possible result of the increased hypoxic stimulus.[172]

6.12.3. Sickle-cell disease

Sickle-cell disease or sickle-cell anaemia is characterized by sickle haemoglobin in the red cell, which is capable of distorting the red cell into a characteristically sickle shape. It is common in 20% of west African negroes and in 10% of West Indians and American negroes. Under severe hypoxia sickling of haemoglobin occurs causing infarctive crises which may be fatal as experienced at high altitude in unpressurized aircraft. It is therefore recommended that negroes from the USA and west Africa should only fly in pressurized planes with simulated altitudes of less than 1500 m unless the absence of sickle-cell disease can be definitely established beforehand.

In a recent publication Singh *et al.*[173] pointed out that a prolonged stay at high altitude significantly lowers the incidence of some of the diseases commonly observed at sea level. This conclusion is based on a study involving 130 700 men stationed on plains between 760 m and sea level and 20 000 men stationed

*This may be an ultraviolet radiation effect (see Table 3.1).

at altitudes between 3692 and 5538 m during the period 1965 to 1972.

When yearly differences in morbidity rates were determined for this period it was found that apart from amoebic hepatitis, goitre and lobar pneumonia, which show a higher incidence, the incidence of infections of bacterial, viral and protozoal origin, diabetes mellitus, hypertension and ischaemic heart disease, asthma and rheumatoid arthritis, gastric disorders, skin diseases, psychiatric ailments and anaemia was significantly lower at high altitude than at sea level.

6.13. OTHER METEOROTROPIC DISEASES

6.13.1. Dental diseases

A great number of meteorologic relationships have been reported but most of them have to be reconfirmed. For example toothache pains seem to be more frequent during drastic weather changes;[174] dental caries shows a seasonal variation in the USA, with maximum incidence in the late winter and early spring and minimum in summer;[175–179] this may be due to the seasonal changes in ultraviolet radiation, hours of sunshine, phosphate and the vitamin D level of the blood serum, which are all decreased in winter; periodontitis pains (an inflammation of the supporting tissues surrounding a tooth) increase after an influx of cold air masses or cold fronts[180] and on föhn* days;[181,182] pulpitis pains (inflammation of the pulpa of a tooth following the beginning of cavitation) show the same increase after atmospheric cooling, perhaps because of meteorologic effects on peripheral nerves which influence the metabolic processes in the pulpa and periodontium (due to changes in cell permeability and capillary flow;[180] haemorrhages (after tooth extraction) are common on cold front days; saliva flow decreases in warm weather and increases at low temperatures; shortage of saliva seems to increase tooth destruction.[183]

Summarizing, the following meteorological factors may be involved in dental diseases:

(a) Decrease in ultraviolet radiation in winter accompanied by lower phosphate, vitamin D and carbonate levels;
(b) The degree of atmospheric cooling;
(c) A meteorological effect on the secretion rate and composition of saliva (e.g. its acidity);
(d) A change in the bacterial activity in the mouth, due to changes in the environmental temperature affecting the mouth temperature during inhalation;
(e) A change in adrenal function during weather stress which affects the capillarity and permeability of tissues.

6.13.2. Diabetes

Diabetes is a clinical syndrome, characterized by the effects of relative or absolute insulin deficiency, which is caused by a dysfunction of the Islands of

*For the definition of föhn p. 190.

Langerhans in the pancreas (see p. 92). A number of environmental factors are responsible for precipitating or affecting this clinical syndrome.[184,185] A few of the major meteorologic relationships are:

(a) *Seasonal effects.* According to Henkel[186] and Chrometzka[187] diabetic coma in Germany is most common from October to April with a maximum occurrence in December and March. Pannhorst and Rieger[188] found that in Germany the onset of diabetes (at least in females) has a small maximum in autumn, in males a maximum in December (a minimum in autumn). They did not believe this to be a true seasonal effect but attributed the increased mortality to infection. However, this would not explain the different periods of maxima for males and females. More recent studies in England by Gamble and Taylor[189] showed for the month of onset of diabetes a significant clustering in March and June with a peak in mid-autumn.

(b) *Short term meteorological effects.* Diabetes patients, transferring from a colder to a warmer climate, occasionally observe a decreased insulin requirement. *Heat stress*, however, causes a significant increase in glucose utilization resulting in hypoglycaemia.

Studies with rats have shown that cold stress in animals with induced diabetes causes early death due to hypothermia as a result of a failure of the oxygen consumption to rise in response to cold. A lowering of blood sugar levels was observed in unfed animals by Brezowski and Hansen[190] in south Germany after the influx of humid warm air masses.

(c) *High altitude effects.* Von Deschwanden[191] observed that, in Switzerland, there was a relatively small number of diabetics but a frequent occurrence of hypoglycaemia in diabetic tourists, who were not aware of the reduced glucose production at altitude and of the increased carbohydrate tolerance. The required insulin doses are thus smaller at high altitude than at sea level, and the alkalosis resulting from high altitude treatments may neutralize the acidosis of diabetics.

(d) *Effects of sea bathing and spas.* In 10–40% of healthy subjects an increase in blood sugar level occurs[192] (after cold seabathing). In diabetics their condition deteriorates as a result of this increase. After warm spa treatment and the drinking of alkaline water, which may neutralize the acidosis, diabetic patients require less insulin.

(e) *Geographical and racial difference in diabetes incidence.* Diabetes is most common and most severe in middle temperature latitudes; it is less frequent in tropical latitudes. Western countries show, according to the W.H.O., a rather uniform diabetes prevalence of 1–3%. In The Netherlands the incidence in Jews is 2–3 times higher than in non-Jews; in women with 3–5 children it is three times higher than in women without children. Certain Indian tribes in Arizona (USA) have a prevalence 30% above the average of the general population. Australian aboriginals have a 19% excess above average. On the other hand in Eskimos in Alaska and Canada very low diabetes prevalence rates (0.3–1.3%) have been recorded.

Most of these differences are due to genetic and racial differences and probably also to differences in food consumption. However Wise[193-195] assumes that perhaps the climatic selection of the diabetic genotype was a major determinant of the global variation in its prevalence.

In this connection it is worth mentioning that in a random group of 24 000 diabetics in The Netherlands most female diabetics were born in March (normal birth maximum is in May). Males did not show this relationship. Also the facts mentioned above seem to suggest that male and female diabetics have a different meteorotropic behaviour.

6.13.3. Föhn disease

Föhn can be defined as an air stream which, during its descent from the mountains, becomes dynamically warm and dry. The name derives from the Latin word *Favonius* which means warm, south wind. It is well known for its biological effects, reported by subjects living in the mountain valleys of Switzerland, Tyrol (Austria) and Bavaria (Germany). In Switzerland it originates when humid air from Italy rises against the mountains of the Swiss Alps. This impact cools the air so much that heavy precipitation occurs on the slopes of the southern Alps. After crossing the mountains the air becomes dry and warm and reaches the northern lower parts of the Alps.

All these specific meteorological conditions are not limited to the Alps but correspond to similar physical occurrences in other regions where they are known by various names, such as 'Santa Anna' (southern California), 'Chinook' (Canada), 'Sirocco' (Mediterranean), 'Zonda' (Argentina), 'Ibe' (Ala-Tau, Caucassus), 'Koebang' and 'Gending' (Java), 'Bohorok' (northern Sumatra), 'Autan' (central France) and 'Tramontana' (northern Italy).

Three phases of föhn weather are distinguished:

(a) the preföhn stage, which lasts until the penetration of the dynamically warm, dry air into the valley;

(b) the actual föhn or valley föhn, which is the period when the warm air has entered the valley and during which period these warm air masses, coupled with gusty winds, effect the population;

(c) the collapsing föhn, which is due to the passage of a following cold front.

The various subjective and objective symptoms in subjects who are sensitive to föhn weather are specially pronounced during the preföhn stage.

According to Halhuber,[196] Von Deschwanden[191] and others there is only a weak correlation between föhn weather and accute appendicitis, gall-bladder stones, renal colics, myocardial infarction, falling blood pressure during anaesthesia etc. However, during the föhn, there is a significant increase in ulcer perforations, of embolism, thrombosis, haemorrhages after operations, bleeding in haemophiliacs, headaches, migraine and the number of suicide attempts.

Some authors also observed a relationship between the time of the occurrence of death and föhn. Significant correlations have been reported also between föhn conditions and human reaction speed (which decreases) and the incidence of traffic accidents. So far it has been very difficult to establish the exact physiological mechanisms involved in the complaints and diseases during föhn weather. One of the following mechanisms may be involved.

(a) Undoubtedly the very dry warm wind during föhn weather affects the thermoregulatory mechanism particularly of subjects with an inefficient thermoregulation mechanism (see pp. 58–62). A serious disturbance of this mechanism could produce a great variety of organic disturbances due to the important role of the hypothalamus in the human brain. It is very likely that serotonin and catecholamines, which play an important role in thermoregulation, are involved in the causation of föhn diseases.

(b) All studies which have tried to find a causal relationship between the changed electrical and electromagnetic conditions of the atmosphere during föhn weather and the above mentioned biological phenomena have so far failed.

(c) An important physiological event during föhn is the change in the capillary resistance and permeability of the membranes. The capillary resistance shows marked variations towards the end of the preföhn phase and usually a decrease rather than an increase. After föhn penetration it shows a tendency to become constant and often even an increase can be observed.[191]

(d) Zimmermann[197] demonstrated a number of meteorotropic relationships between föhn conditions and 17-ketosteroid excretion in 24-h urinary samples. The 17-ketosteroids (17-KS) decrease during the arrival of warm cyclonic conditions and show an increase after cold front passages. In other words the adrenal function and its accompanying hormonal functions are affected (p. 91).

(e) Recently Sulmann *et al.*[198-200] investigated the human reaction to the warm dry wind in Israel (Sharav). They analysed the daily secretion of adrenaline, noradrenaline, serotonin, 5-hydroxyindolacetic acid (5-HIAA), 17-KS, 17-OH, potassium, sodium, creatinine and histamine. In subjects with sleeping disorders, psychological stresses and migraines, Sulman frequently found an increased serotonin and decreased 5-HIAA secretion which was observed days before the influx of the warm, dry desert wind. Similar phenomena can be observed at least qualitatively, in the case of the föhn also, although they are then less pronounced.

Finally Sulman observed exhaustion syndromes of which the symptomatology is accompanied by a strongly reduced catecholamine secretion during the Sharav. As in the case of the föhn the symptoms become more marked from year to year in certain, often elderly, patients. Much more work is required before a definite explanation of the effects of föhn winds can be given; it may reveal the role of serotonin and its metabolites in föhn disease.

6.13.4. Geriatric diseases

According to Brocklehurst[201] in 1969, out of a population of 54 million people in the UK, just under 7 million (i.e., 13%) were aged 65 and over. These comprised 4.3 million females and 2.6 million males. This group of people is often regarded as those who fall into the classification 'geriatric'. The age boundary of 65 has been chosen mainly for administrative convenience as it is the age at which a man may draw his state pension. Around this age many people are beginning to lose their independence, they suffer from an increasing multiplicity of pathological disturbances, and more of them are becoming dependent on those who are younger.

An important factor which must be taken into account in any consideration of disease in old age, is the special habits of old people (as compared with younger ones). One important aspect of these special habits is the lack of mobility. The most obvious effects of the lack of mobility is that it renders many old people housebound so that, perhaps for many years at the end of their lives, they are unable to go out into the open air, or even to be in direct contact with sunlight. This is very serious in view of the important physiological effects of solar radiation (Section 5.2). Lack of mobility leads also to diminished function of many vital processes, partly because of the loss of stimulus from movement and partly because, while remaining in one fairly constant environment, receptor organs are not subject to a series of varying stimuli. Insufficient movement also aggravates most rheumatic complaints, which are common in older people. Elderly people are also more liable to accidents, particularly when weather conditions are hazardous, as with snow and frost and in the long dark nights of winter.

Another hazard to which old people are particularly exposed is that of retirement, which brings a completely new and alien way of life. One feature of retirement in Great Britain and the USA is movement to a different part of the country. The especially favoured retirement areas in England are seaside towns, particularly those on the south coast of England; in the USA the warm southern states (particularly Florida) and California. Whilst such a move may have advantages from the climatic point of view, it creates at the same time, psychological problems because one leaves behind family and friends.

Apart from the above mentioned differences in behaviour between old and young there are a number of basic physiological differences between the groups:

(a) The thermoregulation efficiency of older people is usually rather poor which affects a great number of physiological processes and diseases.

(b) The vascular responses of old age people are changed.

   (i) Fennell and Moore[202] observed, after immersion of the right hand in water of +42 °C, a greater rise of the deep body (core) temperature in the aged; vasodilatation appears to play a small role;

   (ii) Redisch *et al.*[203] observed a significantly lower resting blood flow in healthy old adults (without cardiovascular disease) than in young adults; old adults with cardiovascular complaints have the lowest resting blood flow; young healthy adults receive more blood in the musculature of their extremities;

(iii) Lund and Hellon[204] found that in young male adults sweating started after 15 min of working, in old adults after 30 min, although the number of sweat glands is the same. Sweating in young miners was 1.016 g m^{-2} min^{-1}, in old miners 0.840; during subsequent periods of rest older people sweat more; entering a warm room older males start with a lower temperature but after some time they have a higher temperature than young men;

(iv) Ring *et al.*[205] studied, with a plethysmograph, spontaneous volume changes of the middle finger of a quietly resting subject. In the age group 17–29 years the average change per 10 ml of finger volume is 66 ± 7.0 μl; 30–49 years 44 ± 4.4μl; 50 and over, 32 ± 4.0 μl.

(v) Most elderly people in a cool environment complain about cold feet and hands due to reduced peripheral blood circulation and insufficient closure of the pores in the skin (i.e. too strong cooling). These complaints can be drastically reduced after a series of high altitude treatments (see Section 9.3.1).

(c) The ability to compensate hypoxia decreases with age. This shows up in the following ways: physical capacity is more restricted in old people at high altitude; erythropoesis increases more slowly; respiratory volume per minute increases less; performance of the cerebral cortex decreases; ECG changes occur at lower altitudes; however, young people collapse and suffer from mountain-sickness more frequently than older subjects.

(d) Trembling in cooled subjects, placed in a waterbath (of 35 °C) which is gradually cooled, starts earlier in elderly subjects; their pulse rate and cardiac output per minute increases more quickly; oxygen consumption increases more slowly; the lactic acid level of the blood increases to significantly higher levels; the pores of the extremities, which are insufficiently closed (see above), are responsible for faster rewarming of the skin after intense cooling down.

(e) The capacity of protecting the body against overheating is less in the elderly, they are more susceptible to heatstroke.

In the geriatric age group the two most important causes of death are cardiovascular disease (about 30%) and carcinoma (about 20%). The greatest burden of disease endured by old people comprises stroke, Parkinsonism, arteriosclerotic dementia and various other cerebral syndromes associated with atheroma of the cerebral arteries. These diseases cause a high dependency rate and require hospital and other institutional care. Morbidity in old age is also associated with the degenerative disorders such as osteoarthritis; with chronic infections, particularly chronic bronchitis, asthma and chronic cystitis (inflammation of the bladder); also with effects of malnutrition, especially iron deficiency anaemia, and with various specific disorders which become more common with advancing age such as diabetes, osteoporosis (abnormal porousness of bone) and senile dementia.

Amongst these diseases, which are very frequent in elderly people, rheumatic diseases, stroke, heart diseases, bronchitis and asthma in particular are pronounced meteorotropic diseases which have been reviewed in previous sections.

Apart from these diseases three other, rather common, geriatric diseases should be mentioned: accidental hypothermia, hyperthermia and osteomalacia.

(a) *Accidental hypothermia*: according to Brocklehurst,[201] a survey, by the Royal College of Physicians of London, of 136 patients admitted to British hospitals during three months of a fairly mild winter with temperatures of about 2 °C, suggests a probable overall incidence of patients admitted to hospital, suffering from subnormal temperatures, as being about 9000 during that period. These were mainly old people. While accidental hypothermia in the aged is usually a result of exposure to low temperatures, some is associated with diseases which decrease metabolism, or with drugs which have similar effects. In some patients neither of these two factors are present and the hypothermia seems to be due to an impaired thermoregulation in elderly people.

The clinical symptoms of accidental hypothermia in old people have been well described.[206,207] Patients are either drowsy or unconscious, and do not complain of hypothermia. In the early stage their normal colour may be maintained, but as hypothermia progresses intense vasoconstriction sets in, leading to a blue and white mottled appearance. Respiration is slow; the pulse rate is slow and the skin is strikingly cold to the touch. Investigations have shown a raised blood urea and possibly a raised serum amylase indicating an associated pancreatitis (inflammation of the pancreas). Characteristic changes in the ECG also occur.

The treatment of hypothermic patients should consist of covering them with blankets and bringing them into a warm environment. Rapid rewarming should be prevented. Brocklehurst[201] also suggested warmed intravenous glucose injections or a plasma expandor such as rheomacrodex. The intravenous injection of hydrocortisone at regular intervals is recommended and also an antibiotic. Tri-iodothyronine should be reserved for hypothyroid patients. In general, biochemical disturbances should not be corrected until the process of rewarming is nearly complete. Other conditions which may then require attention are hypoglycaemia, hypoalkalaemia and metabolic acidosis.[201]

(b) *Old-age hyperthermia*: there is evidence that old people are particularly susceptible to hyperthermia and that at high ambient temperatures, death rates among elderly people increase. An analysis of the daily death rate in an institution for elderly people in Australia was carried out[208] throughout six months of the Australian summer and autumn. During that period of time there was no correlation between increased death rate and age, but there was a clear statistical correlation between the temperature of the day at 3.00 pm and the mean number of deaths of elderly people that day.

Studies by Levine[209] suggested that patients who are old and ill are more liable to develop heat stroke.

(c) *Osteomalacia*: is an acquired bone disease, rather common among elderly people, characterized by a decreased or totally absent mineralization of the bone matrix. It is also characterized by vague symptoms such as backache, pains in the legs and general muscular weakness. Blood-calcium, phosphate and vitamin D levels are low.

There is some geographical variation in the incidence so far reported and the question therefore arises as to whether under-exposure to ultraviolet light may be a factor in the etiology. If this is the case then it could result from two different reasons. In the first place the immobility of old people renders them increasingly housebound and thus permanently screened from direct sunlight. In the second place geographical variation may reflect the hours of sunlight in different areas. The highest incidence in Great Britain has been reported from the north rather than the south. Further evidence for the effect of climate as a factor in osteomalacia has been given by Smith *et al.*[210] who compared serum vitamin D levels in groups of elderly women in Michigan and in Puerto Rico; he found major differences.

Many other diseases are observed in aged people, but most of these chronic diseases can be observed in subjects of all age-groups.

6.13.5. Multiple sclerosis

Multiple sclerosis is characterized by an induration, combined with hypertrophy of connective tissue (a sclerosis) in the brain and spinal cord which occurs in patches. As a result a malfunction of motor movements takes place accompanied by paralysis, disappearance of abdominal reflexes, neuralgia of the face nerves causing difficulty in speaking, nystagmus (continuous rolling movement of the eye-ball) etc. The disease occurs particularly among young people between the ages of 20 and 40.

There is some analogy with demyelinating conditions in Swayback disease in lambs, which results from a copper deficiency in soil and food. However, so far no positive evidence has been found that copper or other trace element deficiency is the cause of the breaking up of myelin sheaths (Section 3.4) in humans.

Despite the many research studies in different parts of the world the cause of this disease is unknown. The peculiar geographical distribution and latitude effects of the disease suggest that many factors including meteorological ones, may be involved.

In the USA and Canada the highest incidence occurs in the north eastern states, the lowest in the south western and more in urban than in rural areas. The incidence in Winnipeg (Canada) is six times higher than in New Orleans (southern USA). Also in the USSR the lowest values are found in central Asia and in the south and south west of the USSR, highest values in the northern regions; also in Turkey very low incidence is observed in the hottest southern regions (0.6%), increasing in the cooler northern parts to 3.7%.

A very low incidence occurs in Mexico City (about 1.6/100 000). Practically no cases have been reported from east and west Africa. Japan also has a very low incidence. High incidences are observed in Great Britain and The Netherlands (more than 45/100 000).

In England and Canada no seasonal variation in the attacks has been observed but, in north east Ohio in the USA, Sibley and Foley[211] found that the lowest incidence of attacks occurred in March–June and November–February, the highest during July–October. Several theories have been put forward to explain the various observed relationships.

(a) *Infectious causes.* These are suggested by the increased γ-globulin level in the cerebrospinal fluid in many patients. Miller[212] and others suggested an enteric virus infection as a possible cause in view of the similarity in geographical distribution and other epidemiological features of multiple sclerosis and poliomyelitis. The assumption that multiple sclerosis is caused by a virus was recently supported by studies by Kolodovsky *et al.*[213] and by Henle *et al.*[214] in the School of Medicine in Philadelphia (USA). The virus would be most common near the equator and would cause a higher antibody level. This could explain the decreased incidence of multiple sclerosis in the northern hemisphere from the north to the south.

(b) *Metabolic disturbances.* For example changes in the phospholipid metabolism in the central nervous system may be responsible.

(c) *Trace element disturbances.* This is suggested by the Swayback disease mentioned above.

(d) *Meteorological effects*

(i) Indirect effects. Schwank[215] and others pointed out that climatic changes affect our dietary habits, particularly with respect to the intake of saturated and unsaturated fatty acids in general and animal fats in particular. If this is combined with poor intestinal absorption of unsaturated fatty acids, due to the high gluten content of certain cereal grains, according to Thompson,[215] changes in the plasma linoleate level will trigger a demyelination (Section 3.4.3) of the nerves.

(ii) Direct effects. Both thermal and solar radiation effects may be involved; both change with latitude, season and altitude. They may accelerate or reduce the different physiological processes involved in the causation of multiple sclerosis. For example, if a virus infection is involved both thermal stress and ultraviolet radiation could play an important role in the development of the disease.

An interesting paper by Acheson *et al.*[216] describes possible correlations with solar radiation in relation to latitude. They studied 454 veterans with multiple sclerosis (MS) born in 29 counties in the USA with a population of 300 000 and more. A multiple regression analysis was used for MS incidence, average total annual hours of sunshine and average daily December solar radiation.

The following results were obtained:

(a) There is statistically significant correlation between latitude, respective hours of solar radiation and MS incidence. In the northern hemisphere the incidence decreases from north to south. A similar analysis in the same 29 counties between latitude and ulcerative colitis did not show any statistical correlation, which indicates that the MS correlation is not a statistical freak;

(b) No correlations were found between MS incidence, annual precipitation and average maximum July temperature.

Barlow[217] suggested the possibility that cosmic rays (Section 5.4.2) may be involved as a causative factor. Further analysis of the distribution of MS and cosmic rays on earth seems to contradict this hypothesis.

Finally it is worth mentioning that preliminary studies by Tromp in The Netherlands, using the birth dates of 2000 MS patients suggest that the maximum number of births of MS patients occurs in February and March, the lowest incidence in October (males) and November (females). Only further studies in other countries, using larger volumes of data, can confirm this observation.

Many other, but less important, meteorotropic diseases are known. Several of them have been briefly discussed in Table 6.1 pp. 141–145.

REFERENCES

1. S. W. Tromp (Ed), *Progress in Human Bíometeorology II*, Swets & Zeitlinger, Amsterdam, 1977.
2. C. von Pirquet, *Handbuch der Kindertuberkulose*, Thieme Verlag, Stuttgart, 1933.
3. in M. Rubin, *Diseases of the Chest*, Saunders, Eastbourne, Sussex, 1947, p. 442.
4. S. W. Tromp, *Medical Biometeorology*, Elsevier, Amsterdam, 1963, p. 878.
5. Ref. 4, p. 522.
6. Ref. 1, p. 178.
7. Ref. 1, p. 159.
8. Ref. 1, p. 95.
9. Ref. 1, p. 10.
10. S. W. Tromp and J. J. Bouma, *Monograph Series No. 13*, Biometeorological Research Centre, Leiden, The Netherlands, 1974.
11. R. R. Davies, M. J. Denny and L. M. Newton, *Acta Allerg. (Kbh.)* **18**, 131 (1963).
12. R. R. Davies, *ibid.* **24**, 396 (1969).
13. *idem*, ref. 1, p. 2.
14. S. W. Tromp and J. J. Bouma, *Int. J. Biometeorol.* **9**, 233 (1965).
15. S. W. Tromp, *Monograph Series No. 6*, Biometeorological Research Centre, Leiden, The Netherlands, 1966.
16. *idem*, *Rev. Allergy* **22**, 1027 (1968).
17. S. W. Tromp and J. J. Bouma, *Monograph Series No. 9*, Biometeorological Research Centre, Leiden, The Netherlands, 1968.
18. Ref. 1, p. 23.

19. J. E. C. Schook and C. F. Schüller, *Ned. T. Geneesk.* **118**, 837 (1974).
20. J. Pemberton and C. Goldberg, *Brit. Med. J.* **2**, 567 (1954).
21. M. O. Amdur, W. W. Melvin and P. Drinker, *Lancet*, 758 (1953).
22. Cralley, in ref. 21.
23. N. Goodman, R. E. Lane and S. B. Rampling, *Brit. Med. J.* **2**, 237 (1953).
24. J. T. Boyd, *Proc. Roy. Soc. Med.* **53**, 107 (1960).
25. Martin and Bradley, in ref. 24.
26. P. J. Lawther and J. A. Bonnell, *Proc. 2nd Clean Air Cong.*, Washington DC, Dec. 6–11, 1970.
27. D. D. Reid and R. S. Fairbairn, *Lancet*, 1237 and 1289 (1958).
28. K. De Vries *et al.*, in *Bronchitis II*, N. G. M. Orie and H. J. Sluiter (Eds), Van Gorcum, Assen, The Netherlands, 1964, p. 167.
29. Y. C. Hsidh, R. Frayer and J. C. Ross, *Amer. Rev. Resp. Dis.* **98**, 613 (1968).
30. Ref. 1, p. 38.
31. S. W. Tromp and J. J. Bouma, *Proc. 5th Int. Biometeorol. Congr.*, Montreux, Switzerland, 31 Aug.–6 Sept., 1969, p. 89.
32. S. W. Tromp, *Int. J. Biometeorol.* **7**, 291 (1964).
33. A. J. Lea, *Brit. Med. J.* **1**, 488 (1965).
34. J. R. McKay, *Lancet*, 1393 (1968).
35. J. A. H. Lee, *Lond. Hosp. Gaz., Clin. Sci. Suppl.* **68**, 3 (1966).
36. S. Krasnow, *Int. J. Biometeorol.* **13**, 87 (1969).
37. in E. M. Glaser and J. P. Austin, *ibid.* **13**, 183 (1969).
38. H. Pomp *et al.*, *Med. Trib.* **34a**, 22 and 27 (1973).
39. W. J. J. de Sauvage Nolting, *T. Soc. Geneesk.* **43**, 134 (1965).
40. *idem*, *Geneesk. Gids.* **44**, 178 (1966).
41. *idem*, *Int. J. Biometeorol.* **12**, 293 (1968).
42. D. Stur, *Klin. Wschr.* **65**, 898 (1953).
43. O. Warburg, *ibid.* **5**, 829 (1926).
44. J. A. Campbell and W. Kramer, *Lancet* **1**, 828 (1928).
45. E. G. Sandstroem and G. Michaels, in *Memoirs Univ. Cal. No. 12*, Ch. B. Lipman and E. R. Hendrick (Eds), Univ. Calif. Press, Los Angeles, USA, 1942.
46. H. Brezowski, H. Wrba and H. Rabes, *Naturwiss.*, **52**, 190 (1965).
47. Ref. 1, p. 32.
48. J. Bartels, *Wien. Med. Wschr.* 966 and 2274 (1925).
49. S. G. Beljajen, *Balneologe*, 426 (1936).
50. H. E. Kisch, *Münch. Med. Wschr.* 721 (1908).
51. A. Kolisko, *Handb. f. ärztl. Sachverstand. Tätigkeit.*, 2 (1913).
52. W. Amelung, *Exptl. Med. Surg.* **9**, 238 (1951).
53. H. J. Fladung, Thesis, Frankfurt a/M, 1952.
54. U. Ströder, F. Becker and G. Haas, *Klin. Wschr.*, 312 (1951).
55. F. Wood and O. F. Headley, *Med. Clinics N. Amer.* **19**, 151 (1935).
56. A. M. Master, *J. Amer. Med. Ass.* **109**, 546 (1937).
57. *idem ibid.* **148**, 794 (1952).
58. W. B. Bean, *Amer. Heart J.*, **14**, 684 (1937).
59. *idem*, Bennett and C. A. Mills, *ibid.* **16**, 701 (1938).
60. W. L. Mullins, *Penn. Med. J.* **39**, 322 (1936).
61. S. G. Mintz and L. N. Katz, *Arch. Int. Med.* **80**, 205 (1947).
62. R. Miller, *ibid.*, **83**, 597 (1951)
63. H. J. Hoxie, *Amer. Heart J.* **19**, 475 (1940).
64. Billings, *Amer. J. Med.* **7**, 356 (1949).
65. S. Koller, *Arch. Kreislanfforsch.* **1**, 225 and **2**, 8 (1941).
66. H. E. Heyer, H. C. Teng and W. Barris, *Amer. Heart J.* **45**, 741 (1953).

67. H. C. Teng and H. E. Heyer, *ibid.* **49**, 9 (1955).
68. S. W. Tromp, *Monograph Series No. 3*, Biometeorological Research Centre, Leiden, The Netherlands, 1958.
69. G. E. Burch and N. P. de Pasquale, *Hot Climates, Man and his Heart*, Thomas, Springfield, Illinois, USA, 1962.
70. G. E. Burch and F. D. Giles, *Mod. Conc. Cardiov. Dis.* **39**, 115 (1970).
71. *idem*, in S. W. Tromp, *Pathological Biometeorology*, Swets & Zeitlinger, Amsterdam, The Netherlands, 1977, p. 52.
72. A. Ausari and G. E. Burch, *Arch. Inter. Med.* **123**, 371 (1969).
73. N. P. de Pasquale, *Amer. J. Med. Sci.* **240**, 468 (1961).
74. S. W. Tromp and J. J. Bouma, *Proc. World Asthma Conf.*, Eastbourne, England, 1965, p. 27.
75. M. Millodot, *Nature (Lond.)* **255**, 151 (1975).
76. H. Fisher, Thesis, Leipzig, 1934.
77. Huerkampf, *Klin. Monats. Augenheilk.* **12**, 71 (1952).
78. S. N. Cooper, *J. All India Opthal. Soc.* **12**, 50 (1964).
79. in ref. 4.
80. Hemmes, Thesis, Utrecht, 1959.
81. J. Regli and Stämpfli, *Helv. Physiol. Pharmacol. Acta,* **5**, 44 (1947).
82. R. Lotmar and J. Häfelin, *Arch. Meteor. Geophys. Bioklimatol.* **B7**, 286 (1956).
83. S. Arimatsu, *Tokyo J. Med. Sci.* **65**, 305 (1957).
84. A. Marchionini *et al.*, *Klin. Wschr.* **17**, 663, 736 and 773 (1938).
85. C. Armstrong, *Amer. J. Publ. Hlth.* **40**, 1296 (1950).
86. B. F. Howitt and E. J. Bell, *J. Infect. Dis.* **60**, 113 (1937).
87. W. Zimmermann, *Zentr. Bakteriol.* **167**, 140 (1957).
88. Ref. 1, p. 61.
89. S. W. Tromp and J. J. Bouma, *Proc. 6th Int. Biometeorol. Congr.*, Noordwijk, The Netherlands, 3–9 Sept., 1972.
90. *idem*, *Proc. 6th Int. Congr. on Suicide Prevention*, Amsterdam, The Netherlands, 27–30 Aug., 1973.
91. *idem*, *Monograph Series No. 12*, Biometeorological Research Centre, Leiden, The Netherlands, 1973.
92. A. Fanchamps, *Headache* **1**, 79 (1975).
93. *idem*, *Triangle (Basle)* **15**, 103 (1976).
94. J. W. Lance, *Mechanism and Management of Headache*, Butterworth, London, 1969.
95. J. Pearce, in *Background to Migraine*, A. L. Cochrane (Ed), Heinemann, London, 1971.
96. J. D. Gomersall and A. Stuart, *Int. J. Biometeorol.* **17**, 285 (1973).
97. A. Henschel, J. Brozer and A. Key, *J. Appl. Physiol.* **4**, 340 (1951).
98. H. Knobloch and B. Pasamanick, *Amer. J. Publ. Hlth.* **48**, 1201 (1958).
99. B. De Rudder, *Grundriss einer Meteorobiologie des Menschen*, Springer Verlag, Berlin, 1952.
100. H. J. Tholuck, *Beitr. Gerichtl. Med.* **16**, 121 (1942).
101. H. Rohden, *Arch. Psychol.* **89**, 603 (1933).
102. R. Reiter, *Meteorologie und Elektrizität der Atmosphäre*, Akad. Verlag Geest & Portig, Leipzig, 1960.
103. J. A. Burt, *Practitioner*, **136**, 62 (1936).
104. C. A. Mills, *J. Lab. Clin. Med.* **24**, 53 (1938).
105. B. De Rudder, *Proc. Int. Congr. Rheumatism*, London, 1938.
106. J. S. Lawrence and M. Molyneux, *Int. J. Biometeorol.* **12**, 163 (1965).
107. E. S. Nichol, *J. Lab. Clin. Med.* **21**, 588 (1936).

108. J. R. Paul, *Amer. Clin. Climatol. Ass.* **57**, 172 (1942).
109. S. Cobb and J. Lawrence, *Bull. Rheum. Dis.* **7**, 233 (1957).
110. E. V. McKinley, *Geography of Disease*, Geo. Washington Univ. Press, Washington, USA, 1935.
111. J. Van Breeman, *Amer. J. Phys. Ther.* **6**, 515 (1930).
112. L. Miller, *Münch. Med. Wschr.* 809 (1909).
113. R. S. Fox, *Brit. Med. J.* **1**, 994 (1930).
114. H. Varekamp, *Ned. Tijdschr. Geneesk.* **104**, 862 (1958).
115. *idem*, *Tijdschr. Soc. Geneesk.*, 649 (1958).
116. Anon., *Brit. Med. J.*, 1060 (1933).
117. E. C. Warner, *J. State Med.* **41**, 621 (1933).
118. S. W. Tromp, *Experiments on the Possible Relationship between Soil Resistivity and Dowsing Zones,* Vol. III, Publ. Series, Foundation for the Study of Psychophysics, The Netherlands, 1955.
119. W. S. C. Copeman, *Textbook of Rheumatic Diseases*, Churchill Livingstone, Edinburgh, 1955, pp. 38, 74, 197, 596.
120. S. W. Tromp and J. J. Bouma, *Int. J. Biometeorol.* **10**, 105 (1966).
121. J. Hunter and M. G. Whillans, *Canad. J. Med. Sci.* **29**, 255 (1951).
122. J. Hunter, E. H. Kerr and M. G. Whillans, *ibid.* **30**, 367 (1952).
123. Racker *et al.* in *Muscular Activity*, Hill, Baltimore, USA, 1925.
124. *idem*, in Sacks, *Physiol. Rev.* **21**, 217 (1941).
125. W. Hausmann and H. Haxthausen, *Strahlenther*, **XI**, 1929.
126. H. Jausion, *Les Maladies de Lumière*, Paris, France, 1933.
127. H. F. Blum, *J. Amer. Vet. Med. Ass.* **93**, 185 (1938).
128. *idem*, *Photodynamic Action and Diseases Caused by Light*, Reinhold, New York, USA, 1941.
129. E. Rocha and M. Silva, *Rev. Fac. Med. Vet. Univ. San Paulo*, **1**, 225 (1940).
130. B. M. Kesten and M. Slutkin, *Arch. Dermatol.* **67**, 284 (1953).
131. F. Urbach, *The Biological Effects of Ultra-violet Radiation*, Pergamon, Oxford, UK, 1969.
132. M. Möller, *Der Einflusz des Lichtes auf die Haut.*, Nägele, Stuttgart, West Germany, 1900.
133. B. A. Whistler, *Iowa State College J. Sci.* **14**, 215 (1939).
134. W. F. Wells and H. W. Brown, *Amer. J. Hyg.* **24**, 407 (1936).
135. in D. D. Reid, *Lancet*, 1299 (1958).
136. L. Buchbiner, *J. Bacteriol.* **42**, 353 (1941).
137. Edward, *Lancet*, 644 (1941).
138. A. Marchionini *et al.*, *Klin. Wschr.* **17**, 663, 736, 773 (1938).
139. Ernst, *Deut. Z. Chir.* **238**, 402 (1933).
140. C. Häberlin, *Z. Wissensch. Bäderkunde* **2**, 1001 (1928).
141. K. W. Kalkoff, *Strahlentherapie* **86**, 81 (1951).
142. Ref. 1, p. 89.
143. P. Popchristov and N. Balewska, *Med. Fizkul*, Sofia, Bulgaria, 1966.
144. P. Michailov, S. Borelli and C. Eni-Popescu, *Hautarzt* **18**, 456 (1967).
145. S. Chlebarov and S. Borelli, *Munch. Med. Wschr.* **108**, 589 (1966).
146. C. S. Leithead and A. R. Lind, *Heat Stress and Heat Disorders*, Cassell, London, UK, 1964.
147. M. L. Thompson, *J. Physiol.* **112**, 22 (1951).
148. A. T. Pearson, *Med. J. Australia*, **2**, 968 (1955).
149. E. Sohar and R. Adar in *Proc. UNESCO Symposium on Environmental Physiology and Psychology in Arid Conditions*, Lucknow, India, 1962.
150. *idem*, *Arid Zone Res. (UNESCO, Paris)*, **24**, 55 (1964).

151. K. N. Moss, *Proc. Roy. Soc. (Lond.)*, **B95**, 181 (1923).
152. W. S. S. Ladell, *Lancet*, 836 (1949).
153. Ref. 4, p. 372.
154. M. S. Malhotra, *Nature (Lond.)*, **182**, 1036 (1958).
155. Refs. 149 and 150, p. 63.
156. W. Hancock, A. G. R. Whitehouse and J. S. Haldane, *Proc. Roy. Soc. (Lond.)*, **B105**, 43 (1929).
157. M. B. Schulzberger, H. M. Zimmerman and K. Emerson, *J. Invest. Dermatol.* **7**, 61 and 153 (1946).
158. S. D. Allen and J. P. O'Brien, *Med. J. Australia* **2**, 335 (1944).
159. A. M. Critchley, *Brit. Med. J.* **2**, 208 (1945).
160. R. K. Macpherson, *Univ. Queensland Papers Dept. Physiol.* **1**, 10 (1949).
161. D. H. K. Lee and R. K. Macpherson, *J. Appl. Physiol.* **1**, 60 (1948).
162. E. P. Ellis, *Brit. Med. Bull.* **5**, 13 (1947).
163. B. F. Smith, *Brit. J. Ind. Med.* **15**, 197 (1958).
164. R. A. McCance, J. W. L. Crossfill, C. C. Ungley and E. M. Widdowson, *Med. Res. Council (Lond.) Spec. Rep. No. 291*, 1956.
165. A. C. Burton and O. G. Edholm, *Man in a Cold Environment*, Arnold, London, UK, 1955.
166. K. Lange, D. Weiner and L. J. Boyd, *New England Med. J.* **237**, 383 (1947).
167. C. C. Ungley, G. D. Channell and R. L. Richards, *Brit. J. Surg.* **33**, 17 (1945).
168. M. C. Monge, *Acclimatization in the Andes*, Johns Hopkins, Baltimore, USA, 1948.
169. A. Hurtado, *J. Amer. Med. Ass.* **120**, 12 (1942).
170. *idem et al.*, School of Aviation Med. U.S.A.F., Randolph A.F.B., Texas, USA, Report 56–1, 56 (1956).
171. A. Rotta *et al.*, *J. Appl. Physiol.* **9**, 328 (1956).
172. A. Hurtado, in ref. 4, p. 418.
173. I. Singh *et al.*, *Int. J. Biometeorol.* **21**, 93 (1977).
174. Fries, *Zahnärztl. Wschr.*, 45 (1934).
175. P. G. Anderson *et al.*, *J. Amer. Dent. Ass.* **21**, 1349 (1934).
176. S. F. Erpf, *ibid.* **25**, 681 (1938).
177. B. M. Lathrop, *Amer. J. Ment. Defic.* **46**, 33 (1941).
178. E. C. McBeath and T. F. Zucker, *J. Nutr.* **15**, 547 (1938).
179. C. A. Mills, *J. Dent. Res.* **16**, 417 (1937).
180. D. Schlegel and W. Warmbt. *Dtsch. Zahn-, Mund-, Kieferheigk.* **21**, 122 (1954).
181. B. Rossiwall and E. Waldhart, *Dtsch. Zahnärztl. Z.* **27**, 394 (1972).
182. *idem* and E. Foltin, *Öost. Z. Stomat.* **70**, 219 (1973).
183. Leung, in O. Louridis, N. A. Demetriou and E. Bazopoulou-Kyrkanides, *J. Dent. Res.* **49**, 1136 (1970).
184. Ref. 1, p. 133.
185. Ref. 4, p. 281.
186. G. Henkel, *Z. Klin. Med.* **125**, 52 (1933).
187. F. Chrometzka, *Klin. Wschr.* **38**, 972 (1940).
188. R. Pannhorst and A. Rieger, *Z. Klin. Med.* **134**, 154 (1938).
189. D. R. Gamble and K. W. Taylor, *Brit. Med. J.* **3**, 631 (1964).
190. H. Brezowski and J. B. Hansen, private communication (1960).
191. J. S. von Deschwanden, in *Der Mensch in Klima der Alpen*, J. S. von Deschwanden, K. Schramm and J. C. Thanis (Eds), Verlag Hans Huber, Bern, 1968, p. 189.
192. Catrein, *Z. Klin. Med.* **118**, 688 (1931).
193. Ref. 1, p. 133.
194. P. H. Wise *et al.*, *Nature (Lond.)*, **219**, 1374 (1968).

195. *idem*, *Proc. Asia & Ocean. Congr. Endocr.* Auckland, New Zealand, 1971.
196. M. J. Halhuber, *Med. Met. Hft.* **13**, 112 (1958).
197. W. Zimmermann, in *Lehrbuch der Hygiene*, Fischer Verlag, Stuttgart, West Germany, 1964.
198. F. G. Sulman, *Umschau* **70**, 67 (1970).
199. *idem et al.*, *Int. J. Biometeorol.* **14**, 45 (1970).
200. *idem ibid.* **19**, 202 (1975).
201. Ref. 1, p. 148.
202. W. H. Fennell and R. E. Moore, *Int. J. Biometeorol.* **15**, 325 (1971).
203. Redisch *et al.*, in *Medical and Clinical Aspects of Aging* Vol. 4, H. T. Blumenthal (Ed), Columbia Univ. Press, New York, USA, 1962, p. 453.
204. A. R. Lund and R. F. Hellon, in ref. 203, p. 208.
205. G. C. Ring, G. Isaacs and W. Smith in ref. 203, p. 439.
206. Duguid *et al.*, *Lancet* **2**, 1213 (1961).
207. A. N. Exton-Smith, *Practitioner* **200**, 804 (1968).
208. R. K. MacPherson and F. Offner, *Med. J. Australia*, **1**, 292 (1965).
209. J. A. Levine, *Amer. J. Med.* **47**, 251 (1969).
210. R. W. Smith, J. Rizek, B. Frame and J. Mansour, *Amer. J. Clin. Nutr.* **14**, 98 (1964).
211. W. A. Sibley and J. N. Foley, *Ann. N.Y. Acad. Sci.* **122**, 457 (1965).
212. H. Miller, *Acta Psychiat. Scand.* **35**, Suppl. 147, 55 (1960).
213. U. Kolodovsky *et al.*, *Infect. Immun.* **12**, 1355 (1975).
214. G. Henle *et al.*, *ibid.* **12**, 1367 (1975).
215. in R. H. S. Thompson, *Proc. Roy. Soc. (Lond.)*, **B59**, 269 (1966).
216. E. D. Acheson *et al.*, *Acta Psychiat. Scand.* **35**, Suppl. 147, 132 (1960).
217. J. S. Barlow, *New England J. Med.* **260**, 991 (1959).

CHAPTER 7

METEOROTROPIC EFFECTS OF CLOTHING, NUTRITION, AIR CONDITIONING AND TRAVELLING

7.1. EFFECTS OF CLOTHING ON THE RESISTANCE AGAINST METEOROLOGICAL STIMULI

The three principal factors which determine the effectiveness of clothes in relation to meteorological stimuli, i.e. the material, the shape and the colour of clothing, were briefly discussed on pp. 135–136. Many studies have been carried out by the various armies of the world in connection with the thermal properties of clothing for protection against extreme climatic conditions. Excellent review papers were prepared in England by Renbourn *et al.*,[3] in the USA by Breckenridge and Goldman[4] and by others.

In connection with the importance of the material qualities of clothing the following comments of Renbourn[3] are worth repeating.

'A characteristic feature of cloth is that it is freely permeable to air movement through the low resistance pathways of its interstices. Up to recent years textile fabrics have been derived from animal materials, hair (sheep, angora goat, llama, camel, etc.) and the cocoon of the silkworm, and from vegetable fibres such as cotton and flax. Asbestos, the only fibrous mineral, is sometimes used in the manufacture of fire-proof clothing. Animal fibres are protein (keratin) in nature and burn with difficulty compared to the cellulosic vegetable fibres.

Nowadays a great variety of clothing materials are derived from artificial fibres manufactured from either natural or synthetic chemical substances. Artificial fibres derived from natural sources (known by the generic name of rayons) are manufactured from proteins (e.g. Ardil from peanuts) and from cellulosic materials (viscose and cellulose acetate rayons). The so called synthetic fibres are derived from a number of organic polymers: e.g. Terylene (Dacron) is a polyester, the various nylons are polyamides and Acrilan, Dynel, Orlon, Verel and others are acrylic in nature.

Whereas wool, cotton and linen are derived from short or staple fibres, silk and the artificial fibres are manufactured as long fairly translucent filaments which lend themselves to the production of thin, tight weaves; but if cut into staple lengths (or by the inclusion of loops in the filaments) they can be converted into fairly opaque materials with the appearance and feel of cotton or wool.

Synthetic fibre materials are hard wearing, they are more or less shrink-proof and crease-retaining, they take up very little water and dry quickly, but many have the disadvantage of melting at fairly low temperatures. It is the low water uptake of synthetic fibre materials which is related to their high content of static electricity, an annoyance in underclothes but a potential danger in the presence of explosive mixtures especially in a dry atmosphere. It has been claimed by a number of physicians that the static charge may be of physiological value, e.g. to those suffering from rheumatism. However, this assumption is without any scientific foundation.

In a cold environment the warmth of a cloth is not due to the physical properties of its fibres but mainly to the insulative still air clinging on to the fibres and yarns and the relatively still air in its interstices.

The wool fibre is covered with surface scales and has a crimp or waviness and an elasticity present in few other fibres, natural or artificial. These properties explain the well marked aero-dynamic drag of a woollen material, a property which tends to prevent air moving within or passing through it. A great disadvantage of untreated wool is that it shrinks, a property related to the surface scales.

Cellulosic and particularly protein fibres, either natural or artificial, take up appreciable quantities of water (vapour or liquid) into their substances, a property which may explain why cotton and wool fabrics can absorb up to about 20% and 40% respectively of their weight without feeling damp to the skin, an obvious advantage in underclothes. Such absorption is associated with the liberation of a large quantity of sorption heat. It has been calculated that a wool garment weighing five pounds takes up some three-quarters of a pound of water to give out for an hour or more the equivalent of basal metabolic heat. This may in part explain why it has long been believed that wool maintains an equable temperature around the body.

In warm tropical countries, a skin constantly covered with sweat is liable to prickly heat and fungus infection, hence absorptive and permeable clothing and personal hygiene are particularly important in warm damp areas. Although the minimum clothing compatible with conventions should be worn in the latter, the protection requirements of the jungle area itself must be remembered. Garments should be loose, without constricting belts, and ventilation openings (neck, underarm, sleeves, shorts, etc.) cut on really generous lines. Women's clothing easily lends itself for use in the heat of the tropics, but skin protection may still be required. A thin, fine weave, cotton poplin keeps out biting insects and is not objectively warmer than a thicker but open weave aertex material. The bush jacket has the virtue of combining the shirt with a coat. Starching adds to its appearance and to chimney ventilation, but on the other hand blocks air interstices. Synthetic fibre materials are at present contra-indicated for use in underwear and shirts in any tropical area, but as blends with fine wool, cotton or linen make suitable light-weight suitings.

A flannel binder previously used in the tropics as a protection of the abdomen against cooling is still advised by some tropical practitioners to combat the so called 'chilled abdomen' which is believed (on insufficient evidence) to predispose to diarrhoea and a 'tender belly'. It will, like any other well fitting garment, nevertheless produce in most individuals a band of prickly heat (see p. 184).

Special clothing is also an essential requirement in desert areas, but no longer is it believed that heat disorders arise from the sun's rays penetrating the spine or cranium. As a consequence, the spine pad is almost a museum piece and the sola topee (even with a coloured or aluminium lining), worn only for ceremonial occasions. Any light-weight hat or peaked cap is quite suitable.

A good insulative sole protects the feet from the hot ground, and the addition of a removeable ventilating insole makes for a cooler desert (or jungle) boot. Sandals should be worn whenever possible in all tropical areas as prophylaxis against Tinea pedis (a skin disease between the toes caused by infection of the fungi *Trichophyton* or *Microsporum*).'

For further details the reader is referred to the publications mentioned above.

7.2. EFFECTS OF NUTRITION IN RELATION TO METEOROTROPIC RESISTANCE

It is well known that the kind of food and its chemical composition affect the general metabolism and other important physiological processes in man; therefore they could have a great impact on the resistance of man in relation to his physical environment. As a result of the increasing pollution and poisoning of the atmosphere and hydrosphere the chemical composition of plants, and of animal products, which are consumed by man and animals is changing.

Toxic substances such as arsenic, cadmium, copper, chromium, lead, mercury and nickel etc. have been found in the recently deposited sediments of southern Lake Michigan and will be absorbed by the plants growing on these soils. The same is true for other countries. In other words the resistance of man against his physical environment is strongly affected by his chemical environment.

In a non-polluted environment different environmental thermal conditions require different diets, in order to keep a satisfactory thermal balance in the body; by preventing excessive metabolic processes.

The body's general metabolism is controlled by thyroxine, secreted by the thyroid (p. 70). The secretion of thyroxine is mainly regulated by the anterior lobe of the pituitary by means of the hormone thyrotrophin. Thyroxine mobilizes glycogen from the liver and muscles, where it is stored as an energy producing substance.

Glycogen ($C_6H_{10}O_5$) can be converted into glucose ($C_6H_{12}O_6$) by enzymes; with oxygen it is converted, together with fats, into carbon dioxide, water and heat (1g of glucose produces 4 calories and 1g of fat 9 calories of heat).

Various stress conditions, e.g. meteorological stimuli, may affect the hypothalamic-anterior pituitary system which may lead to excessive thyroxine production and increased metabolism.

Of the three main components of our food, protein, carbohydrates (sugars) and fats the latter two have an energy producing function, the proteins are required only for the building bricks of our body, the cells, tissues, enzymes and hormones. It is evident that the type of food consumed affects our metabolic processes and thus the thermal balance of the body.

Roberts[5] compiled hundreds of published measurements on basal metabolic rate (B.M.R.) of thousands of subjects of different races living in different latitudes. Despite the difficulties in comparing these data it does appear that due to differences in food the metabolic rates of people living near the pole are higher than those for the same race living near the equator. Eskimos and Canadian Indians for example seem to have a higher B.M.R. than North American or Brazilian Indians. Mongolian races in Siberia have a higher B.M.R. than those living near Singapore.

Food also affects the general resistance of the body to infections. In 1946 Hitchings and Falco[6] discovered that mice fed with brown bread are less resistant against pneumococcal infections than those fed with white bread. Studies by McCay and others[5] suggest that undernourished rats are often less prone to bacterial infections. Thus better food does not mean automatically greater resistance against infections.

Food also affects the mental processes of man. According to Mann[5] and Jansen[5] consumption of milk by children and better feeding increases their weight, height, general intelligence and their progress in school. Shortage of thiamine (vitamin B1), which is common in certain rice consuming countries, may give rise to beri-beri diseases, causing neurasthenia (sleep disturbances, irritability, poor memory, emotional instability etc.) and neurologic, cardiac and gastrointestinal complaints. Thiamine is essential for the proper metabolism of carbohydrates, and for the normal functioning of nervous tissue and enzymatic functions.

Summarizing the effects of nutrition,[7,8] it can be stated that in view of the significance of the kind and quantity of food on our metabolism and thermal balance, on the general resistance against infections and on our mental processes, and in view of the influence of weather and climate on each of these physiological processes (as described in previous sections) it is evident that nutrition should be seriously considered in the evaluation of meteorotropic phenomena.

7.3. EFFECTS OF AIR CONDITIONING

Air conditioning can be defined[9] as the control of the temperature, relative humidity and purity of the air in a room or building, in order to create a comfortable environment for the occupants. The main principle in air conditioning is the passing of an air current (at normal or pre-heated temperatures) through a humidifying chamber in which a fine water spray is discharged. Soluble gases and suspended particles are removed and a clean water-saturated air current leaves the chamber.

It is rather surprising that despite the worldwide use of air conditioning equipment, as yet relatively few scientific studies have been made on its biological effects.

In 1932 McConnell *et al.*[10] studied the difference in the incidence of disease among the staff of the Metropolitan Life Insurance Company in New York, USA, who worked partly in a new air-conditioned building and partly in an older office without cooling facilities. The sex and age distribution and the economic status of both groups were about the same. Although the air-conditioned building was considerably more comfortable to work in, there was no significant difference in the incidence of either respiratory diseases or non-respiratory illness between the two groups. Absenteeism was also the same. This result was not confirmed by other research workers. The very constant microclimatic conditions in an air-conditioned room may have favourable biological effects in certain instances (e.g. for asthmatics), but the lack of turbulence in the air, and therefore the lack of physiological stimulation, is not invigorating. Moreover, sunshine is often excluded from air-conditioned buildings. The effect of the lack of sunlight has been discussed on p. 100

The possible effects of the air-conditioning of ships at sea was studied by Smith,[11] between October 1948–September 1952, in ships of the British Royal Navy based in the Middle East and the Persian Gulf. He collected 1855 sickness forms from 104 different warships, of which 40% were from large ships and 49% from smaller ships. The following results were obtained:

(a) Minor sickness increased during hot weather in all classes of ships. Skin diseases in particular showed a pronounced increase with rising environmental temperature;

(b) When the environmental temperature at noon on the upper deck of the ship was 15 °C about 3% of the ship's crew were attending the sick-bay; with 27 °C, 5%; with 32 °C 9%;

(c) A comparison of the incidence of sickness on hot stations in the same ships, before and after the installation of air conditioning, showed that after air conditioning was introduced the number of sick people was reduced by 39%; the incidence of skin diseases and injuries decreased by 43 and 46% respectively;

(d) In the Persian Gulf, with air temperatures above 32 °C, the number attending the sick bay in ships with air conditioning was 7% against 11% in ships without air conditioning; skin diseases were reduced by almost one half in air-conditioned ships.

Whereas air conditioning in ships in very hot climates seems to have a beneficial effect, in temperate climates the effect may be less favourable. Stanley[12] reported that in the S.S. Lurline, travelling between Hawaii and San Francisco during 1951, passengers complained after the ship was air conditioned, that they felt 'as if they were coming down with colds' because of running noses, sore throats and a feeling of stuffiness. Some people felt cold all the time. Several cases of sudden asthma attacks were observed. This may be partly due to a poorly functioning (draughty) air conditioning system, but it could also be caused by the reduced adaptational capacities of subjects living in a constant environment.

Two important physiological problems arise with air conditioning. One is the circulation of bacteria and other internal atmospheric pollutants like tobacco smoke from room to room through the ducts. This may, for example, present problems of cross infection in hospital practice, particularly if recirculated air is used. When air conditioning is used in a medical environment, it is important to realize that ordinary filtration will only remove solid suspended matter. Gaseous pollutants like sulfur dioxide can cause considerable bronchial irritation; therefore in the treatment of smog patients, chemical purification may be essential, especially in highly polluted regions. Simple washing is often beneficial, but this increases the relative humidity which is not always desirable.

Finally a second air-conditioning problem may arise due to duct noise. In badly designed systems, noise may pass from room to room, or along from the plant. This may be extremely disturbing to the occupants. Extreme economy in air conditioning is only likely to be achieved at the expense of efficiency.

Summarizing it can be stated that in hot countries air conditioning improves the comfort, working efficiency and health of the people provided a number of precautionary measures are taken. However, as poorly installed air conditioners can create a draught, and as bacteria and odours can be transported through pipes, which also create noise, in buildings great care is needed in the installation of air-conditioning systems.

7.4. EFFECTS OF TRAVELLING

The rapid displacement in recent years of individuals or groups of individuals, to areas with weather and climatic conditions entirely different from their countries of permanent residence, creates many biometeorological problems. Whereas with careful planning, particularly with elderly people, a change in climate may have very favourable health effects, poorly planned travel may have serious consequences particularly for those not completely healthy.

7.4.1. Sea travel

In northern countries a typical sea climate is often defined as the climate of an area near the sea where the difference in average temperature of summer and winter is less than 15 °C. A typical land climate has differences of more than 20 °C. However, similar temperature conditions to those of a sea climate are also found in most equatorial areas.

The sea climate and weather conditions encountered on a ship depend on a number of factors:

(a) The geographical location (latitude and longitude);
(b) The predominant temperature of sea currents encountered during a voyage;
(c) The proximity of a coastline and the prevailing weather and climatic conditions of the coastal area;
(d) The material, size and construction of the ship, i.e. the microclimatic conditions resulting from a specific architectural design of the ship.

Many factors such as air circulation affected by the general construction of the ship and by the size, shape and number of port-holes of cabins, air-conditioning equipment etc. determine the microclimatic condition, irrespective of external macroclimatic conditions. In reality, both the internal microclimate, determined by the engineering design, and the external macroclimate are closely interrelated and determine the final atmospheric conditions in which the sailor or passenger is forced to live.

A person travelling on a ship experiences different meteorological stimuli, in particular radiation stresses due to changes in infrared (heat) and ultraviolet radiation.[13] The thermal stresses on a moving ship belong to four categories:

(i) The general conductive heat of the atmosphere at the geographical site of the ship;
(ii) The direct and reflected solar radiation heat load (i.e. infrared radiation);
(iii) The conductive and convective heat created by the engines of the ship as changed respectively by an air-conditioning aggregate;
(iv) The transfer of thermal energy in the air through and around a moving ship as a result of air currents.

The significance for human health of air conditioning on ships was discussed in the previous section. During trips in passenger ships in tropical zones passengers often develop minor complaints, particularly if only the dining room or reading room is air conditioned. Under these conditions the human body is continually subjected to great differences in temperature on moving from deck to cabin and dining room. This causes a constant stress on the thermoregulatory mechanism of the body. Whereas the crew may gradually acclimatize to these rapidly changing temperatures (and part of the crew may even not live in air-conditioned quarters) the passengers' sojourn on board is usually too short for them to acclimatize and they may therefore suffer from various diseases, such as common colds, respiratory diseases, rheumatic complaints etc as a result of reduced thermoregulatory capacity.

Apart from thermal stresses ultraviolet radiation effects should not be neglected, the more so as solar radiation near a water surface has stronger effects than on land.

The high humidity of the air is also an important factor on ships. In a warm climate it seriously affects our thermoregulation and general thermal comfort. The effect of increased humidity on the nasal mucosa and growth of bacteria was discussed earlier on p. 165. The high aerosol content of the air above the sea, which contains iodine and magnesium and sodium chlorides, could have physiological effects on persons travelling on ships.

Solar radiation and ozone oxidize the iodine ions in sea water droplets into volatile iodine pentoxide. Recent studies by Balogh in Hungary suggest that an increased aerosol-iodine content of the air decreases the heat production of animals and the I^{132} uptake by thyroid gland, whereas a low content was found to be associated with a higher metabolism.

Continuous oral inhalation of high concentrations of magnesium chloride which dissolves in the saliva, could result in this reaching the stomach together with the food. The possible effect on the stomach and the intestines has never yet been studied, however it is very likely that it would affect the digestive processes in the intestines.

The high salt content of sea air affects the cells of the nasal membranes and lungs and often has favourable effects in subjects suffering from respiratory complaints.

In addition to the effects mentioned above, the biological rhythm disturbances (due to rapid east-west movements of ships) should also be mentioned. However these effects are particularly pronounced in the case of air travel and will be discussed in that context in Section 7.4.2(b).

Few data exist on the actual morbidity frequency of different diseases observed amongst sailors and passengers on ships of different sizes and navigating in different parts of the world. The following data are based on the statistics of the C.I.R.M. (International Radio Medical Centre) in Rome (for the period 1935–1963) supplied by the former director Professor G. Guida. According to these data 11 073 cases of various diseases were treated during the period, of which the majority seem to belong to the group of infectious diseases (24%) followed by diseases of the digestive organs (18%), of the genito-urinary tract (10%), oto-rhino-laryngological diseases (6.6%) and respiratory diseases (4.9%). However, these frequencies probably do not represent the actual morbidity frequencies. They indicate only the frequency pattern of the more serious cases for which the ship surgeons requested the radio-telephone aid of C.I.R.M.

7.4.2. Air travel

Apart from the psychological affects two major biological effects are experienced during air travel, a high altitude effect and a biological rhythm effect.

(a) *High altitude effect*: In most commercial aircraft the simulated altitude in the cabins surpasses the 1500 m altitude boundary, particularly in the USSR, Latin America, India and Pakistan. Considering the physiological and clinical effects of altitudes above 1500 m (see Section 5.3), similar effects must be expected in passengers and flying personnel during commercial flights. After a stay of several hours at this reduced pressure improved ventilation and vital capacity of the lungs can be expected. Increased peripheral bloodflow, increased sensitivity of the autonomic nervous system, stimulation of the hormone production by the adrenal gland and of the blood producing mechanisms and improvement of the overall thermoregulation efficiency will occur. Unfavourable effects are increased sensitivity to drugs and toxic substances, the explosive development of latent infectious diseases (e.g. apendicitis, common colds, influenza etc.), toothache due to dental root infection, etc. Provided the simulated altitudes are below 1800 m an aeroplane trip is very beneficial for passengers without latent infectious or dental diseases. Heart patients also usually do not experience any discomfort at this altitude if they remain seated. However, in the case of pregnant women, particularly those in their 3rd month of pregnancy, flight at high altitudes is not recommended in view of possible effects on the foetus (Section 8.2.1, p. 220).

(b) *Biological rhythm effect:* In recent years it has been well established that most processes in the biological world, and many phenomena in the non-living world, have a rhythmic nature with specific amplitudes, wavelengths and frequencies.[14–33] In humans the regular heartbeat and respiration are good examples, but hormonal processes, body fluid excretions etc. also take place in a rhythmical manner. The physiological mechanisms involved in most of the biological rhythms are known. The rhythms are usually caused by changes in illumination, temperature, atmospheric pressure or in other meteorological parameters of our environment. This group of environmentally determined rhythms is known as exogenous rhythms. However, there are several rhythmic phenomena with an internal cause. These are known as endogenous rhythms. In addition to the effects of the hypothalamus (thermoregulatory centre), the pituitary and the rhinencephalon (see pp. 70–72), the epiphysis or pineal gland (also known as the photo-periodic brain) probably plays a key role in many endogenous rhythms. It regulates the timing of many physiological events, and is rich in noradrenaline, histamine, and, in particular, serotinin, the level of which shows striking diurnal changes, being highest at 1 p.m. and lowest at 11 p.m. Therefore the pineal gland, which is richly innervated by sympathetic nerve terminals having swellings containing noradrenaline, is also affected by environmental stimuli (particularly light), and is an important component of the endogenous time measuring system. However, many endogenous rhythmic processes still have unknown causes.

Due to these various exogenous and endogenous stimuli all hormonal- and neural-controlled processes in the human body vary continuously in a periodic, semi-periodic or fluctuating manner. It is thus evident that if one changes one's environment by a rapid move to the east or the west, then the various rhythms will become out of phase with the local time, and it will be some time before the body is adapted again. As all these internal physiological processes also affect our moods and behaviour, it is logical that if one flies from Europe to the USA or vice versa, one's internal time mechanisms become completely out of balance. As a result the body demands sleep during working hours and food at times no food is served. These effects are even more severe during long lasting space flights around the earth, or to the moon, and may become very serious if inter-planetary travel becomes possible. Readers interested in obtaining further details about biological rhythms and their effects should consult the references at the end of this chapter,[14–33] and two recent reviews.[34,35]

7.4.3. Car travel

A passenger travelling in a car for many hours experiences a number of biological effects apart from the usual psychological stresses which drivers experience in modern traffic. The biometeorological stresses are:

(a) *Radiation stresses.* If the windows of the car cannot be protected against the sun solar heat radiation (infrared rays) will reach the passenger; however, ultraviolet light from the sun is mostly absorbed by the window glass before reaching the passenger.

(b) *Cooling stress* (temperature-wind effect) occurs if the car is driven with great speed and the wind in the open air is very strong; the effect is usually small if the car is constructed satisfactorily and no air currents can enter the car.

(c) *Pollution stress*, created by the inhalation of even a small percentage of atmospheric contamination (e.g. exhaust gases, carbon monoxide, carbon dioxide, sulfur dioxide, hydrocarbons, etc.), takes place if the passenger must stay for a long time in the closed space of the car.

(d) *Electric stresses* may occur if a metal car is not a perfect Faraday cage, in other words, most electromagnetic waves are able to penetrate; the potential gradient is practically zero; the natural ratio of positive and negative ions changes; usually an excess of positive ions exists because of passengers smoking, together with increased humidity of the passenger space due to breathing and the presence of pieces of metal in the car, which affect the ion content of an air stream.

The possible biological effects of each of these stresses have been described in previous sections.

REFERENCES

1. S. W. Tromp, *Medical Biometeorology*, Elsevier, Amsterdam, The Netherlands, 1963.

2. *idem* (Ed), *Progress in Human Biometeorology, II*, Swets & Zeitlinger, Amsterdam, The Netherlands, 1977.
3. Renbourn *et al*., in ref. 1, p. 426.
4. Breckenridge and Goldman, in ref. 2, p. 194.
5. in B. C. P. Jansen, *Moderne Voedingsleer*, Wyt, Rotterdam, The Netherlands, 1959.
6. G. H. Hitchings and E. A. Falco, *Science* **104**, 568 (1946).
7. E. V. McCollum and H. Simmonds, *Newer Knowledge of Nutrition*, Macmillan, New York, USA, 1929.
8. C. C. Furnas and S. M. Furnas, *Story of Man and his Food*, New Home Library, New York, USA, 1943.
9. T. Bedford, *Basic Principles of Ventilation and Heating*, Lewis, London, UK, 1948.
10. W. J. McConnell, H. H. Fellows and M. G. Stephens, *Publ. 14th Ann. Conf. Life Office Management Ass*., Chicago, USA, 1937.
11. F. E. Smith, *Brit. J. Ind. Med.* **15**, 197 (1958).
12. L. L. Stanley, private communication, 1951.
13. S. W. Tromp, *Some Aspects of Nautical Biometeorology*, Centro di Studi e Recherche per l'Assistenza Sanitaria e Sociale dei Maritimi, Rome, Italy, 1965.
14. J. Aschoff (Ed), *Circadian Clocks,* North-Holland, Amsterdam, The Netherlands, 1965.
15. *idem, Oecologia* **3**, 125 (1969).
16. *idem*, F. Ceresa and F. Halberg, *Chronobiologia* **1**, 1 (1974).
17. F. A. Brown Jr., J. W. Hastings and J. D. Palmer, *The Biological Clock,* Academic Press, New York, USA, 1970.
18. E. Büning, *Die Physiologische Uhr*, Springer Verlag, Berlin, Germany, 1963.
19. R. T. W. L. Conroy and J. N. Mills, *Human Circadian Rhythms*, Churchill, London, UK, 1970.
20. J. L. Cloudsley-Thompson, *Rhythmic Activity in Animal Physiology and Behaviour*, Academic Press, New York, USA, 1961.
21. E. R. Dewey, *Cycles*, Hawthorn Books, New York, USA, 1971.
22. J. Harker, *The Physiology of Diurnal Rhythms*, University Press, Cambridge, UK, 1964.
23. K. E. Klein, H. Brüner, H. Holtmann, H. Rehme, J. Stolze, W. D. Steinhof and H. M. Wegemann, *Aerospace Med.* **4**, 125 (1970).
25. C. S. Pittendrigh, in *Life Sciences and Space Research, Vol. 5*, North-Holland, Amsterdam, The Netherlands, 1967 p. 122.
26. H. A. Reimann, *Periodic Disease*, Davis, Philadelphia, USA, 1963.
27. A. Reinberg and J. Ghata, *Rythmes et Cycles Biologiques*, Presse Universitaire, Paris, France, 1957.
28. L. E. Scheving, F. Halberg and J. E. Pauly, *Chronobiology*, Thieme Verlag, Stuttgart, West Germany, 1974.
29. P. V. Siegel, S. J. Gerathewohl and S. R. Mohler, *Science* **164**, 1249 (1969).
30. A. Sollberger, *Rep. 7th Conf. Soc. Biol. Rhythm*, Panminerva Medica, Turin, Italy, 1962, p. 138.
31. *idem*, *Biological Rhythm Research*, Elsevier, Amsterdam, The Netherlands, 1965.
32. H. von Mayersbach, *The Cellular Aspects of Biorhythms*, Springer Verlag, Berlin, Germany, 1967.
33. R. Wever, *Kybernetik* **1**, 139 (1962).
34. in ref. 2, p. 223.
35. S. W. Tromp (Ed), *Progress in Human Biometeorology, IB,* Swets & Zeitlinger, Amsterdam, The Netherlands, 1974.

CHAPTER 8

MISCELLANEOUS METEOROTROPIC EFFECTS

8.1. EFFECT OF MONTH OF BIRTH ON MENTAL DEVELOPMENT AND PRONENESS TO DISEASE

The scientific study of the possible effects of meteorological conditions during the month of conception on the mental development of the embryo, and on proneness to disease after maturity has been reached, is seriously hampered because the subject is emotionally heavily charged. Since the beginning of human civilization astrologers believed in cosmic influences. They believe that man's development and future events are determined by planetary and other cosmic conditions prevailing during the month and day of his birth. However, the phenomena which will be described in this section are not concerned with astrology and the month or day of birth, but with the month or day of conception and with foetal development during the following months. It is not cosmic effects in the sense of astrology but the meteorological conditions which prevail during the month or day of conception and the following foetal development which will be discussed.

Physiologically speaking there are two important dates in embryonic development at which meteorological stimuli may affect the future physiological functions of the developing embryo. These are the date of conception and the date, some three months later, when the cerebral cortex is becoming organized and the nervous system is developing. As both of these dates are only known approximately, the month and date of birth, which are known exactly, are used for statistical purposes. In the following pages the effect of the month of birth on various aspects of subsequent development is discussed, but the processes responsible are those influenced by the meteorological stimuli during the month of conception nine months earlier.

In view of the many seasonal effects on the basic physiological mechanisms and diseases of humans, some effect of the month of birth (or conception) on mental development and proneness to disease could be expected on purely theoretical grounds. Kevan[2] summarized the many reasonably scientific papers which suggest a statistically significant relationship between month of birth and mental development. However statistics can never prove a causal relationship, but do at least stimulate further research in this field. Some of Kevan's major findings, after extensive literature research, can be summarized as follows.

8.1.1. Effect on mental development

Research concerning the effects of month of birth upon mental development can be classified according to the relation of month of birth to intelligence, personality or human behaviour. The theories that have been proposed in order to explain these relations may be divided according to whether they stress pre-natal or post-natal factors. Pre-natal or endogenous theories suggest that the atmospheric environment may affect the growth of the foetus during gestation. Post-natal or exogenous theories suppose that the weather and climate affect the child's early experience in such a way as to determine the future development of that child. In general, the advocates of post-natal theories have been the investigators of intelligence and month of birth, whereas those who have studied the relationship of mental illness to month of birth have tended to propose pre-natal theories.

The first serious studies on intelligence in relation to month of birth were made by Fairgrieve.[3] He discovered that the boys who were born in late spring and early summer had a tendency to get lower intelligence test scores than those boys who were born in late autumn and early winter.

In 1931 Pintner[4] re-awakened the academic community's interest in this subject. Throughout the next decade Pintner and his associates[5] undertook many studies on the relationship between season of birth and intelligence. The data collected from tens of thousands of 'normal' subjects revealed that there is a seasonal relationship with intelligence; subjects born during the warmer months of the year tended to have slightly higher intelligence levels than those who were born during the colder periods. However, the observed differences were usually not statistically significant. Later studies[6–14] showed somewhat similar trends, but not all the observed relationships were statistically significant. These studies need to be continued with considerably larger groups of subjects both in schools and universities, before a firm conclusion can be reached (for further details and references see Kevan's paper[2]). Gibson[15] observed, in a population near Oxford (Great Britain), an association between IQ and ABO blood groups. As preliminary studies at Leiden (The Netherlands) suggest a relationship between month of birth and certain blood groups, the observations of Gibson may support the findings reviewed by Kevan.[2]

Studies by Armstrong and Tyler[16] and others suggest that the IQ of young children can be affected by a disturbed amino acid metabolism having an unfavourable effect on the function of the brain, due to a disturbed oxidation of the amino acid phenylalanine to tyrosine. Meteorological stimuli affect the thyroid function and general metabolism. Therefore on theoretical grounds a meteorological disturbance of amino acid and metabolism is not excluded.

Between 1920 and 1929 other workers[17–21] observed a greater speed of calculation and a greater accuracy in solving arithmetical problems, after a few weeks of thalassotherapy, but a decrease in attentiveness. Their studies have not been repeated in recent years.

The relationship between month of birth and personality was particularly studied by Huntington.[22] After the analysis of a very large number of data he concluded that people born during the coldest months of the year show a very great variability in aptitude compared with those born during the warmest months of the year, who are the least likely to succeed in any field. Forlano and Ehrlich[23] observed that, in New York City college, freshmen who were born during the summer had a tendency to be more introverted and to suffer more from feelings of inferiority than students born at other times of the year. Davies[7] and Farley,[24] were unable to confirm these relationships. Recent studies undertaken in The Netherlands,[25] which require further confirmation, suggest a possible relationship between the month of birth and partner choice in marriage and the probability of divorce.

8.1.2. Effect on proneness to disease

Many studies seem to suggest that the month of birth of a person could have a strong effect on his future proneness to disease. This has been demonstrated on statistical grounds in the case of schizophrenia by workers in several countries.[26–34] In the case of cancer, workers in the USA, and in The Netherlands,[29, 35–37] observed a similar proneness to the disease in relation to the month of birth.

In the case of schizophrenia it was found by these independent research workers, using data from more than 30 000 cases, that schizophrenics are born most frequently in January, February and least frequently in June, July, and August. The differences deviate significantly from the normal birth-rate peaks in the countries concerned. It has been assumed that the high winter peak is due to conditions six months earlier, i.e. June, July, August, when the embryo is three months old and the cerebral cortex of the unborn child is becoming organised. Any disturbance in the metabolic processes in these summer months (e.g. due to excessive thermal stress) could create a serious injury in the developing cortex of the foetus.

In the case of cancer the highest number of birth dates occur in the period December–March, the lowest around June–July (studies based on more than 15 000 cases). This seasonal variation seems to be independent of the cancer site. Data collected by Keogh[38] in Australia in the southern hemisphere and analysed by De Sauvage Nolting,[29] show a similar peak, but during the summer months of the northern hemisphere, i.e. the winter period of the southern hemisphere. Cancer cases amongst Australian residents who were immigrants from the northern hemisphere showed a northern hemisphere pattern. It can be calculated that if all children of the northern hemisphere were born in the summer then seven to eleven per cent fewer cancer cases would be anticipated.

In view of the practical importance of these findings similar relationships were studied,[39] in The Netherlands, in subjects attempting suicide in the cities of Amsterdam, Rotterdam, The Hague and Utrecht during the period 1954–1970. The data from about 23 620 cases of diabetes and about 2000 cases of patients suffering from multiple sclerosis were also analysed.[40] The following preliminary findings were obtained:

(a) *Suicide attempts.* In The Hague the maximum number of births of normal subjects occurred in the month of May (1958–1968) whereas the birth dates of those who attempted suicide occurred most frequently in February and March. In fact the whole birth date pattern of people attempting suicide during this period was completely different from the birth date distribution of normal subjects.

(b) *Diabetics.* The normal distribution of births in The Netherlands (1956–1968) showed a maximum in May. However, the maximum number of birth dates of diabetics is found in March. The differences observed in female diabetics are statistically significant, but not so for males.[40]

(c) *Multiple sclerosis.* The monthly birth date distribution in both males and females is different from the normal distribution, the maximum occurring in February and March. The period January–March has a higher birth date incidence than April–June. The lowest incidence occurs in October (males) and November (females). However the differences are not significant at the 0.01 level.

The results of a study carried out by Schook and Schüller[41] in Hilversum, The Netherlands, on the relationship between month of birth and asthma were discussed in Section 5.3.4.

8.2. METEOROTROPISM OF CONGENITAL MALFORMATIONS

Hippocrates, in his book *Airs, Waters and Places*, stressed repeatedly that the developing embryo is influenced by the condition of the mother and by the climatic conditions in which she lives. However, it is only in recent years that the subject has been studied scientifically by biometeorologists.

Since 1950 more than 50 publications have appeared on the possible influence of environmental factors (both chemical and physical) on congenital malformations in animals and man (see Table 8.1). Despite many empirical data and a number of experimental studies little is still known about the deeper mechanisms involved.

8.2.1. Direct meteorological effects

At least five different meteorotropic mechanisms have been suggested to explain the observed weather relationships. Thermal stresses, high altitude, seasonal influences, noise, and water pollution.

TABLE 8.1
Medical-geographical relationships of congenital malformations

Kind of malformation	Reported environmental relationships
General	Malformations develop during pregnancy after virus infections (e.g., rubeola), maternal diabetes (due to insulin), etc. In England incidence 6/1000. According to Petersen[44] in the meteorologically quiescent southern USA less foetal malformations occur than in the North (Mississippi/Maryland ratio 1:6). In the northern USA highest incidences are in areas with many atmospheric depressions, high rainfall and strong cooling.
Anencephalus (absence of the vault of the skull; the brains are degenerated or missing)	Probably due to intra-uterine environmental factors e.g. increased α-foetoprotein after the 15th week of pregnancy in the amniotic fluids produced by the foetus liver, reaching a maximum during the 30th week.[45] According to Knox[46] a high protein diet (cheese, meat etc.) protects against malformation. Incidence in north and west England 7–8/1000, south east England 3/1000; in Africa, south America and Japan 1/1000; in USA incidence about 3/1000 decreasing from east to west and from north to south. Seasonal variations. Particularly evident if the month not of birth, but of conception is used.[47] In Birmingham and Scotland the highest frequency was observed in October–March, minimum April–September;[48,49] MacMahon *et al.*[50,51] found, in Rhode Island (USA), similar seasonal effects; De Groot[52] found in The Netherlands highest incidence in December–May; Frezal *et al.*[53] did not find a seasonal effect in France in 402 cases; Leck and Record[54] observed low incidence of April–August only from 1939–1958, from 1961 weak effects; Silberg *et al.*[55] found similar seasonal effects in various malformations, but no clear effect is anencephalus; Elwood[56,57] observed in Belfast maximum in October, minimum in February–March. Long term changes were observed in Birmingham (decrease from 1940–1947), Scotland (increase from 1939–1958 from 2.58–2.86/1000 Boston (decrease from 1875–1965 from 2.04–0.65/1000; in between short rises from 1900–1904 and 1930–1934); Netherlands (decreases from 1951–1968 from 1.47–0.84/1000 with small rises in 1954, 1958, 1962, 1968).
Cleft lip (cleft involving the lip only) and cleft palate (fissure of the median line of the palate, i.e. roof of the mouth)	Possible causes: Chemical (drugs during pregnancy e.g. thalidomide), rubeola and rubella (measles) virus during pregnancy, maternal diabetes (use of insulin) etc. Mainly occurring in males. Highest incidence in Japanese (2.1/1000), followed by Caucasians, low in American negroes (0.4/1000), in England 1/1000; in children of epileptic mothers (using anti-convulsant drugs) cleft lip incidence 29 × more frequent than in non-epileptics. Probably caused by anti-folic acid action of drugs. Highest incidence in November–March.
Hip dislocation (congenital) (dislocated hip at birth)	Male/female ratio 1:7, often associated with breech/birth. High frequency amongst Laplanders and some American Indian Tribes; in England 4/1000; in USA maximum incidence January/March; in southern hemisphere (Victoria, Australia) maximum incidence

TABLE 8.1—contd

Kind of malformation	Reported environmental relationships
	February–June, lowest July–January; in northern hemisphere maximum incidence September–February with peak from end October–middle January minimum June–August.
Pyloric stenosis (narrowing of the pylorus i.e. the junction of the stomach and duodenum)	Five times more common in males than in females; frequent in north west Europe (1–3/1000 births), less common in Chinese, Jews and black Africans. In England 3/1000. Seasonal variations: Some studies suggest incidence more frequent in children born in summer. Other studies did not confirm this finding.
Spina bifida (cleft of vertebral column with meningeal protrusion)	Same etiological factors as anencephalus; male/female ratio = 0.85. According to Petersen[44] high in north east and north west USA. According to McKeown and Record[48] no seasonal variations of spina bifida and hydrocephalus.
Talipes equinovarus (type of clubfoot, present in the foetus and often disappearing during the eighth week of foetal life)	Male/female ratio 2:1; high incidence in Polynesians (about 5/1000 births); incidence rate in England 1/1000.

(a) *Thermal stresses.* The observation that the maximum incidence of hip dislocations occurs during the cold winter months suggests a thermal or reduced solar radiation effect. Edwards[42,43] demonstrated in Sydney that hyperthermia of the pregnant guinea-pig, in an incubator at 43 °C for 1 h daily during the 18th–25th day of gestation, creates in the young animals a high incidence of the *talipes* (club foot like) deformities of the hind- and fore-limbs. Also abnormalities of the spinal cord and brains were observed. Similar experiments with rats produced increased foetal resorptions, growth retardation etc. Edwards also succeeded in producing arthrogryposis multiplex congenita experimentally in new-born guinea pigs. Arthrogryposis (a persistent flexure of a joint of a child or animal) is characterized by muscular atrophy and immobility of joints. The pregnant animals were subjected, for one hour daily, to a temperature of 43 °C in an incubator between days 4 and 67 of pregnancy. Arthrogryposis was observed in the young from mothers exposed between the 26th and 46th day of pregnancy. Arthrogryposis of the neck, fore limbs, thorax and abdomen followed maternal hyperthermia between days 26 and 31. In the hind limbs it occurred as a result of hyperthermia between days 30 and 46.

The fact that the thermal stress should take place at a certain stage of gestation indicates clearly that a similarity in average temperature in different countries with a different incidence of an effect cannot disprove a thermal hypothesis. Metabolic and thermal disturbances in the mother during the 3rd month of pregnancy, when the brain and nervous system are formed in the human embryo, may be responsible for the proneness to malformations.

(b) *High altitude.* Lilienfeld and Pasamanick[58] pointed out that many malformations, including convulsive seizures in children, are due to anoxia during pre-natal and post-natal periods, which may result in brain damage. Experimental studies[59,60] suggest that reduced oxygen pressure, under natural or simulated high altitude conditions (observed at 7000 m altitude), could create malformations of the vertebral column. According to Degenhardt[61] ingestion of vitamin E could reduce the number of malformations.

The degree of malformation depended on the rate of oxygen deficiency and the day of pregnancy that the oxygen deficiency occurred. Werthemann *et al.*[62] and Naujoks,[62] observed changes in the eye lens and retina of pregnant laboratory animals and chickens which were subjected to oxygen deficiency during the beginning of pregnancy.

(c) *Seasonal influences.* A number of seasonal meteorological parameters seem to be responsible for the monthly seasonal changes in incidence of congenital malformations. Particularly changes in solar radiation and thermal stresses may be involved, which seem to be responsible for metabolic disturbances of pregnant mothers.

Considering the different effects of meteorological stimuli on thyroid functions (see p. 106) a relationship between weather, metabolic disturbances and malformations seems to be possible. The significance of the thyroid is suggested by studies[63-65] in struma areas. A higher incidence of malformations is observed in these areas. Langman and Van Faasen[63] studied the effect of extirpation of the thyroid in white rats, before pregnancy, on embryo development. The number of embryos decreased and different anomalies occurred, such as abnormal ossification, cleft palate, eye lens degenerations etc.

The importance of α-foetoprotein produced by the liver (see Table 8.1, anencephalus) of the foetus also suggests that meteorological factors may create malformations through serious disturbance of its liver function.

A third meteorotropic effect may be due to the influence of weather and climate on insulin production. Experiments[66] indicate that injection of insulin in animals creates several malformations. Also the higher incidence of malformations in children of pregnant diabetic mothers suggests the influence of insulin metabolism on the occurrence of malformations.

(d) *Effects of noise.* Sontag *et al.*[67] pointed out that emotional stages of pregnant females could affect the physiological-endocrinological state of the mother and probably indirectly of the foetus; e.g. anxiety could affect the thyroid function and blood sugar levels which could influence the development of malformations. Sontag and Wallace[68] observed in the foetus an immediate increase in heart rate and foetal movement after sound stimulation of the mother. Most likely the stimuli create a maternal sensory response which is mediated through the endocrine-humoral mechanisms. Therefore it should not be excluded that strong sound stimuli in our physical environment and emotional stresses could affect the development of malformations.

(e) *Other factors*. Undoubtedly many other environmental factors may be involved in the causation of malformations, for example pollution of drinking water. Studies by several workers[69–71] suggest high incidence of malformations due to water containing lead compounds, which seem to be able to penetrate the placenta in rats. Soft water (i.e. low in calcium carbonate) is richer in lead compounds. This could explain the higher incidence of anencephalus in rural areas with usually soft water, at least in The Netherlands.[70,71] Also the high lead content of certain rivers (e.g. Rhine water) may influence the incidence, particularly in very warm regions where evaporation increases the concentration of pollutants.

8.2.2. Indirect meteorological effects on malformations

These could be expected because weather and climate determine the type of food consumed, particularly in the not heavily industrialized countries. According to Knox[72] high protein diets reduce the development of malformations. The importance of the food consumption of the mother is also stressed by Renwick in England[73–76] who suggested a relationship between anencephalus and spina bifida and the consumption of winter-stored potatoes which have been infested by the potato fungus *Phytophtora infestans*. The fungus develops particularly under mild, wet weather conditions. Renwick observed a significant correlation between potato blight incidence and incidence of anencephalus and spina bifida and in particular between the incidence of malformations and the blight severity in the year prior to that in which the teratogenic effect occurs. The effect would be caused, according to Martin[77] and Renwick,[73–76] by certain toxic substances in the potato tuber e.g. *solanidine* ($C_{27}H_{43}NO$), a steroid alkaloid which is toxic to man. Light, particularly ultraviolet radiation, increases the concentration of the toxin. The medical-geographical studies of Renwick[73–76] were partly supported by the experimental studies of Poswillo *et al.*[78] Although pregnant Wistar rats fed with 'blighted' potato tubers did not show any abnormalities, the same experiments with certain monkeys (*Callithrix jacchus*) showed an increase in malformations.

This brief summary of the various studies on the possible effects of weather and climate on congenital malformations clearly shows the extremely complex nature of the problem. Extensive research studies on the meteorotropic aspects of congenital malformations are warranted because the solution of these problems may have far reaching sociological consequences once the deeper mechanisms involved are better understood.

8.3. EFFECTS OF THERMAL STRESS AND ALTITUDE ON THE RESPONSE TO DRUGS

After a drug is administered its passage into the blood and its therapeutic action are determined by physical, physiological and biochemical processes that include absorption, binding to protein molecules, transfer across cell membranes, interactions with cell receptors, metabolism etc. Both specific and non-specific enzymes play an important role in many of these processes. Depending on the genetic structure of the subject the enzyme processes will be different.

Since 1926 several authors[79–85] have pointed out that apart from genetic factors meteorological factors are also involved in the different response to drugs, which could explain that the same chemical substance, in the same concentration, administered to the same subject may create an entirely different response under different weather, climate and altitude conditions. The study of these drug effects belong to the science of Pharmacological Biometeorology, which was briefly summarized in 1963.[86] Since 1963 several review papers have been published on this topic.[87–90] The interested reader is referred to these publications and in particular to their extensive lists of references.

The most important meteorological factors which could affect the response to drugs are thermal stresses (combined temperature, air movement and humidity effects), ultraviolet radiation (in particular at high altitude) and reduced partial oxygen pressure (and all accompanying physiological processes due to altitude affects above 1500 m). All these meteorological stresses affect the membrane permeability and the rate of drug absorption in the human body. Both changes explain the many observed meteorological effects on drug action, some of which (altitude affects) were described on p. 111. In Table 8.2 a brief compilation is given of some observations made in animals and man, in connection with meteorological and in particular thermal effects on the action of drugs.

[a]Amphetamine: drug causing exhilaration and increased muscular efficiency. Atropine sulphate: parasympatholytic drug used for dilatation of the pupil and for hepatic colics. Chlorpromazine: used in the drug Largactyl, it has a sedating effect on anxiety psychoses and other psychiatric diseases and is used as antihistamine. Digitalis: drug used for patients with congestive heart failure. Epinephrine is a cardiac stimulant. Insulin: important substance secreted by Islands of Langerhans (see pp. 91, 188) and important for carbohydrate metabolism. Mepazine: sedative in psychoses and severe neuroses. Meprobamate: tranquilizer. Morphine sulphate: analgesic drug suppressing pain. Pilocarpine: drug causing constriction of the pupil. Phenothiazine: neuroleptic drug used for psychiatric patients. Trifluopherazine: drug against psychoses. Tuberculin: used in intradermal tests for sensitivity to tubercle bacilli.
For references see Tromp from p. 224.

TABLE 8.2
Most important effects of meteorological stimuli on the action of drugs[91]

Drugs used[a]	Type of stress	Observed changes in drug action
Amphetamine	Meteorological stress or crowding of animals.	High mortality even in small doses.
Atropine sulfate	Meteorological stress.	Some days children require twice the normal dose before the eye reaction occurs.
Digitalis	High altitude (Sierra Nevada, USA, 3000 m).	In San Francisco 29% of pigeons and cats started vomiting with 15 mg dose; at 3000 m 63%; 90 mg doses killed 20% of pigeons at 3000 m, 0% in San Francisco; with 105 mg 67% killed at 3000 m, 0% at sea level.
Epinephrine and Pilocarpine	Heat stress.	More rapid increase of blood pressure than under normal conditions.
Glucose	Cold stress.	In rats increase in hyperglycaemic action of orally administered glucose.
Insulin	Cold stress.	Increased need for insulin and inhibition of its hypoglycaemic action.
Meprobamate	Thermal stress (10 °C and 43 °C).	800 mg dose at 10 °C causes lower core temperature than in placebo group, moderate elevation at 43 °C.
Morphine sulfate	Cold fronts.	Lowest mortality in mice during meteorologically stable days; if front passage closely precedes injection, toxicity highest; if front passage took place long before injection toxicity low; during cold front highest death rate with alkaline diets; after cold front passage death rate highest with acid diets.
Phenothiazines	Heat stress.	Sometimes oral temperatures decrease.
Chlorpromazine (Ph-derivative)	Cold or heat stress.	Hypersensitivity to freezing, to cold water washing or swimming (of old patients), and to hot humid weather (after 300 mg dose) leading to fatal hyperpyrexia, particularly after humid heat wave.
		Anaesthesia of horses in humid hot climate requires lower doses than in cold climate.
Trifluopherazine (Ph-derivative)	Heat stress (33 °C).	Severe hyperthermia after large doses of drug.
Mepazine (Ph-derivative)	Heat wave (40 °C).	Development of heat stroke and hyperthermia.
Tuberculin	Weather fronts	Daily changes in sensitivity.
	Seasonal effects	Greater sensitivity March–April than Autumn.

8.4. EFFECT ON SEX RATIO

In western Europe the ratio males/females amongst newborn babies is about 105/100, i.e. 5% more boys. The ratio is largest at the 3rd month of pregnancy and this ratio gets gradually smaller till the 9th month due to increased death of male foetuses. In other words 20–30% more male than female foetuses exist during the first days of pregnancy of which gradually more male than female foetuses die during the following 9 months. After a war, even in countries not involved in the war, the ratio males/females increases, i.e. the male surplus becomes larger. According to certain authors[92] this may be related to a higher intercourse frequency after wars.

Several causes of changes in the sex ratio have been suggested:

(a) Studies by Schilling *et al.*[93] showed a relationship between the size of sperm cells, their motility and the sex ratio. Spermatozoa of bulls and rams were separated by sedimentation into a bottom and top fraction. Sperm cells of the bottom fraction were larger, their heads 2.4% longer and 5.8% broader as compared with the top fraction. The tails of the spermatozoa of both fractions were the same. The spermatozoa of the bottom fractions moved more slowly due to their greater size. According to Schilling[94] the smaller and faster moving spermatozoa (according to Haag and Werthessen[95] the latter are characterized by a low sulfhydryl content which favours fertility) represent (male determining) Y-spermatozoa. They have a greater fertilization capacity than the larger, slower ones representing (the female determining) X-spermatozoa. Fertilization with the larger spermatozoa would favour the development of female embryos, the smaller ones favour male development. Therefore external factors affecting the ratio of these two spermatozoa types could affect the sex ratio.

(b) In 1925 changes in the sex ratio in animals were reported by Riddle and Fisher[96] in pigeons. In three species of pigeons kept on the same diet throughout the year they observed in the early spring an increased thyrotrophin production and thyroid activity together with an excess of male offspring, whereas during the summer, a period of less activity and smaller sized thyroid, there was an excess of female pigeons. Similar observations on changing sex ratios through the year were made previously by Adler[97] in frogs and by Whitman.[97] Pucek[98] observed great changes in the sex ratio of young, immature specimens of *Sorex araneus* (common shrew) captured at the Bialowicza National Park in Poland. The highest sex ratio (average of 10 years) occurs in June (male/female ratio 1.16 against 1.079 of the whole material), the lowest in August (0.88).

(c) Scholte[99] observed in The Netherlands that the sex ratio of the children of females, who had been irradiated with 270 rads changed giving 6% more females. Paternal irradiation increased the number of males by almost 5%.

(d) Lyster[100] assumed that there is a direct correlation in Great Britain between water hardness and an increase in the sex ratio which may be related to the presence of certain trace elements (lithium, magnesium, etc.) in drinking water.

(e) An interesting hypothesis was put forward by Petersen.[101] He assumed that there is a correlation between sex and the weather condition during the period of conception. Conception during cold periods would give a statistical predominance of males, whereas females would be more common if conception took place during warm periods. An increased metabolic rate of the ovum during fertilization would increase a physico-chemical tendency towards maleness. During cold stress thyrotrophin production increases which (at least in the case of pigeons, see above) would increase the proportion of male offspring, perhaps due to an increased spermatozoa production and a dominance of smaller and faster moving Y-spermatozoa (experiments of Schilling *et al.*[93]). It is evident that this hypothesis needs very careful checking.

(f) Pfaundler[102] and De Rudder[103] observed a change in the ratio after birth between the number of deaths of male and female children. The male surplus at birth gradually disappears due to greater male mortality during the early part of life. Also during summer the mortality sex ratio (males/females) decreases, i.e. more female babies die in summer.

8.5. EFFECT ON MORTALITY

The very high mortality from arteriosclerotic heart diseases and stroke, and the effects of weather and climate on these diseases, was discussed in Section 6.5. Some of the more general studies on the effects of weather and climate on mortality can be summarized as follows.

In Germany, Spann[104,105] studied the total number of deaths (40 417) in the city of Munich during the years 1950–1953, excluding deaths due to accidents, homicide, suicide, etc. In 39 980 cases the exact hour of death was known. The following interesting correlations were found:

(a) Mortality increased during the night and reached a maximum in the early hours of the morning; it decreased during the day and reached a minimum in the early evening;
(b) The mortality rate was higher in winter than in summer;
(c) Mortality was highest during Ungeheuer's weather phase 4, i.e. in periods when there is an influx of new air masses and a steep fall in barometric pressure.

In Great Britain Boyd[106] studied the relationship between respiratory mortality and meteorological conditions during the winter months of 1947–1954, in the Greater London area and in the rural districts of East Anglia. The observed correlations were essentially the same in both areas. A close association was found between weekly deaths from respiratory diseases, in people of 45 years and over, with temperature and humidity particularly if the mortality data were correlated with weather conditions in the preceding week. It was found that mortality increased with falling temperatures. In the London area at temperatures below 0 °C, bronchitis mortality during heavy fog was three times greater than with light fog. This influence of fog on mortality remains the same only up to a temperature of 2 °C. In other words, fog can only be correlated with bronchitis mortality when the temperature is very low.

In the USA, Kutschenreuter[107] studied the relationships between total mortality and temperature in the New York area for the period 1949–1959. The following observations were made:

(a) In the age group 45 and older, mortality was significantly highest on Mondays and lowest on Sundays; in younger age groups this relationship was either not very pronounced or even lacking. This weekly periodicity in mortality is probably a sociological correlation;

(b) During the influenza pandemic (October 1957–March 1958) no correlation was found between mortality, temperature and precipitation;

(c) In other periods, in New York, Los Angeles (Cal.) and Cincinnati (Ohio) the total monthly mortality curve showed an inverse relationship to temperatures: i.e. the highest mortality occurred in the coldest months and vice versa; the colder the winter the higher the mortality; a sudden increase in mortality may nevertheless also occur during very hot summers; the mortality-temperature relationship was only observed for those above the age of 24 and below that of one year; from 1–24 years no death-temperature relationship was observed, perhaps partly due to the relatively low mortality figures for this age group. There was no significant difference between these various temperature/mortality relationships for the white and non-white segments of the population, except that in the case of the whites maximum mortality occurs in the 65 and over group, in the non-whites in the age-group of 45–64;

(d) A statistically significant correlation was found in the New York area from January to June, between days of passage of active front and high mortality; but in October, after the hot weather of the New York summer, the passage of fronts (accompanied by cooler weather) caused a reduction in mortality. This relationship is different from that for western Europe, where the front-mortality relationship peaks in the autumn;

(e) In 1950, Reiter[108,109] studied the statistical relationships in Bavaria between 52 238 mortality cases and days of great disturbance in electromagnetic long waves in the atmosphere. Reiter observed an increase in mortality of from 11–30% on such days as compared with quiet days. Males showed a better correlation than females. However, as days of strong disturbances of the electromagnetic field are usually related to other weather changes (fronts, etc.), it seems more likely that the observed correlations are not causal and that the electromagnetic long waves are only characteristic biometeorological indicators.

Extensive studies on the seasonal effects on human mortality in particular in Japan, were carried out by Momiyama.[110] As she later reported[111] the seasonal variation of total mortality since 1900 in Japan revealed the existence of some quantitative regularities in human mortality:

(a) Deaths occurred most frequently in the hot months (summer concentration) late in the 19th century;
(b) The summer peak steadily flattened, while on the other hand mortality gradually increased in winter in the first decades of the current century;
(c) The winter peak has become higher in contrast to the disappearance of the summer time maximum in the past decades.

According to the seasonal mortality calendars[110] deaths occurred most frequently in summer in the last decades of the 19th century in Japan. Even old age ailments took a heavy toll twice a year, i.e. in summer and winter. A change began to appear in the 1920s, and a complete turnover occurred in the postwar pattern of seasonal mortality calendars: most of the diseases caused frequent deaths in the cold months, whereas dysentery alone occurred in the hot months.

Further studies by Momiyama[112,113] revealed similar regularities in the vital statistics of 18 countries. In the west European countries, the present mortality peak is seen in winter, though chronological changes differ from those of Japan. On the other hand in the United States and the Scandinavian countries, signs of 'deseasonalization' of mortality are seen: apparently the influence of seasonal changes is disappearing. Further details are given in Momiyama's original publications.[110–113]

8.6. EFFECT ON REACTION TIME, PERFORMANCE CAPABILITY AND ACCIDENTS (ROAD AND INDUSTRIAL)

Many studies have been made[114–119] of performance capability under extreme heat stress, e.g. studies in coal mines, heavy industries and weaving mills. In England MacWorth[118] demonstrated the effect of thermal stress on experienced wireless operators. Hot and moist air seriously impaired accuracy. Similar observations were made by Pepler[119] in Singapore.

Several empirical and experimental studies have been carried out on reaction time. Reiter[120–122] tested 53 000 visitors to a Traffic Exhibition in Munich for their reaction speed. A statistically significant decrease (up to 20%) was observed on days with considerable disturbance of certain electromagnetic waves in the atmosphere. However, this relationship is not a causal one, it is only an indicator for certain changes in particular weather patterns. Mücher and Ungeheuer[123] repeated these studies during another traffic exhibition in Munich around 1960.

In Poland during the period 1963–1970 Maczynski[124] compared the daily changes in the optical reaction time of 10 identical subjects (total of 40 000 observations) with changes in the weather and the number of traffic accidents on the same day (total of 6677 traffic accidents). Reaction time measurements were carried out in the same room and at the same hour each day. All three phenomena, optical reaction time, weather and traffic accidents, proved to be inter-related. The relationships were statistically significant. It was found that seasonal changes in reaction time occurred with variations of up to 20 ms. The longer reaction time was observed in June, the shortest in December.

Apart from indirect meteorological effects on road accidents due to cloudiness, rain, snow or glazed frost, which reduce the light intensity and field of vision of the driver and which make roads very slippery, considerable evidence has been collected suggesting direct effects of weather and climate on the physiology, of car drivers and of industrial workers, which under certain circumstances could trigger road and industrial accidents.

As it has been shown that weather and climate affect the reaction time of a subject, it is understandable that meteorological stimuli would also affect industrial accidents. Reiter[120–122] studied 362 000 industrial accidents which occurred in Bavaria, Württemberg and Hessen (Germany) in the period of 1950–1952. A statistical analysis demonstrated a mathematically significant relationship between electromagnetic disturbances in the atmosphere and industrial accidents. On days of considerable disturbance of the long wave electromagnetic radiation the number of accidents increased by 20–25%.

A similar analysis was carried out by Reiter[120–122] of 21 000 traffic accidents which occurred in Bavaria between July 1949 and June 1950. It was found that a statistically significant increase in the number of traffic accidents occurred during days or periods of electromagnetic disturbances.

Considerably more research is needed to clarify these observed relationships.

8.7. SPORT PERFORMANCE

Considering the effects, described in Section 8.6, of the weather and climate on performance capability it is logical to assume that meteorological factors could also affect performances at sports quite apart from the direct effect of thermal conditions on muscular performance. Apart from thermal stresses sport performance is also influenced by altitude. The thermal aspects were studied by Eagan *et al.*[125,126] The altitude effects were particularly studied during the Olympic Games in Mexico City in 1968, and the results have been summarized by Jokl *et al.*[127,128]

8.7.1. Thermal stress effects

In most sporting activities, the physiological stress results primarily from an increased heat production which tends to exceed heat loss; this hyperthermal stress is usually manifested by an increased deep body temperature. But in some cases, the sport endeavour requires performance in the cold where, in spite of increased heat production, the circumstances of the event may cause risk of local cold injury. This threat to the peripheral tissues of the body results from hypothermal stress.

Eagan[126] summarized the principal concepts as follows:

'Sport performance is typically assessed in terms of points per contest, distance jumped, or distance run per unit time. Skill, efficiency and metabolic power are all involved. In events, such as distance running, likely to produce thermal strain, duration and intensity of effort are of prime importance. Skill and efficiency are not unimportant but are less selective in this type of endeavour than is metabolic power. It is the ability to sustain high energy output that most marks the superior performer. The constituent attributes are a high maximum oxygen uptake, the capability to work at a high percentage of maximum oxygen uptake for the duration of the event, and the efficient dissipation of the incidental, but obligatory, heat produced.'

An example is performance in the marathon (42.2 km) where the rate of work is directly related to the speed of running which, in turn, is limited by the level of endurance fitness. It is seen that the power output of a top runner equals about 1500 W, enough power to heat a fair-sized room. This power output, which is 14 times the resting value of 105 W, is maintained for over 2 h. Total energy release is largely independent of the rate of work though there is an apparent 12% increase in total kcal from the slowest to the fastest runner. This would indicate that the fitter, and hence the faster, the runner, the less thermally efficient he is. Åtrand showed that the maximum oxygen consumption for men and women running at the same speed was independent of both sex and level of fitness.

A high level of physical fitness reduces the strain of heat stress. A major part of this advantage is the greater capability of the trained person to sweat, which results in a lower body temperature similar to what occurs with heat acclimatization alone.

Men exercising under conditions of high thermal stress due to high power output tend to drink only 50 to 70% of the amount of water lost in sweat, hence they tend to suffer from dehydration (p. 89). This involuntary dehydration is not of great practical consequence until the water shortage reaches about 3% of body weight in fit persons. But under conditions of sustained endurance performance man cannot depend upon physiological clues to maintain his water balance. If a man deliberately drinks to maintain his body weight within 3% of normal he is relatively hyperhydrated. If he does not, he is hypohydrated and suffers the consequences in reduced performance.

8.7.2. Effects of high altitude

The effect of heavy exercise at altitude depends on the duration of the sporting event. For example the short track races at Mexico City were run conspicuously fast and the long races conspicuously slowly. The extraordinarily good results achieved at Mexico City in the sprints, the hurdles, the long and the triple jump were due to a variety of influences, among them the facilitation of reflexes at altitude. The poor performances in the endurance events seem to be due to the reduced partial oxygen pressure.

According to Jokl *et al.*[127,128] the acute clinical manifestations after exercise at altitude have not been sufficiently studied. During the Olympic Games in Mexico City, several hundred collapses occurred among the highly selected and optimally trained participants. Although many of these collapses were severe, there were no fatalities. The collapses at Mexico City divided themselves into four diagnostically distinct entities: effort migraine, sudden primary loss of consciousness, shock, and powerless attacks: each of them had been known to occur also at sea level. However, their frequency at Mexico City was unprecedentedly high.

(a) *Effort migraine* occurs particularly after 400 m races. Its chief manifestations are the appearance of scintillating scotomas (i.e. appearance of dark, vanishing, cloudy patches before the eyes), followed by nausea, vomiting and severe throbbing headache.

(b) *Sudden primary loss of consciousness* occurs during or after exhausting long-lasting athletic performances. The unconsciousness may last from a few minutes to one hour.

(c) *Shock* was characterized by cyanosis of lips and nose, not often encountered at sea level, and a sharp drop in arterial blood pressure in several cases to a level of 66 mm Hg (systolic). If the upright posture was assumed prematurely, the subjects fainted. Several swimmers collapsed in this way during the flag ceremonies.

(d) *Powerless attacks* take place after great emotional excitement. One of the Olympic champions collapsed when he was told the results of his world-record long jump.

Apart from the direct clinical manifestations the following three aspects of sport performance at altitude must be considered.

(a) *Influence of altitude adaptation*: the clinical manifestations after heavy exercise at high altitude are considerably less if the subject is adapted to high altitude after a three weeks light training above 1500 m altitude or after low pressure climatic chamber treatments (see Section 9.3.1).

(b) *Effects of drugs*: they could affect the sport performances in two ways. As shown in Section 8.3 certain drugs could have a very toxic effect if consumed at high altitude. On the other hand diuretics,[129] metenolanacetate,[130] dianabol[131] and other anabolic substances (anabolism is a constructive metabolism during which an organism converts simple compounds into higher organized substances) seem to improve the performances of athletes at altitudes above 1500 m.

(c) *Effects of altitude and exercise on infections*: it is known (see Section 5.3.1 (vii)) that if a subject moves to altitudes above 1500 m certain infections caught at sea-level may suddenly appear explosively at altitude, e.g. respiratory and gastro-intestinal infections, appendicitis etc. Also the resistance against viral infections is lowered after strenuous physical exertion.

All these factors should be taken into consideration in the organization of sport events in hot areas or at high altitude.

8.8. RESTLESSNESS OF SCHOOL CHILDREN

A few studies have been made on the restlessness of school children as a function of meteorological conditions. Studies by Koch *et al.*[132] indicated that, in Leipzig during the period November 1964–January 1965, a better mental performance during school tests was observed with young children during low atmospheric pressure conditions. Ungeheuer[133] observed greater restlessness of children during sudden changes in the weather. Unrest before thunderstorms is probably not due to changes in electric conditions in the atmosphere but due to the accompanying thermal stresses (heat and humidity) which affect the inefficient thermoregulatory system of young children.

Although the results of animal studies cannot be applied automatically to man, studies[134] during 4½ years in Zürich (Switzerland) in which the daily average locomotor activity of ten mice living under standard conditions were compared with temperature and humidity conditions outdoors, indicated highest motility values during spring and summer, lowest in autumn. In each season the motility was highest during the transition to undisturbed weather, lowest during the transition to weather phase 5 of Ungeheuer (cold front). The mice were not exposed to humidity and temperature outdoors, but were only affected by the natural changes in daylight during the season.

Studies by Rieger[135] (1970) indicated a circadian cycle of motor activity in 26 subjects who lived for 23 days in an underground isolation chamber in Erling-Andechs (Bavaria, Germany). The total circadian cycle was 24.99 h, the average activity time was 16.56 h, average rest-time 8.43 h. The fact that the whole period is almost 1 h longer than the usual 24-h period suggests an endogenous origin for this periodicity.

8.9. REPRODUCTION IN HUMANS

The only scientific analysis of the relation of reproduction to climate was carried out by Macfarlane[136] in Australia. His final conclusions can be summarized as follows:

'Human conception rates change with the season and the greatest fluctuation is in the warm temperate to subtropical latitudes. There is less seasonality among populations in the subarctic and equatorial regions. Latitude and altitude affect the general pattern; and conception seasonality is not the same in all places at a given latitude.

Island populations show a smaller fluctuation than mid-continental areas except Japan, which has a high spring maximum. Of the cool temperate continental areas there is a summer depression of conception in the United States, while in Europe and Australia there is a high summer rate of conception and a winter minimum.

Seasonal conception rate patterns found in the 1930s changed little in the 1960s, except that the rates in under-developed areas have moved towards the distribution of the more urbanized communities. In spite of increasing control of the microhabitat of man there is little reduction of seasonality and in some places the fluctuation has increased.

Negroes in the United States have a greater seasonality of conception than whites, but in New Zealand the Maoris show less seasonal change than Europeans. Race as such appears to have little effect on seasonal variability of conception.

The average maxima and minima of conception in cool temperate latitudes occur at mean monthly temperatures of 16 °C (maximum) and 3 or 11 °C (minimum). In the warmer temperate zones, maximum conception rates occur at 13.6 °C and the minimum at 23 °C. In the tropics maxima are at 26 °C and minima at 28 °C (though near the equator the mean temperatures may be reversed). Temperature as such cannot explain all conception seasonality, but humidity in the hotter regions (above 25 °C) is important in depressing conception.'

According to Macfarlane[136] the possible neuroendocrine mechanism involved is the following: an increased release of thyroxin and of cortisol resulting from the cooling of the body in a cold climate. Both substances increase metabolism and may be associated with reduced gonadotrophic activity. Most human populations use clothing, heating and shelter to prevent any undue cooling of the body, so this is not likely to be a serious factor in modifying conception in the large groups registering seasonal changes of conception. In Japan, however, where buildings have been in the past lightly constructed and where indoor heating is minimal, cold may have contributed more to the low winter rate of conception, through hormone responses to the environment.

When the temperature rises above 30 °C there is a reduction in energy turnover together with lowered thyroxin and cortisol output. Pennycuik[137] and Macfarlane[138] pointed out that the effect of the chronic exposure of rats with adequate nutrition to a hot environment is the reduction of the overall level of pituitary activity (a mild hypopituitarism with slow growth, low metabolism, poor fertility and reduced lactation). Associated with this there is a reduction in ovulation as well as in implantation of spermatozoa and foetuses. The number of foetuses resorbed in the early stages is also increased as a result of exposure to high temperature, even when all nutrient needs have been adequately supplied. These effects could be due to lowered pituitary outputs of follicle stimulating and later of luteinizing hormone. When the high temperatures become excessive, cortisol is likely to be released and this in turn leads to foetal resorption. This cortisol release is more likely when there is rapid cooling in autumn, or during the high contrasts of warm and cold in spring. However, whether such mechanisms are at work in man is uncertain. Considerably more research is required to explain the observed weather relationships with reproduction.

REFERENCES

1. S. W. Tromp (Ed), *Progress in Biometeorology II*, Swets & Zeitlinger, Amsterdam, 1977.
2. Kevan, in Ref. 1, p. 160.
3. M. M. Fairgrieve, *Proc. Roy. Soc. (Edin.)* **41**, 150 (1921).
4. R. Pintner, *J. Appl. Psychol.* **15**, 149 (1931).
5. *idem* and G. Forlano, *J. Educ. Psychol.* **24**, 561 (1933).
6. H. N. Fiaklin and R. O. Beckman, *J. Genet. Psychol.* **52**, 203 (1938).
7. A. D. M. Davies, *Brit. J. Psychol.* **55**, 475 (1964).
8. A. M. MacMeeken, *The Intelligence of a Representative Group of Scottish Children*, Univ. Lond. Press, London, UK, 1939.
9. O. C. Held, *J. Genet. Psychol.* **57**, 211 (1940).
10. A. B. Fitt, *Educ. Res. Serv. (N.Z. Coun. Educ. Res.)*, 17 (1941).
11. C. A. Mills, *Human Biol.* **13**, 378 (1941).
12. J. A. F. Roberts, *Brit. Med. J.* **1**, 320 (1944).
13. F. S. Dunlop, *Cancer J. Psychol.* **1**, 87 (1947).
14. H. C. Gordon and B. J. Novak, *Science* **112**, 62 (1950).
15. J. B. Gibson, *Nature (London)* **246**, 498 (1973).
16. M. D. Armstrong and F. H. Tyler, *J. Clin. Invest.* **34**, 76 (1955).
17. B. Berliner, *Veröff. Zentral Anst. Balneol.* **2**, 1 (1920).
18. *idem*, *Veröff. Med. Verw.* **15**, 3 (1920).
19. A. Langelüddeke, *Veröff. Zentral Anst. Balneol.* **2**, 57 (1925).
20. M. Muchow, *Z. Paedagog. Psychol.* **27**, 18 (1926).
21. Petrasch, *Z. Angew. Psychol.* **34**, 47 (1929).
22. E. Huntington, *Season of Birth: Its Relation to Human Abilities*, Wiley, New York, USA, 1938.
23. G. Forlano and V. Z. Ehrlich, *J. Genet. Psychol.* **32**, 1 (1941).
24. F. H. Farley, *Brit. J. Psychol.* **59**, 281 (1968).
25. P. P. A. M. Kop and B. A. Heutz, *J. Interdiscipl. Cycle Res.* **5**, 18 (1974).
26. M. Tramer, *Arch. Neurol. Psychiat.* **24**, 17 (1929).
27. T. Lang, *Arch. Rassenbiol.* **25**, 42 (1931).
28. W. J. J. de Sauvage Nolting, *Ned. T. Geneesk.* **79**, 528 (1934).
29. *idem, Int. J. Biometeorol.* **12**, 293 (1968).
30. H. Henne, Thesis, Univ. Tübingen, Germany, 1949.
31. H. Knobloch and B. Pasamanick, *Amer. J. Publ. Hlth.* **49**, 1201 (1958).
32. N. J. Bradley and R. J. Lucerno, *Amer. J. Psychiat.* **115**, 343 (1958).
33. H. Barry and H. Barry Jr., *Arch. Gen. Psychiat.* **5**, 1292 (1961).
34. P. Dalen, *Acta Psychiat. Scand.* **203**, 55 (1968).
35. D. Stur, *Klin. Wschr.* **65**, 898 (1953).
36. F. M. Allan, *Lancet* **1**, 439 (1964).
37. B. K. S. Dijkstra, *J. Nat. Cancer Inst.* **31**, 511 (1963).
38. Keogh, private communication.
39. S. W. Tromp and J. J. Bouma, *Monograph Series Biometeorological Research Centre, No. 12*, Leiden, The Netherlands, 1973.
40. S. W. Tromp, *Proc. 6th Int. Biometeorol. Congr.*, Noordwijk, The Netherlands, 3–9 Sept., 1972.
41. J. E. C. Schook and C. F. Schüller, *Ned. T. Geneesk.* **118**, 837 (1974).
42. M. J. Edwards, *Arch. Path.* **184**, 42 (1967).
43. *idem*, *J. Path.* **104**, 221 (1971).
44. W. Petersen, *J. Obstet. Gynec.* **28**, 70 (1934).
45. Brock, in C. O. Carter, *Proc. Roy. Soc. Med.* **69**, 38 (1976).
46. E. G. Knox, *Brit. J. Prev. Soc. Med.* **26**, 219 (1972).

47. J. Overbeke, Thesis, Univ. Utrecht, Netherlands, 1971.
48. T. McKeown and R. G. Record, *Lancet*, 192 (1951).
49. J. H. Edwards, *Brit. J. Prev. Soc. Med.* **12**, 115 (1958).
50. MacMahon *et al., ibid.* **5**, 254 (1951).
51. *idem ibid.* **7**, 211 (1953).
52. M. J. W. De Groot, *Huisarts en Wetenschap (Netherlands)*, **8**, 121 (1965).
53. J. Frezal, J. Kelly, M. L. Guillemot and M. Lamy, *Amer. J. Hum. Genet.* **16**, 336 (1964).
54. I. Leck and R. G. Record, *Brit. J. Prev. Soc. Med.* **20**, 67 (1966).
55. S. L. Silberg, F. R. Watson and J. C. Martin, *Canad. J. Publ. Hlth.* **59**, 239 (1968).
56. J. H. Elwood, *Brit. J. Prev. Soc. Med.* **24**, 78 (1970).
57. *idem*, *Develop. Med. Child. Neurol.* **12**, 582 (1970).
58. A. M. Lilienfeld and B. Pasamanick, *J. Amer. Med. Ass.* **155**, 719 (1954).
59. G. Badtke, K. H. Degenhardt and O. Lund, *Z. Anat. Entivickl. Gesch.* **121**, 71 (1959).
60. A. Rutt and K. H. Degenhardt, *Arch. Orthop. Unfall. Chir.* **51**, 120 (1959).
61. K. H. Degenhardt, *Z. Menschl. Vererb. Konst. Lehre*, 1959.
62. A. Werthemann, M. Reiniger and N. Thoelen, *Schwerz. Z. Allg. Path.* **13**, 6 (1950).
63. J. Langman and F. Van Faasen, *Ned. T. Geneesk.* **99**, 2119 (1955).
64. A. Polman, *T. Soc. Geneesk. (Netherlands)*, 393 (1947).
65. *idem*, *Genetica*, **25**, 29 (1951).
66. W. Landauer, *Extrait Arch. Anat. d'Hist. et d'Embr. Normales et Expérimentales*, 155 (1962).
67. L. W. Sontag, W. G. Steele and M. Lewis, *Human Develop.* **12**, 1 (1969).
68. L. W. Sontag and R. F. Wallace, *Child Develop.* **6**, 253 (1953).
69. J. R. Paul, *Clinical Epidemiology*, Univ. Chicago Press, Chicago, 1966.
70. J. Fedrick, *Nature (London)*, **227**, 126 (1970).
71. *idem*, *Ann. Human Genet. (London)*, **34**, 31 (1970).
72. E. C. Knox, *Brit. J. Prev. Soc. Med.* **26**, 219 (1972).
73. J. H. Renwick, *ibid.* **26**, 67 (1972).
74. *idem*, *Lancet* 336 and 967 (1972).
75. *idem ibid.* 96 (1973).
76. *idem*, *Brit. Med. J.* 172 (1973).
77. Martin, in refs. 75 and 76.
78. D. E. Poswillo, D. Sopher and S. J. Mitchell, *Nature (London)* **239**, 462 (1972).
79. E. Findersen, *Bioklim. Beibl.* **10**, 23 (1943).
80. D. I. Macht, *Amer. J. Pharm.* **106**, 135 (1934).
81. J. D. Marshall, *Hygiene* **57**, 484 (1959).
82. A. J. Nedzel, *J. Lab. Clin. Med.* **23**, 1063 (1938).
83. K. Ossoinig, *Mschr. Kinderheilk.* **31**, 371 (1926).
84. W. F. Petersen, *The Patient and the Weather, Vol. II*, Edwards, Ann Arbor, Michigan, USA, 1935.
85. F. I. Sargent and A. J. Nedzel, *Biokl. Beibl.* **6**, 26 (1939).
86. S. W. Tromp, *Medical Biometeorology*, Elsevier, Amsterdam, The Netherlands, 1963, p. 585.
87. F. A. Fuhrman, *Arid Zone Res. Rep. (UNESCO, Paris)* **22**, 223 (1963).
88. K. J. Furman, in *Survey of Human Biometeorology*, F. Sargent II. and S. W. Tromp (Eds), W.M.O. Tech. Note Ser. 65, 1964.
89. W. H. Weihe, in *Proc. Symp. Pharmacol. and Thermoreg., San Francisco, 1972*, Karger Verlag, Basle, 1973, p. 409.
90. *idem*, *Ther. d. Gegenw.* **104**, 1179 (1965).

91. in ref. 1, p. 371.
92. J. I. Goodwin, *Lancet* **1**, 652 (1971).
93. E. Schilling *et al., Z. Tierzucht. Tierzuchtungsbiel.* **83**, 331 (1967).
94. E. Schilling, *Zuchthyg. Fortpflauz.-Stör. Besam. Haustiere* **1**, 5 (1966).
95. F. M. Haag and N. T. Werthessen, *Fertil. Steril.* **7**, 516 (1956).
96. O. Riddle and W. S. Fisher, *Amer. J. Physiol.* **72**, 464 (1925).
97. L. Adler, *Arch. Exp. Path. Pharmakol. Naunyn-Schiedeberg's* **86**, 159 (1920).
98. Z. Pucek, *Acta Thermol.* **3**, 43 (1959).
99. P. J. L. Scholte, Thesis, Univ. Leiden, The Netherlands, 1973.
100. W. R. Lyster, *Science J.*, Nov. 1970. p. 61.
101. W. F. Petersen, *Man, Weather and Sun*, Thomas, Springfield, Illinois, USA, 1947.
102. Pfaundler, *Z. Kinderheilk.* **64**, 1 (1943).
103. B. De Rudder, *Grundriss einer Meteorobiologie des Menschen*, Springer Verlag, Heidelberg, West Germany, 1952.
104. W. Spann, *Dtsch. Med. Wschr.* **82**, 251 (1957).
105. *idem*, *Germ. Med. Monthly* **2**, 116 (1957).
106. J. T. Boyd, *Proc. Roy. Soc. B* **53**, 107 (1960).
107. P. H. Kutschenreuter, *Trans. N.Y. Acad. Sci. Ser. 2*, **22**, 126 (1959).
108. R. Reiter, *Arch. Meteor. Geophys. Bioklimatol., Ser. B. IV*, **3**, 327 (1953).
109. *idem*, *Strahlenther.* **89**, 682 (1953).
110. M. Momiyama, *Papers Meteor. Geophys.* **14**, 190 (1963).
111. *idem*, in *Progress in Human Biometeorology, Part IB*, S. W. Tromp (Ed), Swets & Zeitlinger, Amsterdam, The Netherlands, 1974, p. 521.
112. *idem*, *J. Meteor. Soc. Japan*, **47**, 466 (1969).
113. *idem*, *Geog. Rev. Japan*, **42**, 1 (1969).
114. H. C. Weston, *Rep. Ind. Fatigue Res. Board*, 20 (1922).
115. H. M. Vernon *et al.*, *ibid.*, 39 (1927).
116. K. N. Moss and J. L. Halls, *Brit. J. Ind. Med.* **90**, 62 (1935).
117. T. Bedford, *Report to Dept. Hlth. India*, 1946.
118. N. H. MacWorth, *Brit. J. Ind. Med.* **3**, 143 (1946).
119. R. D. Pepler, *Rep. Clim. Effic. Sub-cttee., R.N. Pers. Res. Cttee. T.R.U.* 3/57 (1953).
120. R. Reiter, *Med.-Meteor. Hfte.* **9**, 35 (1954).
121. *idem*, *Munch. Med. Wschr.* **17**, 479 (1954).
122. *idem, ibid.* **18**, 526 (1954).
123. H. Mücher and H. Ungeheuer, *Perceptual Motor Skills* **17**, 163 (1961).
124. B. Maczynski, *Int. J. Biomeorol., Suppl.* **16**, 75 (1972).
125. C. J. Eagan, *Fed. Proc.* **26**, 775 (1967).
126. *idem* in ref. 1, p. 282.
127. E. Jokl and P. Jokl, *Exercise and Altitude*, Karger Verlag, Basle, Switzerland, 1968.
128. *idem* and D. C. Seaton, *Int. J. Biometeor.* **13**, 309 (1969).
129. I. Singh *et al.*, *New England J. Med.* **280**, 175 (1969).
130. E. Albrecht and H. Albrecht, *Pflügers Arch. Ges. Physiol.* **293**, 1 (1967).
131. M. Steinbach, *Sportarzt. Sportmed.* **19**, 485 (1968).
132. I. Koch *et al.*, *Arch. Phys. Ther. (Lpz.)* **18**, 435 (1966).
133. H. Ungeheuer, *Ber. Dtsch. Wetterdienstes, Bad Kissingen* **16**, 3 (1955).
134. Stille, H. Brezowski and W. H. Weihe, *Artzneimittel-Forsch.* **18**, 892 (1968).
135. P. Rieger, *Int. Z. Ang. Physiol.* **28**, 332 (1970).
136. W. V. Macfarlane, in ref. 111, p. 557.
137. P. R. Pennycuik, *Austral. J. Biol. Sci.* **17**, 245 (1964).
138. W. V. Macfarlane, *Int. J. Biometeorol.* **14**, 167 (1970).

CHAPTER 9

THERAPEUTIC APPLICATIONS OF BIOMETEOROLOGICAL EFFECTS

The various meteorotropic phenomena described in the previous sections, and the study of the different physiological mechanisms involved, have made it possible to develop therapies for many meteorotropic diseases.

The different methods which have been suggested can be grouped together under three headings: General Climato-Therapy, Aerosol Therapy and Climatic Chamber Therapy (Low and High Pressure Therapy respectively).

9.1. GENERAL CLIMATOTHERAPY

In the previous sections we have tried to demonstrate the profound influence of the weather and climate both on healthy humans and on the diseased person. As each climate has its specific meteorotropic effects, the obvious course is to select certain areas with very specific climatic conditions for the treatment of certain diseases. Such areas are known as climatic health resorts, also known as Kurort (in German-speaking countries) or Spa (in English-speaking countries). Four main areas have been selected for climatotherapy: areas along the coast (the treatment is known as Thalassotherapy); areas surrounding thermal springs (the treatments are called Balneotherapy, the area of treatment is specifically known as a Spa); and mountainous, not windy areas above 1500 m altitude where the treatment is known as High Altitude Therapy; a fourth possibility of climatotherapy is the transfer of large groups of people (or single persons) from unfavourable areas to areas with favourable weather and climate conditions.

9.1.1. Thalassotherapy*

The most important meteorological factors of a sea climate which may affect the physiological processes in man are discussed on p. 26. In view of the

*The name derives from the Greek word '*thalassa*' = sea.

climatic characteristics of coastal areas, certain patients should be advised not to live near the sea coast, whereas it may be beneficial to others. Apart from a climatic effect, sea bathing has also a balneological effect which will be discussed in Section 9.1.2. The meteorological effects are mainly due to the air turbulence and cooling near the coast and to the excessive solar radiation the biological effects of which are discussed on p. 99.

Air turbulence and cooling are greater near the coast. Patients suffering from diseases associated with a dysfunction of the thermoregulation mechanism, such as bronchial asthma and various rheumatic diseases, should preferably not undergo thalassotherapy. The cold stress, due to cooling air and in particular, cold sea-water bathing, increases diuresis and the corticosteroid level in urine. The latter factor may enhance the general resistance of the body to infections[1] but for patients suffering from kidney diseases the increased diuresis may have unfavourable effects. It is evident that in warm countries with little air turbulence a sea climate may benefit asthmatics and rheumatics.

Whereas these cooling effects are harmful to certain diseases, they invigorate the thermoregulation mechanism of healthy persons and strengthen their resistance to infectious diseases.

Excessive ultraviolet radiation near sea water may be harmful to patients suffering from an excessive secretion of gastric acid, but it seems to stimulate the appetite of healthy persons spending a holiday near the coast; the time required for food to pass through the intestines appears to be shorter. According to Ellinger,[2] in hyperthyroid persons exposure to ultraviolet may cause an exacerbation of their hyperthyroidism. The effect of excessive solar radiation on the pituitary along the coast of Jugoslavia was demonstrated by Milin[3] (see p. 105). The beneficial effect of ultraviolet radiation on blood-pressure, haemoglobin content, calcium and phosphate levels of the serum and on protein metabolism, and its destructive effect on micro-organisms are favourable features for many patients living in a sunny sea climate.

According to several workers[4–7] a 40–50% increase in basal metabolism occurs in people visiting the coast for some weeks.

Owing to these various climatic stress conditions, some people, especially those with a non-stable autonomic nervous system, may be thrown out of balance during the first week of their stay at a sea-side resort. It can be shown[8] that two to three weeks are generally required at the coast for complete acclimatization and that some patients (particularly children) may need as much as 5–6 weeks. This may explain why a short holiday at the coast can be detrimental, whereas an extended stay is decidedly curative in some cases. According to Häberlin[4] it is also very important to consider the typology of the patient (see Section 5.13.4) because the same thalassotherapy may have opposite results in different types of human subjects.

The application of thalassotherapy dates back to ancient Greek and Roman times. During the Middle Ages, and up to about 1700, however, thalassotherapy was practically unknown. It was mainly due to the English physician

Richard Russell (1700–1771) that the first extensive study was made of the therapeutic effect of sea water and a sea climate. Two English physicians, John Cookley Lettsome (1744–1815) and John Latham (1761–1843) continued his work during the late 18th century. This led to the building of the first thalassotherapeutic sanatorium in 1796, with 30 beds, in south-eastern England at Margate. As the physicians pointed out 'it was for the relief of the scrophulous poor of London and of all England'.

A number of therapeutic effects of health resorts on the coast have been reported. The following summarizes briefly the most important findings.

(a) *Blood diseases.* Anaemia, which is characterized by a decrease in the erythrocyte and haemoglobin content of the blood, is often successfully treated at the sea coast.[9–13] Häberlin[10] even reported an increase in haemoglobin of from 30–70% in anaemic patients. The average number of erythrocytes may increase by half a million. Conradi[13] says that an abnormal leucocyte level is also quickly restored as a rule. The blood sedimentation rates are reduced.[12] Successful treatment of anaemia is observed only in its temporary forms. In the case of pernicious anaemia, leucaemia etc. a sea climate apparently has an aggravating effect.[14,15] This view is supported by various reports suggesting a higher incidence of these diseases near the coast.

(b) *Diabetes*. In Warnemünde (Germany) Catrein[16] observed an increase in the blood sugar level of 10–40% in healthy persons after a period of sea bathing. Therefore, most workers in the field of thalassotherapy do not advise diabetics to go in for cold sea bathing. On the other hand, wind-protected sun bathing usually lowers the blood sugar level.

(c) *Goitre.* Thalassotherapy is a highly successful treatment for simple goitre, but according to Ellinger[2] hyperthyroidism is apt to be exacerbated by over-exposure to sunlight in cases of exophthalmic goitre.

(d) *Heart diseases.* Considering the influence of strong cooling on various heart diseases, in general thalassotherapy cannot be recommended for heart patients unless the bathing takes place in swimming pools of warm sea water protected from the wind. However, as ultraviolet radiation often has a favourable effect on the blood pressure of hypertensic subjects a holiday on the coast may with due discretion be of benefit even to heart patients, particularly if psychological stress is the main cause of the heart trouble. A slow and gradual adaptation to cold brief sea-water bathing may have a very stimulating effect under these specific conditions. Berg and Zeplin[17] state that the blood pressure of hypertensives adapted to the cold can be seen to drop by 10–20 mm Hg in 20 minutes after a short spell (5–8 min) of sea bathing.

(e) *Nervous and mental diseases.* The majority of nervous patients should be advised to avoid a marine climate. In many instances a forest climate may have a soothing effect on the patient. Nicolas,[18] Leblanc[19] and Curschmann[20] noticed that patients suffering from migraine responded well to sea air. Epileptics and patients suffering from chorea infectiosa (or *St*

Vitus' dance, a nervous disease with involuntary irregular movements, particularly common in young girls aged between 5 and 15, and due to a disturbance of the corpus striatum in the brain) should be advised not to spend a long holiday at sea resorts because of the strong cooling effects and intensive sunlight which, on the other hand, have a favourable (i.e. quieting) effect on schizophrenics.

According to Curschmann,[20] Laignel-Lavestine[21] and Krauel,[22] nervous patients with an overstimulated parasympathetic nervous system are favourably affected by sea climate. Its influence on the mental efficiency of school children was briefly described on p. 216

Although part of the beneficial effects of a sea climate is entirely due to psychological factors (lack of responsibilities, gaiety of beach life, etc.), a number of direct meteorotropic influences on mental and nervous diseases cannot be denied. It is this observation which made Huntington[23,24] believe that people living near the ocean are more productive and energetic than populations living in the centre of continents. He suggested that this could account for the desire of coastal nations all through history to explore new parts of the world.

(f) *Respiratory diseases*. Most physicians seem to agree that lung tuberculosis and serious cases of chronic bronchitis should not be treated on the coast. Asthmatics (bronchial asthma) living near the coast in windy areas, with much fog and considerable daily fluctuations in environmental cooling, often improve considerably after moving a few miles inland. Sea bathing, on days when there is little or no wind, is reported to have favourable effects on asthmatics because the bathing in cold water trains the thermoregulation system and stimulates the orthosympathetic nervous system. Also if bronchial asthma is partly due to allergens a sea climate may be beneficial because of the smaller number of allergens present in sea air. The same applies to hay-fever and allergic rhinitis.

Chronic nose and throat catarrhs usually improve rapidly near the coast or during a sea voyage due to the presence of salt aerosols in the air.

Owing to the greater cooling effects near the coast, the thermoregulation mechanism of a normal person is trained during thalassotherapy, and this may be one of the reasons why people living near the sea suffer less from the common cold.

(g) *Skin diseases*. Häberlin and Roeloffs[25] reported favourable effects of a sea climate on skin tuberculosis (lupus vulgaris). Häberlin, and others[25–27] are of the opinion that abdominal tuberculosis and osteotuberculosis can also be successfully treated at seaside sanatoria in western Europe, particularly during the winter season.

Thalassotherapy improves the blood circulation in the skin and, therefore, its elasticity. After 6 weeks the colouring of the skin after cold application is considerably more pronounced than shortly after arrival at the sea coast.

According to Döhring[28] and Häberlin[25–27] various forms of eczema (eczema pruriginosum,* a chronic skin disease, marked by small, pale, papules and intense itching), allergic dermatoses, neurodermatitis, psoriasis, furunculosis, etc. (see Section 6.10), respond to thalassotherapy. Considerably more research has to be done to confirm the reports of the Kurort and Spa physicians who are often biased in their interpretations of the results obtained.

9.1.2. Balneotherapy†

The therapeutic effects of Balneotherapy can be classified into three groups of factors:

(a) *Influence of the local spa climate*, also known as balneo-climatotherapeutic treatment.

(b) *Influence of hydrological treatment*, also known as balneological treatment. This consists of two different treatments:

 (i) bathing in thermal waters, i.e. external balneological treatment; this treatment, which consists of a thermal, a physico-chemical and a hydrostatic stress on the skin, is often combined with massage, aerosol inhalation, etc.; mud-baths are sometimes used instead of water;

 (ii) drinking thermal waters, i.e. internal balneological treatment.

(c) *Influence of psychological factors.* The influence of these factors, which comprise the effects of a holiday without any responsibilities, the healthy life (the patient goes to bed very early and wakes up rather late, in the afternoons he takes another nap), the gaiety of many spas with concerts and so forth, is so important that without considerable research it is impossible to determine whether an observed improvement is really due to the actual balneological treatment, either hydrological or meteorological or both. Despite the often rather unscientifiic approach of many spas, the interest of the general population and of physicians in spa-treatment is enormous, particularly in countries like Belgium, France, Germany, Austria, Switzerland, Italy, Poland, Czechoslovakia, Yugoslavia, Roumania and the USSR.

9.1.2.1. Climatotherapeutic effects

A patient visiting a spa for a short period is only subjected to the daily changes in the local weather conditions, but if he stays for a considerable time, an important factor is the average daily meteorological conditions, i.e. the local microclimate of the spa. The daily weather and microclimatic conditions of a spa are closely related to its general location. In mountainous areas the general weather conditions and the biological effects of a mountain climate (see Sec-

*Also known as prurigo.

†The name derives from the Latin word '*balneum*' = bath. For references see *Medical Biometeorology*, pp. 955–956.

tion 1.5.1) will prevail, near the coast that of a maritime climate (see Section 1.5.3); inland it may be a forest climate (see Section 1.5.2) and so forth. As practically all the climatic zones can be found in many large countries, a scientific study of the biometeorological characteristics of the various spas in such a country would enable a physician to find a favourable one for almost every meteorotropic disease.

9.1.2.2. Hydrological effects

As indicated above balneotherapeutic treatments consist of an external and internal balneological treatment. The external hydrological treatment of a subject has three physiological effects:

(a) *Influence of thermal stimuli*: it has been pointed out that, according to Zimmerman[29] and others, an increased 17-ketosteroid secretion can be observed after very warm thermal baths and after sauna baths; heat stress in thermal baths also affects the oxygen exchange, general circulation, cardiac output, blood pressure, pulse rate, peripheral resistance, dehydration, etc.[30] The effects of such a thermal bath treatment will depend on the hour of the day (i.e. the biological rhythm), sex and age of the patient, his personality pattern, the season and the degree of acclimatization.

(b) *Influence of physico-chemical properties of the water*: experiments by Buettner[31] and others have shown that the outer horny layer of the skin, the *stratum corneum*, is not as impermeable as one may be inclined to believe. Experiments in heavy water have shown that a foot or arm placed for one hour in 100 ml of water may absorp 2 ml. A small portion remains in the epidermis but the major part enters the system and is finally excreted in the urine. Not only water, but also water vapour can penetrate the skin. The water and vapour intake increases with increased temperature and activity of the body. A high concentration of salt, exceeding 5% of sodium chloride, causes a decreased water uptake. With 10–15% of sodium chloride, the transfer ceases. Stronger solutions draw water out of the skin.

In normal sea water, used during the thalassotherapy, the total salt content is about 33–37g l^{-1} (33 in northern polar seas, 34 in the Antarctic, 36 in the tropical Pacific and Indian Oceans, 37 in the tropical Atlantic Ocean). The various salt percentages are the following:

Sodium chloride	77.7
Magnesium chloride	10.8
Magnesium sulfate	4.7
Calcium sulfate	3.6
Potassium sulfate	2.6
Calcium carbonate	0.34
Magnesium bromide	0.22

A thermal spring in the well-known Kurort Baden (Switzerland) for example produces 4.5 g of salts l^{-1} at a temperature of 48°C, comprising about 2 g of sodium chloride, apart from many other salts of calcium, magnesium, potassium, lithium, strontium, etc. In German law, only springs with more than 1 g of salts l^{-1} of water or with temperatures above 20°C can legitimately be called 'curative springs' (Heilquellen). Other curative springs should be called 'mineral springs'.

In view of the fluid penetration into the skin, as shown by trace elements which penetrate into the body, the use of hormonal products in bath salts is dangerous. Particular chemical compounds, which have a solvent effect on fatty substances of the skin, can penetrate more easily. According to various authors particularly carbon dioxide, hydrogen sulfide, arsenic acid (H_3AsO_4), thiosulfuric acid ($H_2S_2O_3$), chlorides, bicarbonates and various salts of heavy metals (such as iron, copper, cobalt, manganese) belong to this group of compounds. A hypertonic thermal water will also cause a shrinkage of the body cells by dehydrating the cells. According to Terrier[32] thermal water containing 0.45–0.48 ml l^{-1} of hydrogen sulfide would cause a penetration of sulphur into the blood circulation.

According to Goldzieher[33] there seems to be a close relationship between the adrenal cortex, liver function and sulfur metabolism. This could be one of the reasons for the supposedly favourable effects of thermal sulfur springs on certain rheumatic diseases. A similar explanation has been given for the favourable effects of lithium waters which would assist in dissolving uric acid compounds in *podagra* (gouty pains in the big toe).

The penetration of thermal water through the skin will affect also the diuresis and functions of the kidney; most likely the penetration of solutions through the skin affects also the peripheral nervous system, but no accurate data are available.

Thermal water will also affect the acidity of the skin.

(c) *Influence of hydrostatic pressure*: it has been suggested that the hydrostatic pressure in a bath may also have important physiological effects. The circumference of the abdomen during thermal baths may be reduced by 3–6 cm. The pressure on the extremities affects the blood flow and function of the heart.

The problem of the reduced weight of a body floating in water has been studied, notably in space medicine (Section 5.9). The floating condition is described as a hypodynamic environment, the study of the physiological changes during prolonged immersion of a body as a result of weightlessness is called hypodynamics. The biological effects of the immersion of the body in water have been described already but so far only the effects of external hydrological treatment have been discussed. However, during the treatment at spas, patients are also advised to drink the mineral water.

This internal hydrological treatment may considerably affect the final outcome of the climatological treatment.

Drinking thermal waters may admit two kinds of substances to the stomach, fluids and gases.

(a) *Gases*. An important gas in mineral springs is radon which is present in the atmosphere in very small quantities but which may occur in considerable quantities in certain thermal springs. The drinking of such water or the inhalation of aerosols prepared with this water is known as 'radon therapy', 'emanotherapy' or 'radium emanation therapy'. The aerosols usually contain less than 80 Mache units (MU) l^{-1} of air, the fluids 800 MU l^{-1}. Engelmann,[33] who studied the matter in 1920 at the Physiological Institute of the University of Frankfurt, suggests that the radon radiation stimulates cell functioning and, in particular, the oxidizing processes. Thermal waters rich in carbon dioxide (i.e. at least 1000 mgl^{-1} CO_2) apparently stimulate gastric secretion and the digestion of food.

(b) *Fluids*. The fluids which are drunk are usually complex mixtures of various salts. Both the chemical composition and pH of these fluids may considerably affect the stomach and intestines.

Alkaline waters (preferably taken before meals) seem to bring alleviation to ulcer patients suffering from excessive secretion of gastric acid. They should not, however, be taken regularly by healthy persons. Changes in the alkalinity of the stomach also affect the blood serum, and diabetics have sometimes been advised to take alkaline water as it inhibits pancreatic secretion and neutralizes acidosis.

Acid waters (usually rich in bicarbonate and calcium) appear to increase the calcium content of the blood of people with a low calcium level, but do not affect a healthy person. Thermal waters rich in sodium sulfate and magnesium chloride or sulfate cause accelerated digestion; magnesium chloride waters stimulate bile secretion by the liver and assist both the relaxation of Oddi's sphincter (the muscle closing the common bile duct) and the tonic contraction of the gall bladder. Bile salts facilitate the digestive action of all pancreatic enzymes; they promote the absorption of vitamins D and K and the physiological processes affected by these vitamins.

Arsenic waters (i.e. waters containing more than 1 mg of arsenic l^{-1}, the toxic dose being 10 mg arsenic) have been recommended because arsenic tends to reduce the activity of the thyroid in cases of hyperthyroidism; it has a regulating effect on the cornification of skin and nails; it seems to stimulate the formation of erythrocytes and reduces the formation of leucocytes. It has therefore been recommended for leukaemia patients and for cases of secondary and pernicious anaemia.

Sulfur waters (containing more than 1 mg l^{-1} of sulfur, usually as hydrogen sulfide) have been recommended for rheumatic diseases, itching skin diseases, etc.

Summarizing it can be stated that internal hydrological treatment may bring about objective physiological changes in the human body, whether favourable or unfavourable; but, owing to the psychological factors involved in such a treatment and the lack of systematic research to date, it has been very difficult to determine the exact physiological processes which occur after different balneological treatments. Although the hydrological aspects of thalasso- and balneotherapy actually do not belong to the study of biometeorology it was considered to be useful to discuss this topic because millions of people in western Europe receive this kind of treatment each year.

9.1.2.3. Effects of balneotherapy

Some of the major results reported in balneotherapy may be summarized as:

(a) *Rheumatic diseases.* Warm thermal baths, mud-baths or sand-baths, surrounding the body and protecting it from local differences in the infrared radiations of the environment, considerably improve the disturbed thermoregulation mechanism of rheumatics. Drinking alkaline waters, which neutralize uric acid in the body, has proved to be beneficial for certain forms of rheumatism. Sulfur waters seem to stimulate the adrenal cortex and are therefore recommended for the treatment of various rheumatic diseases.

(b) *Diabetes.* During periods of acidosis the drinking of alkaline waters is apparently very helpful and the patients seem to require less insulin to control their metabolism.

(c) *Skin diseases.* Various skin diseases (e.g. eczemas, psoriasis, etc.) can be alleviated by balneological treatment.

(d) *Hyperthyroidism.* Arsenic waters have been recommended for reducing thyroid activity.

(e) *Hypertension.* Thermal baths have proved in many instances to reduce the diastolic blood pressure of temporary cases of hypertension.

Extensive research is required before any of these benefits can be generally accepted.

9.1.3. High altitude therapy

In Sections 5.3 and 6.12 it was pointed out that apart from the causation of chronic mountain sickness and of sickle-cell disease many diseases are favourably affected by high altitude treatment above 1500 m and some of the effects on diabetes on allergy and on skin diseases were described. The principal application of high altitude therapy however is the treatment of respiratory diseases, in particular of asthma and bronchitis, which is described in Section 9.3.

9.1.4. Travelling to favourable climates

Increasingly each summer during the last ten years, millions of people in western Europe move from the colder northern countries to coastal areas of southern Europe (in particular to Yugoslavia, Greece, Italy, Spain and Portugal and also to the Canary Islands). In winter time thousands of people visit the high altitude areas in Switzerland and Austria, in recent years the Carpathian Mountains in Poland and Roumania also, for winter sport activities. Also in the USA and Canada many tourists move in winter time to the Rocky Mountain areas for skiing.

In view of the many drastic physiological changes which occur during each of the above mentioned climatic and altitude changes it is evident that visitors should be careful not to start very strenuous activities immediately upon arrival.

For example heart patients should not exceed an altitude of 1200 m without previous and careful adaptation. They should also be careful in visiting very hot countries. Young healthy people visiting ski resorts above 1500 m, should adapt for at least three days before starting vigorous exercise at this altitude. Even healthy tourists visiting a mountainous area are recommended to stay in a hotel below 1500 m and to use a ski-lift or car to go to 1800 or 2000 m during the day for skiing or walking. They should return in the evening to the lower altitude because most people experience disturbed sleep and unpleasant dreaming if they sleep above 1500 m. Such a winter sport schedule, in a sheltered area, is also highly recommended for asthmatics. If they go to higher altitudes each day for a month their disturbed thermoregulation will be normalized and their respiratory and adrenal functions will be greatly improved. Staying for ten or more days in the Alps above 1500 m will result in a considerable increase in the haemoglobin content of their blood on returning to sea level. The blood is then able to bind more oxygen after breathing and this results in greater fitness. Also the increased ultraviolet radiation at high altitude, with a different wavelength distribution from that at sea level produces many beneficial effects as well as some harmful ones (e.g. see Table 5.1, ulcer and hyperthyroid patients).

Finally the favourable effects on high blood pressure experienced when subjects suffering from hypertension move to warmer areas (not too hot) must be mentioned. However, moving to a warm country also affects the physicochemical state of the blood; the level of γ-globulin and other antibodies will be reduced. As a result, visitors returning from a holiday in a warm area may be rather prone to infectious diseases, but preventive consumption of large doses of vitamin C the day of return will usually stop common colds and other respiratory infections.

9.2. AEROSOL THERAPY

The structure and size of aerosol particles and their physical and biological properties were discussed on p. 115. The present section briefly reviews some of the major clinical applications of aerosols which have been reported.[34–41] These clinical applications, known as aerosol therapy, are particularly applied in France, Germany and most East European countries. This therapeutic method has gained considerable importance in modern medicine and found its formal recognition in the founding of the International Society for Aerosols in Medicine in 1970.

In Table 9.1 the principle applications of aerosol therapy are compiled and Table 9.2 lists the types of medication used. Aerosol inhalation for the deposition of drugs in the respiratory tract accounts for the most frequent application of aerosol therapy. Inhalation of aerosols from saline solutions or mineral waters is also well known. It is particularly popular in the European inhalatoria where this treatment is frequently administered as group therapy.

Several methods have been developed to generate aerosols to be used for

TABLE 9.1
Principal applications of aerosols in medicine[a]

Applications	Examples
I Therapy.	
(a) Inhalation of drugs or saline solutions in broncho-pulmonary and upper respiratory conditions.	Sympathomimetic bronchodilators and spasmolytics to effect bronchodilation and to relieve bronchospasm; enzymes, mucolytic agents and detergents to liquify obstructing secretions and to improve bronchial hygiene; antibiotics in bacterial infection; decongestants and antiphlogistic agents in mucosal congestion and inflammation; foam-inhibiting agents in pulmonary oedema; inhalation of aerosols from saline solutions or mineral waters as practiced especially in European spas.
(b) Inhalation of systemic drugs when other routes of administration are undesirable or impractical.	
(c) Local application of medicated aerosols for non-respiratory conditions.	Certain skin diseases and certain conditions of the vagina and uterus.
II Prophylaxis.	Mass immunization by inhalation of vaccine; pre- and post-operative prophylaxis by inhalation of medicated aerosols to prevent or reduce complications during and after surgery; humidification of dry air.
III Diagnostics.	Provocation of expectoration for bacterial and cytological studies; inhalation of radioactively labelled or radio-opaque substances for diagnostic purposes; inhalation of pharmacodynamic aerosols such as Aludrin for pulmonary function tests.
IV Disinfection.	Disinfection of rooms by dispersion of bactericidal aerosols.
V Anaesthesia.	Local anaesthesia by inhalation for insertion of tracheal or bronchial catheters.

[a]Copied from a paper by Wehner.[39]

therapeutic purposes.[41] These aerosol generators are classified into two groups: gas/liquid jet atomizers and ultrasonic nebulizers.

In jet atomization an instrument removes non-respirable droplets exceeding a certain size by centrifugal or gravitational forces. This takes place prior to application of the aerosol for inhalation treatment. According to Dirnagl[41] a concentration of 10–15 mg of solution remains in droplets, of up to 4 μm diameter, per litre of aerosol. In the case of ultrasonic nebulizers the ultrasonic waves may produce droplets of 2 μm and concentrations of 100–200 mg l^{-1}. It is only due to the very small size of the particles that the deepest spots in the respiratory track can be reached, locations which otherwise are difficult to treat. Most artificially generated aerosols are electrostatically charged. Therefore the term electro-aerosol therapy is often used.

TABLE 9.2
Types of medication used in aerosol inhalation therapy[a]

Action of drug[b]	Kind of drug[b]
Bronchodilation, spasmolysis	Epinephrine and certain of its derivates; isoproterenol; racemic *n*-isopropylnorepinephrine sulfate; demethylphenylethylenimine; papaverine.
Mucosal decongestion	Phenylephrine and cyclopentamine as sympathomimetic vasoconstrictors; certain antihistamines, e.g. acetic l-dimethylphenylimminothiazolidine.
Mucolysis	Acetylcysteine; deoxyribonuclease.
Moistening, thinning, wetting	Tyloxapol; glycerine; propylene glycol; triethylene glycol; $NaHCO_3$; sodium lauryl sulfate.
Antibacterial action	Antibiotics such as tetracyclin, chloramphenicol, neomycin, bacitracin.
Antifoaming action	Ethyl or octyl alcohol, ethylhexanol.
Anti-inflammatory action	Topical steroids; 1,4-dimethyl-7-isopropylazulene; certain hypo- and isotonic mineral waters or solutions.
Expectoration	Certain hypertonic mineral waters or solutions; certain ethereal oils.

Other drugs used in aerosol inhalation therapy: vitamin C; pantothenic acid (a vitamin of the B complex); hyaluronidase; trypsin; anticholinergic agents; hormone extracts from the posterior lobe of the pituitary gland; H_2O_2.

[a]Copied from a paper by Wehner.[39] Table 9.2 summarizes the types and action of drugs frequently used in inhalation therapy. Wehner[39] pointed out that drugs are often combined for optimal effects, the combination being dependent on the individual case. While, for example, epinephrine is both a bronchodilator and vasoconstrictor, the bronchodilator isoproterenol is often combined with such vasoconstrictors as phenylephrine and cyclopentamine for decongestant action. Inhalation of bronchodilators, decongestants and mucolytics often precedes or is combined with that of other drugs such as topical antibiotics, steroids etc. to improve access to the target tissue for the latter drugs. Some drugs combine several desired effects, such as epinephrine, or certain ethereal oils which act as expectorants and disinfectants. In each case of drug inhalation the dose, rate of absorption and side effects including the possibility of allergic reactions, has to be carefully considered.

[b]This list is neither complete nor does it necessarily imply an endorsement of the listed agents by the author. Only the chief action for which the drug is used is mentioned. Collateral and side effects are not considered.

Inhalation therapy is mainly used for chronic respiratory diseases. In general secretolytic, bronchospasmolytic, antiphlogistic (diminishing inflammation) and antibiotic drugs are used in aerosol therapy. Mucostasis is one of the central problems in chronic respiratory diseases. Pathological mucus is a very good culture medium for pathogenic bacteria, it sometimes obturates parts of the bronchial tree with consecutive atelectasis (non-expansion of the lungs, causing anoxia and often death). Death during an asthmatic attack, for example, is often caused by obstruction of a great deal of the bronchi by mucus. Even distilled water, when inhaled, has a secretolytic effect by increasing the water content of airway secretions.

Aerosol therapy is a very complex interdisciplinary field of study. As pointed out by Wehner[39] it involves engineering, aerosol physics, medicine, physiology, pharmacology and biochemistry. Aerosols, in concentrations generally used for medical purposes, are rather unstable systems. Their behaviour is often not readily predictable due to intangibles for which, in the theoretical calculations, certain assumptions have to be made which may or may not come close to conditions actually encountered by the individual aerosol particles. Considerable complications for obviously important aerosol deposition calculations are introduced by the complexity of the respiratory tract. This tract resembles a tree whose trunk, the trachea, has a diameter, of 1.6 cm and which eventually splits into 1 to 5×10^7 end-pieces, the alveolar sacs, with a diameter of 0.04 cm each. During passage from the trachea to the alveoli at a flow rate of 1 l s^{-1}, the velocity of the aerosol particle, being a function of the cumulative cross-sectional area of the tract, changes from approximately 400 cm s^{-1} to practically zero in the alveolar sacs. Andersen[42] in Denmark in particular studied on an experimental basis the penetration mechanism of aerosols in the respiratory tract. For further details see *Progress in Biometeorology*, 1977, Part II.[38]

9.3. CLIMATIC CHAMBER THERAPY

Both low pressure and high pressure climatic chambers are used to simulate high altitude and deep water conditions. The methods are also described as hypobaric and hyperbaric treatment.

9.3.1. Low pressure or hypobaric climatic chamber therapy

Low pressure climatic chamber therapy simulates different altitudes (see Fig. 9.1). A pressure of approximately 716 mm Hg in the chamber corresponds roughly to an altitude in the mountains of 500 m above sea level; 675 mm equals 1000 m; 635 mm equals 1500 m; 595 mm equals 2000 m; 560 mm equals 2500 m; and 525 mm equals 3000 m.

As pointed out on p. 109 it is only about 1500 m altitude that important physiological changes take place. Many German (particularly Nückel[35]), Austrian, Swiss and French physicians have applied high altitude low pressure therapy to various diseases but mainly to respiratory diseases. Their patients

Fig. 9.1 Low pressure climatic chamber as developed and used by the Biometeorological Research Centre, Leiden, The Netherlands, since 1961

are brought to high altitude sanatoria where they usually stay for a year or longer. In recent years also in the east European countries and particularly in the USSR this method has been applied to respiratory and other diseases. However, in most published reports the follow-up studies were insufficient. Very accurate long term studies, using climatic chambers, were carried out by Bouma and Tromp at Leiden, The Netherlands, mainly between 1961 and 1970.[43–46] The following therapeutic applications were found.

(a) *Asthma.* It was pointed out earlier that asthmatics have an inefficient thermoregulatory mechanism and adrenal function; their blood pressure is usually low and the threshold for the contraction of the bronchi, after atmospheric cooling, is much lower than in normal healthy subjects. Each of these basically different physiological functions can be improved or even normalized (in young subjects below the age of 40) after a long series (70 or more) of one hour natural or simulated high altitude treatments above the critical altitude level of 1500 m.

In the climatic chambers temperatures of 10–15 °C and 40–50% relative humidity are recommended to improve the comfort of patients if they have to sit with six or more patients in the same chamber. Usually patients sit for one hour in the chamber at a reduced pressure equivalent to an altitude above 1500 m (preferably 2000–2500 m).

The rise, after several sessions, to 2000 m, takes only two to three minutes, the return to sea level may take 10 min or more depending on the ear troubles experienced by the patients. For the first two or three five-day weeks a one-hour session is held. In following weeks it is reduced to four, three and two sessions per week till the thermoregulation, adrenal and respiratory functions are completely normalized.[43–46] In Table 9.3 an

TABLE 9.3
Changes in the respiratory function of a girl of 14 years of age, (height 1.74 m) after 48 low pressure climate chamber treatments at Leiden, The Netherlands[38]

Date	Number of climatic chamber treatments	Vital cap. in ml	Forced expiratory vol. in ml	FEV as % VC	Forced inspiratory vol. as % VC	Maximal breathing cap. litres
4.2.64	0	3290	2350	71	91	70.5
23.6.64	13	3900	3350	86	89	100.5
1.9.64	24	4080	3380	83	89	101.4
17.11.64	33	4170	3636	89	100	108.9
11.1.65	38	4280	4180	98	99	125.4
21.4.65	48	4470	4290	96	95	128.7
8.9.65	Since 24.4 no cl. ch. treatm.	4420	4230	96	89	126.9
Normal values		4250		>70		108.3

example is given of the type of respiratory changes which occur after a series of simulated high altitude treatments.

(b) *Rhinitis.* Most rhinitis patients have a poor thermoregulation mechanism which explains their sudden attacks during drastic changes in their thermal environment. A long series of low pressure climatic chamber treatments could improve their thermoregulation mechanism and reduce their subjective complaints.

(c) *Arthritis.* Certain forms of arthritis may be improved if regular low pressure climatic chamber treatments are given at simulated altitudes of 2000–2500 m and at temperatures of 30–35 °C. The observed improvements are probably due to one (or all) of the following three factors:

(i) Arthritic patients have a poor thermoregulation mechanism which explains their great sensitivity to changes in their thermal environment; regular high altitude treatments will improve this mechanism and make an arthritic patient more resistant against atmosphere cooling;

(ii) Rheumatic patients usually have an insufficient adrenal function, high altitude treatments improve this condition;

(iii) An important factor in the causation of arthritic pains seems to be the sudden increase in viscosity of the lubricating synovial fluids in the joints (Section 6.9). Increased excretion of hexosamine in the urine takes place during high temperature in the chamber.

(d) *Raynaud disease and other vascular disorders.* The improved peripheral blood circulation and a better balanced autonomic nervous system as a result of simulated high altitude treatments above 1500 m (Section 5.3), may improve serious peripheral vascular disorders which may otherwise cause cyanosis of the extremities, and in extreme cases may lead to gangrene. Observations at Leiden, The Netherlands, suggest that regular treatments at simulated altitudes of 2000–2500 m and temperatures of 35 °C could reduce or stop the vasospasms and the following cyanosis.[43–46] Also the general thermoregulation efficiency of these subjects, which is usually poor, is improved after a series of treatments. It is very likely that peripheral disorders in geriatric patients and serious phantom pains in amputated limbs could also be reduced or partly cured by similar climatic chamber treatments.

(e) *Eczema.* Certain resistant forms of eczema, which do not react or react only temporarily to medication, are often improved or cured if the patient is treated at very high altitudes above 2500 m and at low temperatures (below 10°C). Examples are certain forms of neurodermatitis and psoriasis. The combined effect on peripheral blood circulation, general thermoregulation and the improved equilibrium between the autonomic nervous systems may be responsible for the observed improvements.

(f) *Migraine.* Observations[43–46] at Leiden, The Netherlands, suggest that serious cases of migraine, which could not be explained anatomically, neurologically or psychosomatically, and where no encephalographic indications can be found, could be improved or disappear after 20 or more low pressure climatic chamber treatments above 2000 m altitude and at temperatures below 10°C. Also headaches remaining after a previous brain concussion usually disappear after cold treatments at lower altitudes but above 1500 m. The physiological mechanism involved may be twofold:
  (i) a better balanced autonomic nervous system as a result of the high altitude treatments;
  (ii) due to the reduced partial oxygen pressure the pH of the various body fluids is affected.

(g) *Whooping cough.* Several German physicians have reported the favourable effects of high altitude treatment of whooping cough.

(h) *Anaemia.* It is well known that a long stay at high altitude raises the erythrocyte and haemoglobin content of the blood. In certain forms of anaemia a long stay at high altitude or in a climatic chamber (for over 2 weeks) could have similar effects. This therapy has been applied successfully particularly in the USSR.

(i) *Schizophrenia.* Clinical studies reported from the USSR in high altitude regions seem to indicate considerable improvements in the behaviour of schizophrenics. This may be related to their improved thermoregulation and other high altitude effects. As pointed out in Section 6.8 schizophrenics are characterized by a very inefficient thermoregulation mechanism.

(j) *Heart diseases.* Many patients suffering from myocardial diseases are very sensitive to excessive heat stress or sudden cooling due to their inefficient thermoregulatory mechanism. In areas with extreme heat stresses heart patients, just before or during an attack, can be relieved considerably by putting them in a cooled climatic chamber. Training of the thermoregulation efficiency of heart patients is possible by bringing them slowly above simulated altitudes of 1500 m and by gradually lowering the environmental temperature in the chamber.

Several other applications of hypobaric treatment have been suggested: for example the testing of subjects who decide to move permanently to other countries with extreme climates; the treatment of certain eye diseases due to lack of capillary blood circulation; the improvement of memory of elderly people by improving capillary blood circulation in the brain etc.; finally sport performance in general and particularly at altitude could be improved by previous hypobaric training.

Although several of the reported findings have to be confirmed in further studies the many well established data sufficiently indicate the great clinical significance of low pressure climatic chamber treatments in medicine.

9.3.2. High pressure or hyperbaric climatic chamber therapy

The first use of compressed air for technical and industrial purposes was made by Triger in 1841 with a caisson for deepening the bed of the Loire river. This technique has been used regularly in bridge building since that time, although accidents (caisson or decompression disease) and even death have frequently been reported because the high pressure air bubbles enter the blood stream during too rapid a decompression.* Apart from bridge building, increased atmospheric pressure has been used in mining when areas of quicksand have to be by-passed. Another application of caissons of increasing importance is in deep-sea investigations. The work of Cousteau and Sealab II are examples of this method.[47]

In 1956 Boerema and co-workers[48,49] started experiments on the medical applications of a high-pressure chamber in Amsterdam. The main goal was to determine the value of increased atmospheric pressure in cardiac surgery. The idea of using oxygen under these circumstances was closely related to the practice of hypothermia in which, by lowering the body temperature to 28 °C, the oxygen consumption was reduced to about 50% of its level at 37 °C. A period of complete circulatory arrest for seven to eight minutes was well

*Manifestations of decompression sickness (commonly known as 'the bends') include: the joint pains of bends; the discrete and often confusing neurologic 'stroke like' pictures; severe, life-threatening gas pulmonary embolism and vasomotor collapse with shock. Treatment requires emergency recompression and oxygen breathing in the hyperbaric chamber to reduce intra-vascular bubble size and to eliminate bubbles.

tolerated and it was possible to perform small intra-cardiac repairs in this way. However, the change of body temperature to lower levels frequently resulted in arrhythmias and is not without considerable risk in clinical application. In order to increase the amount of time available for intra-cardiac repair. Boerema suggested that the body should be loaded with a maximum amount of oxygen (2–3 atm) before circulatory arrest.

The first cases of palliative surgical treatment in a hyperbaric chamber of patients with cyanotic heart disease were reported by Boerema[49] and co-workers in 1962 and Bernhard and Tank[50] in 1963. Later on Bernhard *et al.*[51,52] reported 135 cases treated in this way and Meyne[53,54] reported the treatment of 58 cases.

Apart from cardiac surgery hyperbaric therapy is used in the following cases:

(a) *Gas gangrene or Clostridial myonecrosis.* A rapidly spreading necrotizing inflammation, caused by anaerobic bacteria, e.g. species of *Clostridium.* The results of hyperbaric treatment as described by Boerema and Brummelkamp[48] are impressive. Meyne[47] pointed out that the gain of oxygen under increased pressure causes not only a decrease of mortality, but also an increase in the salvage of limbs. There is experimental evidence that oxygen under increased pressure has an effect on alpha toxin production and also a lethal effect on *Clostridium welchii* bacilli. Besides the effects of oxygen under increased pressure, there can be little doubt that reduction of the gas bubbles in the tissue results in a better tissue vascularization and, in patients with a lowered cardiac output, improvement is very likely to occur.

(b) *Vascular occlusion.* According to Meyne[47] there is good experimental evidence that during occlusion of the femoral artery there is a positive correlation between the oxygen tension in the ischaemic area and the arterial oxygen tension. There is a very strong impression that the acute cases especially benefit from hyperbaric oxygen. Acute vascular occlusion, during arterial spasm, impending crush syndrome* and skinflaps† with impaired circulation, are good indications for the therapy.

(c) *Carbon monoxide poisoning.* Carbon monoxide is toxic even at a very low concentration. It has a very great affinity for haemoglobin and patients dying in the acute phase of intoxication die from hypoxia. Meyne[47] pointed out that oxygen administration at two atmospheres reduces the amount of tissue hypoxia, mainly by increasing the dissolved oxygen in the plasma; furthermore high oxygen tensions increase the rate of carbon monoxide elimination. Oxygen treatment should be started as soon as possible after intoxication.

(d) *Burns.* Meyne[47] and Hart[55,56] reported very successful treatment of serious burns provided the oxygenation is applied within 24 h after the accident.

(e) *Acute deafness.* Yanagita[57] in Japan reported a successful treatment of acute deafness by hyperbaric oxygenation. Treatment should start within

*Crush syndrome is a renal failure following severe crushing of a limb.

†Skin flaps are thin masses of tissue containing little subcutaneous tissue.

days after the occurrence. Often the deafness had already disappeared after only a few treatments.

(f) *Chronic Osteomyelitis.* This inflammation of bones is caused by pus producing micro-organisms and can be arrested to the extent of 60–90%. According to Depenbusch, Thompson and Hart[56] the treatment must be coupled with the excision of existing pieces of dead bone and the use of appropriate antibiotics. Hyperbaric treatment takes place at 2.4 atm for 1.5 h daily, six days a week for a period of 30 days. The mechanism involved seems to be stimulation of osteogenesis and of fibroblastic activity with increased collagen production.

Good results have also been obtained in the case of osteoradio necrosis (necrosis of bone as a result of excessive exposure to radiation). The increased fibrinoblastic activity and collagen production seem to stimulate the closing of nonhealing wounds. Finally good results have been reported in certain cases of stroke. However most of the reported results have to be reconfirmed by others. Only the applications in cardiac surgery, gas gangrene and carbon monoxide poisoning seem to be well established facts. In recent years hyperbaric treatments are also common in the USA, in particular in the USAF Hyperbaric Centre at the USAF School of Aerospace Medicine at Brooks, Texas, where the first hyperbaric chamber was installed in 1960. In a recent report it was stated that in the period 1940–1959, 17 000 cases of altitude decompression sickness occurred in the US Air Force.

REFERENCES

1. W. Zimmermann, *Z. Bakteriol.* **167**, 140 (1957).
2. F. Ellinger, *Medical Radiation Biology*, Thomas, Springfield, Illinois, USA, 1957.
3. R. Milin, *Verh. Anat. Ges. (54 Versammlung in Freiburg/Br.)* **22**, 191 (1959).
4. C. Häberlin, in S. W. Tromp, *Medical Biometeorology*, Elsevier, Amsterdam, The Netherlands, 1963, p. 945.
5. F. Müller, *Lehrang über Heilwirkung der Dtsch. Ostseebäder* **6**, 12 (1923).
6. Gauvain, in C. Häberlin, *Veröff. Zentralanst. Balneol.* **2**, 312 (1915).
7. H. Kleinschmidt, *ibid.* **2**, 20 (1923).
8. H. Jungmann, *Med.-Meteor. Hfte.* **1**, 7 (1950).
9. M. Bracher, *Strahlentherapie* **40**, 650 (1931).
10. C. Häberlin, *Berl. Klin. Wschr.* **800**, 2301 (1908).
11. E. Wolff, Thesis, Univ. Rostock, East Germany, 1931.
12. K. Scheel, Thesis, Univ. Rostock, East Germany, 1931.
13. Conradi, *Folia Haematol.* **17**, 105 (1913).
14. H. Curschmann, *Wien Klin. Wschr.* 13 (1928).
15. *idem*, *Arch. Phys. Therap.* **2**, 73 (1950).
16. Catrein, *Z. Klin. Med.* **118**, 688 (1931).
17. J. Berg, *Z. Klin. Med.* **120**, 64 (1932).
18. Nicolas, *Winterkuren an der See*, Hamburg, West Germany, 1907.
19. Leblanc, *Veröff. Zentralanst. Balneol.* **2**, 73 (1925).
20. H. Curschmann, *Verh. Ges. Dtsch. Naturforsch. Ärzte* **93**, 159 (1934).
21. Laignel-Lavestine, *4ième Congr. de Thalasso Thérap.* Arcachon, 1925.
22. G. Krauel, *Med.-Meteorol. Hfte.* **1**, 66 (1949).

23. E. Huntington, *Proc. Amer. Ass. Ment. Def.* **4292**, 116 (1937).
24. *idem*, *Civilization and Climate*, Yale Univ. Press, New York, USA, 1948.
25. C. Häberlin and F. Roeloffs, *Die Meeresheilkunde in Deutschland*, Griese Verlag, Hamburg, West Germany, 1938.
26. C. Häberlin, *Strahlentherapie* **28**, 286 (1928).
27. *idem*, *Z. Wissenschaft. Bäderheilk.* **2**, 1001 (1928).
28. P. Döhring, *Med.-Meteorol. Hfte.* **1**, 5 (1949).
29. W. Zimmermann, *Arch. Physik. Therap.* **7**, 405 (1955).
30. H. C. Bazett, J. C. Scott, M. E. Maxfield and M. D. Blithe, *Amer. J. Physiol.* **119**, 93 (1937).
31. K. J. K. Buettner, *J. Appl. Physiol.* **14**, 261 (1959).
32. Terrier, private communication, 1938.
33. in C. Häberlin and G. Goeters, *Grundlagen der Meeresheilkunde*, Thieme Verlag, Stuttgart, West Germany, 1954.
34. K. Bisa, *Z. Aerosol. Forsch. Therap.* **4**, 288 (1956).
35. H. Nückel, *Aerosol Therapie*, Schattauer Verlag, Stuttgart, West Germany, 1957.
36. K. Dirnagl, *Z. Angew. Bäder-Klimaheilk.* **16**, 536 (1969).
37. *idem*, *Dtsch. Med. Wschr.* **22**, 245 (1971).
38. S. W. Tromp (Ed), *Progress in Human Biometeorology II*, Swets & Zeitlinger, Amsterdam, The Netherlands, 1977.
39. A. P. Wehner, in ref. 38, p. 233.
40. D. Cop, I. Ruzdic, T. Durrigl and M. Majsec, *Arch. Interamer. Rheumat.* **3**, 426 (1960).
41. K. Dirnagl. in ref. 38, p. 238.
42. I. Andersen, *Biometeorology* **5**, (II), 229 (1972).
43. S. W. Tromp, *Monograph Series Biometeorological Research Centre No. 8*, Leiden, The Netherlands, 1964.
44. *idem* and J. J. Bouma, *ibid.*, No. 9, 1968.
45. *idem ibid.*, No. 13, 1974.
46. *idem*, *Physik. Med.* **5**, 23 (1976).
47. N. G. Meyne, in ref. 38, p. 249.
48. I. Boerema and W. H. Brummelkamp, *Ned. T. Geneesk. (The Netherlands)* **104**, 2548 (1960).
49. I. Boerema *et al.*, *Surgery* **52**, 796 (1962).
50. W. F. Bernhard and E. G. Tank, *ibid.* **54**, 203 (1963).
51. I. W. Brown Jr. and B. D. Cox (Eds), *Hyperbaric Medicine*, Proc. Nat. Acad. Sci., Nat. Res. Council Pubn. 1404, Washington, USA, 1966.
52. W. F. Bernhard *et al.*, in ref. 51, p. 345.
53. N. G. Meyne, *Hyperbaric Oxygen and its Clinical Value*, Thomas, Springfield, Illinois, USA, 1970.
54. *idem* and H. M. Mellink, in *Hyperbaric Medicine*, J. Wada and T. Iwa (Eds), Igakushoin, Tokyo, Japan, 1970, p. 335.
55. G. B. Hart, *Proc. 5th Int. Hyperbaric Congr. USA* 1977.
56. Depenbusch, Thompson and Hart, private communication, 1977.
57. N. Yanagita, *Proc. 5th Int. Hyperbaric Congr., USA*, 1977.

CHAPTER 10

ARCHITECTURAL AND URBAN APPLICATIONS (INCLUDING SELECTION OF HOSPITAL SITES)

10.1 GENERAL METEOROTROPIC CONSIDERATIONS

It is more and more realized that the architectural design of houses, buildings, factories, hospitals and sanatoria, considerably affects the comfort and general health of people working in these buildings. However, the general layout of a town, the distance and height of buildings etc. are also extremely important. In Section 2.9 some of the major physical factors which are affected by the plant environment of buildings and in and around cities were reviewed. One of the leading experts in this field is Givoni[1] from Israel who is using small scale models in wind tunnels to predict the best architectural design and urban layout. The mathematical formulae of Section 4.6 are also being applied to predict the degree of comfort which can be expected in various designs of building.

The vital importance of an efficient thermoregulatory system in humans has been stressed repeatedly. In most developing and highly developed countries both architectural construction and urban lay-out considerably affect the direct meteorological environment of large population groups and therefore also the thermoregulation efficiency, which, in its turn, affects the physico-chemical properties of the blood of these population groups, particularly elderly people and people suffering from respiratory, rheumatic and heart diseases. These people represent a very large proportion of the present day population in larger cities. Due to the rapidly growing urban population all over the world, of which already 30% live in settlements of 5000 or more persons and 18% in cities of more than 100 000, it is essential to consider those biometeorological aspects of our present day architectural and urban planning which are either completely neglected, or insufficiently considered, by most architects, engineers and urbanists responsible for present day large scale building projects. This is a very serious problem particularly in Europe where, contrary to USA practice, errors in these projects will last for many years, and usually can never be corrected owing to lack of funds. Consequently one or two future generations may have to suffer for the architectural and urban errors of the present day planners.

One of the principal reasons why architectural and urban planners often neglect to introduce certain biometeorological measures in their building projects is the fact that mainly financial considerations seem to determine the final decisions of governments, municipal boards or private building companies. As a result in modern buildings unsatisfactory thermal conditions often prevail, particularly near outside walls and glass windows. The same is true for humidity conditions and air movement, which can be most uncomfortable in ill-designed air-conditioned houses or buildings. Natural lighting conditions are often inadequate or spectral distributions are introduced which are entirely different from natural sunlight, and this can have important effects on the natural biological rhythms and endocrinal functions of man (see p. 211). There is even a tendency in certain countries to exclude natural sunlight (e.g. in schools) by omitting all windows from a building. Excessive sound waves (noise), in and outside buildings are also often not sufficiently reduced by proper insulation or other measures which reduce the noise level.

In the areas between buildings too much air movement and, in particular, strong air turbulence, is responsible for considerable variations in the cooling index at different points in an urban area both in the horizontal and the vertical planes. All of these factors must be considered in any new lay-out of large complexes of high buildings. However, even if these factors are considered, the meteorological data used for the planning of the comfort conditions are usually based on recordings from meteorological stations far from the centre of a city. Another important factor which determines the degree of comfort experienced by the people living in modern urban areas is the degree of air pollution. Complexes of high buildings, with narrow streets and a lack of air circulation, may accumulate a high concentration of air pollutants.

All these biometeorological aspects of our modern living quarters have a profound influence on the health and general well being of man and his working efficiency. This is particularly true for elderly people, who have gradually become a high proportion of the present day population due to the continuous increase of the average life expectancy of man. Apart from the elderly, three groups of younger persons, who suffer for a considerable part of their life from meteorotropic diseases which are difficult to cure, must be mentioned. These are the sufferers from respiratory diseases like asthma and bronchitis, rheumatic diseases (in particular rheumatoid arthritis) and certain heart diseases. Patients suffering from these three types of disease represent very large population groups. In western Europe at least one, but probably two or three out of every 100 inhabitants suffer from asthma or asthmatic bronchitis. In the USA it has been estimated that at least 22 million working days are lost per year due to asthmatic workers, representing a yearly financial loss of about 450 million dollars.

In the case of rheumatoid arthritis a survey in 1966 in the USA revealed that about 13 million Americans suffer from some form of arthritis causing an annual disability of 186 million days of restricted activity, including 57 million days that are spent in bed and 12 million lost from work, representing a yearly economic loss of 220 million dollars.

A third important meteorotropic disease is the group of heart diseases, in particular myocardial infarction and angina pectoris. Biometeorological studies in different parts of the world have shown that sudden extreme thermal stresses, either cold or heat, are very important triggers for heart diseases which together with cancer are the major components in the total mortality figures of European and American countries.

Considering the importance of the human thermoregulatory mechanisms it is evident that all architectural constructions and urban lay-outs which may have a disturbing influence on the efficiency of the hypothalamic function, should be rejected.

It is often not sufficiently realized that the microclimate in the lower 200 m of the atmosphere, in other words the interface between ground and air, known as the planetary boundary layer, is very much controlled by the topography and physical properties of the buildings and the spaces between them. These near-surface conditions are of vital importance for the various energy budgets of the atmosphere.

For architectural constructions and urban lay-outs to have favourable effects on the hypothalamic thermoregulatory functions of the people living in those areas, the following factors at least must be considered: windspeed, air turbulence, heat islands, air pollution and noise (particularly from traffic). A summary of these problems was given recently by Landsberg[2] in the USA and by Chandler[3] in Great Britain.

10.2. WIND EFFECTS

The wind in the boundary layer is seriously affected by friction with the earth's surface, which is very great in urban areas. Mean horizontal winds are usually reduced. In the London area the difference between the average wind speeds in the centre and in surrounding rural areas is about 13% for speeds in excess of 1.5 m s^{-1}.[3] Investigations in London[4] and New York[5] have shown that light winds are sometimes accelerated as they move into towns, whilst strong winds are usually decelerated. There are also differences between day and night in urban areas. By night, owing to frequent thermal stability, winds tend to be less turbulent and the mean horizontal wind increases sharply with height. During the day when most people go to their work, often very strong air turbulence is experienced, particularly between high buildings separated by narrow streets causing a tunnel effect. In factories particularly draught plays an important role. Many factory workers must work in draughty spaces for several hours a day and move around between very cold and very hot spaces, causing serious thermoregulatory disturbance particularly in older age groups.

10.3. TEMPERATURE EFFECTS

During periods with little regional wind the centre of a city is considerably warmer than the surrounding rural areas. The difference usually is 0.5–2.0 °C but can be 10 °C or higher. Such heat islands in cities occur particularly by night with clear skies and calm air, i.e. during high pressure conditions.[4,6] In western Europe and the USA the most intense heat islands occur in summer and autumn. The heat island strength is particularly dependent on urban development density and not so much on city size.[7] The favourable and unfavourable effects of the air-conditioning of houses and buildings on the health and comfort of man during periods of unfavourable weather or climatic conditions, were discussed in Section 7.3, p. 207.

10.4. AIR POLLUTION EFFECTS

In addition to wind and heat control the prevention of air pollution should also play an important role in urban planning, the more so as bronchitics and patients suffering from bronchitic asthma (common in the elderly) suffer badly from polluted air. Polluted air in cities also reduces the intensity of ultraviolet radiation with its many favourable effects on various physiological processes in man. The air pollution patterns in towns are very complex as sources of pollution are randomly sized, they vary in amount, type, temperature and height of emission and occur in complex topographical and urban lay-outs. If adequate spaces are left between buildings to allow for sufficient vegetation, plants can be used successfully as air pollutant removers (p. 51), at the same time restoring the oxygen level of the air. If polluted air, containing for example 150 parts of ozone per thousand million parts of air, passes over a forest of trees 4–5 m tall for one hour, only 60–90 parts of ozone will remain, the rest being absorbed by plants. After 8 h the filtering would leave only 30 parts. The taller the trees the better their filtering effect.

Plants work also as air conditioners (p. 52) by modifying the temperature, air flow and moisture content of the atmosphere. They can also be used as air cleansers (p. 52) by removing air-borne dirt, sand, dust, odours, fumes etc. from the air,[8] or as windbreaks.

10.5. EFFECTS OF NOISE

Plants are also very important for reducing traffic noise. Their screening of moving vehicles has also a psychologically pleasing effect. Heavy street traffic may cause a noise level of 70 decibels (dB). Proper planting can reduce this by 60%. As noise has not only psychological but also endocrinological effects

(Section 5.12) proper landscape planning may reduce traffic noise considerably and thus improve the health of people living in noisy areas.[9,10]

10.6. PRACTICAL PROPOSALS FOR THE DESIGN AND SITING OF BUILDINGS

(a) The spacing between, and the height and shape of future buildings should be such that the wind effects, particularly turbulent ones, at ground level and near balconies, are reduced to a minimum in order to lower serious biological cooling effects on the inhabitants, in particular on the elderly and on those suffering from meteorotropic diseases. In hot tropical countries cooling effects should be increased.

(b) Sufficiently large green patches should be preserved between buildings to allow sufficient exchange of fresh air and the penetration of radiation, in addition to taking advantage of the psychological aspects of such a lay-out. It would also enable the urbanists to create large areas covered with plants and trees with their function of air pollution removers, air cleansers, wind-breaks and noise reducers.

(c) Supermarkets and other shopping centres should be well protected from wind and rain, preferably heated during very cold weather and cooled during the hot season, but, at the same time, sufficient sunlight should enter the streets at least in the cold countries. In warm southern countries sunlight should be considerably reduced in shopping centres.

(d) (i) The size of windows and their location in buildings in the cold countries should be such that a maximum quantity of sunlight enters the apartments. This is necessary in view of the important phsyiological effects of natural sunlight on a number of endocrinological functions and the natural daily biological rhythm in man. In warm countries the incoming sunlight must be drastically reduced because of the heat radiation.

(ii) This measure should be combined with a number of better construction methods, such as: double windows; thermally well-insulated walls, floors and ceilings (a loss of heat near a wall of only 0.03–0.045 cal s^{-1} creates a sensation of cold, which feeling increases if the loss reaches a value of 0.1–0.2 cal s^{-1}); adequate and equally distributed space heating; prevention of air currents (so common in imperfectly constructed air-conditioning systems), etc., i.e. measures which would enable one to sit or work near large windows or near outside walls without being subjected to radiation cooling, which seriously affects the thermoregulation mechanism.

Similar precautionary architectural measures must be taken with buildings housing cattle, sheep and swine. In particular fat animals such as pigs, which cannot perspire or have difficulty in doing so, cannot support thermal stresses. Cows, in general, support low temperatures better than hot ones because of their great metabolic heat production and inefficient perspiration. As discussed in Section 11.2.2, high cow-shed temperatures may reduce milk production by more than 50%, a common feature in hot developing countries. But also in the colder European countries the hot and humid microclimatic conditions in cattle sheds may resemble the microclimates of tropical countries.[11] Not only the walls but also the floors of cattle sheds should not be too cold in northern countries. Intense body cooling, if the animal lies on the floor, can be prevented by putting a layer of straw on the floor. This reduces the heat loss of a cow towards the floor from 150–40 kcal m^{-2} h^{-1}.[11] On the other hand, in tropical countries the floors should not be made of highly-insulating materials, as this prevents sufficient body cooling.

Finally very briefly a new architectural aspect must be mentioned; the possible effect of geophysical stimuli from the soil on animals and on man in particular.

Several studies have shown the indirect effect of trace element deficiencies in the soil on animals and man, through food growing on these soils and water percolating through such soils. However, direct geophysical effects of soils on living organisms have also been observed in a number of localities both in The Netherlands and elsewhere but these observations require further confirmation.

In many sedimentary areas certain zones occur which are characterized by a sudden, strongly reduced electrical resistivity of the soil due to geological discontinuities in the sub-soil, such as rising sand or peat layers under a clay cover, changes in the porosity of sand layers, faults etc. Such zones are also characterized by measurable changes in infrared soil radiation. Many empirical studies[12,13] have shown that if such zones underlie a house, those sleeping or working for many hours a day above such zones may have rheumatic and other complaints which disappear once the subjects move outside the geophysical zone.

Three kinds of experiments support the assumption of biological effects of such zones.

(a) Petschke,[14,15] a German physician, observed differences in blood sedimentation rate (BSR) patterns if the test tubes were placed outside or above such geophysical zones. Experiments by Tromp and Van Veen[16] at the bloodbank in Leiden have shown that with constant temperatures of 24 °C differences in the BSR pattern of the same blood can be observed if the tests are carried out at ground level, in the basement or on the top floor of the concrete building.

(b) Beran[17] in Vienna, found similar differences in blood coagulation speed depending on the location of the blood sample (in the laboratory or in a cellar). Further studies have shown that temperature variation could not explain the observed differences.

(c) Jenny *et al.*,[18–20] between 1933 and 1945 at Suhr (in Switzerland) kept mice in cages six metres long which were kept partly above a pronounced geophysical zone (strongly reduced soil resistivity from 3200 to 1500 Ω) and partly outside of it. He observed that mice preferred to sleep outside the zone (6434 mice slept outside, 1626 inside the zone); mice treated with carcinogenic tar developed up to 30% more carcinomas if placed in the zone than if placed outside of it.

These well-controlled experiments suggest that unknown direct (probably radiation) geophysical effects of the soil on the living organisms occur, which could be partly responsible for the concentration of cancer cases and other diseases in certain areas in the world as observed during medical geographical studies. If these observations can be confirmed in future studies it will be necessary to consider soil conditions during the selection of the site of a private house or building.

What has been said about the precautionary measures to be taken in the siting and construction of buildings applies particularly to hospitals and sanatoria. These should not be built at windy locations with much air turbulence, they should be built in non-polluted areas without excessive rainfall and with a great number of sun hours yearly. Air conditioning should be built in such a way that patients do not feel a draught above their beds or if they sit in their rooms; but even more important are the preventive measures to be taken to ensure that the circulating air does not carry bacteria, viruses or odours to the rooms of the patients from other parts of the hospital.

REFERENCES

1. B. Givoni, *Man. Climate and Architecture*, Elsevier, Amsterdam, The Netherlands, 1969.
2. H. E. Landsberg, *Tech. Note BN 741*, Inst. Fluid Dynam. Appl. Maths., Univ. Maryland, USA, 1972.
3. T. J. Chandler, *Proc. A.C.I.Conf., The Hague, The Netherlands, 25—26 November 1971*, p. 98.
4. *idem*, *The Climate of London*, Hutchinson, London, UK, 1965.
5. Bornstein, *J. Appl. Meteor.* **7**, 575 (1968).
6. F. L. Ludwig, *Pub. No. SF-70-9* Amer. Soc. Heat. Refrig. Air-cond. Eng., 1970, p. 40.
7. T. J. Chandler, *Weather* **19**, 170 (1964).
8. G. Robinette, *Horticulture* **46**, 26 (1968).
9. M. Kampfer, *Vegetationskunde (Bad Godesberg, Germany)*, 61 (1967).
10. W. Hess and E. Kürsteiner, *Wohltätiger Wald. Schweiz. Forst., Verein*, Zurich, Switzerland, 1962, p. 56.
11. S. W. Tromp, *Proc. A.C.I. Congr., The Hague, The Netherlands, 1970*, p. 6.
12. *idem*, *Psychical Physics*, Elsevier, Amsterdam, The Netherlands, 1949.
13. *idem*, *Monograph Series, No. 2, Biometeorological Research Centre*, Leiden, The Netherlands, 1955.
14. H. Petschke, *Medizinische* **39(I)**, 1263 (1954).
15. *idem ibid.* **52(II)**, 1759 (1954).
16. S. W. Tromp and T. Van Veen, *J. Interdiscipl. Cycle Res.* **7**, 1 (1976).
17. G. Beran, *ibid.* **3**, 207 (1972).
18. E. Jenny, A. Oehler and H. Stauffer, *Schweiz. Med. Wschr.* **24**, 572 (1936).
19. E. Jenny, *Gesund. Wohlf.* **1**, 1 (1947).
20. *idem* and A. Oehler, *Schweiz. Med. Wschr.* **67**, 33 (1937).

CHAPTER 11

INFLUENCE OF THE WEATHER AND CLIMATE ON ANIMALS AND ITS PRACTICAL ASPECTS

11.1. PHYSIOLOGICAL EFFECTS OF METEOROLOGICAL STIMULI

In view of the great economic importance of meteorological effects on animals many meteorotropic studies have been made. Some of the major handbooks in this field are listed in the references. Discussion will be limited to the nine most important groups of animals: cattle, sheep, pigs, wild ungulates, ruminants, birds in general and domestic chicken in particular, fish, insects and arthropods in general. For each group the economic importance of biometeorological studies will be reviewed and, as far as information is available, for each of the groups thermal, light and high altitude effects will be discussed.

11.1.1. Physiological effects of the weather and climate on cattle

(a) *Thermal effects*. In Section 5.1, the various thermal effects on the higher mammalia (in particular man) have been extensively discussed. The physiological processes involved in balancing heat production and heat loss in animals are regulated by the thermoregulatory mechanisms in the body (p. 58) which are located in the hypothalamus in the brain (see Figs 3.1 and 3.2). These mechanisms differ for different animals. They are vital for the survival of living organisms and it is therefore not surprising that the study of these mechanisms represents a major part of the studies of animal biometeorology. A dysfunction of these mechanisms seriously affects the health of animals, it may cause or trigger animal diseases, it influences milk production, hair and wool growth, it affects the reproduction and the quality of cattle and their resistance against extreme climates. Both heat and cold affect a great number of an animal's physiological mechanisms. The most important ones can be grouped together under different headings such as effects on cellular metabolism, enzyme systems, endocrine functions, water and electrolyte balance, vitamin metabolism, cardiovascular functions, growth of animals and several other important mechanisms. It would exceed the purpose of this publication to discuss each of these processes in detail (for further details see Reference 1).

Whereas in man excessive heat in the body is reduced by stimuli from the principal heat regulatory centre in the brain, the hypothalamus, through the opening of the peripheral pores in the skin and by evaporation of both sensible (sweat) and insensible perspiration (see Section 3.3), in cattle and other mammals in addition to this evaporative thermoregulation respiratory thermoregulation also plays an important role, particularly in animals with no, or few, sweat glands.

In this latter case the increase in the volume of air exhaled per unit of time (the respiratory ventilation) is known as panting. Some authors speak only of panting if the mouth of the heat stressed animal is opened. However, in ox and sheep for example, there may be a marked increase in respiratory ventilation with the mouth remaining closed. Respiratory ventilation of the ox may increase 10-fold, that of sheep 12-fold, that of the rabbit 15-fold, that of the dog 23-fold. Thermal stimulation of the thermoreceptors in the skin or of the hypothalamus controls the mechanism of panting.

It is well known that animals do not grow and produce as well in very hot climates as they do in temperate or cold climates. With a sudden onset of hot weather, broilers and other meat animals grow more slowly, hens produce fewer eggs, egg size decreases, and the eggshells become thinner and more subject to breakage.

Heat stress, particularly when combined with high humidity, also influences the vitamin requirements and the food consumption, which is generally reduced. Therefore, with an increase in environmental temperature, the levels of those nutrients required in definite daily amounts, such as the vitamins, must be increased in terms of percentages of the diet. In an experiment conducted with broiler chickens at the College of Agriculture in the Philippines, a 10% increase in growth was achieved by feeding the broilers during the cooler night-time hours as compared to the growth of broilers fed conventionally during the hot day-time hours.[1,2]

A considerable amount of work has been carried out on the heat balance of cows by Findlay *et al.*[3] in Scotland. They were able to show that cows are better adapted to low than to high temperatures, the comfort zone for cattle being between 0 and 15 °C. The main reasons for ill-adaptation of cattle to hot environments are the following:

(i) The high level of heat production: in a non-lactating, non-gestating cow it amounts to 10 000 kcal day^{-1}, in a cow yielding 20 kg of milk per day it amounts to 20 000 kcal day^{-1};

(ii) Poor sweating facilities of cows: the skin moisture evaporation reaches a maximum at only 17 °C ambient temperature.

Pigs seem to be more affected by summer temperatures than other farm animals; a pig is a non-sweating animal and, in addition, with increasing weight the fatty covering acts as a barrier to the loss of body heat. With rising environmental temperature the respiration rate in swine rises rapidly.

Cattle of northern latitudes feel uncomfortable when exposed to sun without wind. The grazing time of temperate breeds of cattle is very much less than that of tropical breeds. When temperate breeds of cattle rest, they never do so in the sun, whereas tropical breeds never lie down in the shade.

The difference in heat regulation between tropical and temperate cattle shows up particularly in the respiratory frequency which increases steeply in temperate cattle with rising environmental temperature (especially above 15 °C), whereas the increase is more gradual in tropical cattle. This is partly due to the fact that the surface area to body weight ratio of tropical cattle is about 12% greater than of temperate cattle. The heart rate usually increases steeply above 25 °C (in temperate cattle); the rectal temperature above 30 °C. This holds only under conditions of low humidity. With great humidity the rise starts earlier and is more precipitous. Thermal stress also affects the blood composition of cattle. At air temperatures of 18–40 °C there is a large increase in the creatinine content of the blood; the carbon dioxide combining capacity, the ascorbic acid and cholesterol levels in the blood are reduced to less than half the levels at 10 °C. Unlike man, there are no disturbances in the water, electrolyte balance or colloid concentration. This is probably due to the fact that in man, above 27 °C, the evaporation loss rises exponentially with increasing temperature, whereas there is no increase in cattle.[4] Heat stress also affects the reproduction rate of animals (Section 11.3).

The thermal effects of infrared solar radiation on the hypothalamus of cattle were studied particularly by Hutchinson,[5] Finch,[6] Dawson,[7] Brown[8] and their colleagues. The following paragraphs are based on their findings and those of other workers.[3,9,10] The thermal effect on the sensible heat exchange between an animal and its environment is not defined physiologically by the solar radiant energy absorbed but by the heat exchange at the base of the coat. A white, yellow or red coat with a smooth and glossy texture considerably minimizes the effects of solar radiation.

The heat exchange at the base of the animal coat can be assessed in terms of the effect of insolation on heat production and its dissipation by sweating, respiration and by sensible exchanges. The exchange is affected by the thermal resistance of the coat, the convective heat transfer and the penetration depth and transmission of the solar radiation in the coat. Different formulae have been developed to calculate and predict each of these factors. All these problems have been investigated, particularly by Cena, Findlay *et al.*, Hutchinson *et al.*, Monteith, Thwaites and others (see Ref. 1). Some of their major conclusions are summarized below.

Both mathematical and empirical studies have shown that the coats of arid zone animals should be very pale in colour, relatively deep, have a high bulk density and perhaps should be glossy. Although little information on the characteristics of the coats of arid zone mammals is available, they do not seem to conform to all these requirements. The most common coat colours of wild animals are various shades of fawn, red and brown. Many appear relatively pale, and some species such as the impala (*Aepyceros melampus*) have very pale or even white hair on the underside of the body. The absorbance of the coats of a few wild species has been measured. According to Dawson[7] and Brown,[8] in the red kangaroo it may vary from 0.7 for fur on the back to less than 0.5 for the palest hip fur. Finch[6] found absorbances of 0.75 for the eland and 0.65 for the hartebeest. Because some brown coats may have a heat load as high as black coats due to the effect of penetration as described above, the advantages afforded by the various shades of buff and brown are uncertain. The advantage of a low absorbance is not the only factor of evolutionary significance influencing the coat colour. In some populations of red kangaroos, most females have a relatively dark 'blue' coat while that of the males is a pale red. In African mammals, populations of the same species from different locations may vary in colour; variations in colour due to protective camouflage are common, but few animals have very dark coats.

The coats of most mammals that cannot avoid solar radiation in warm or hot arid regions appear to be shallow. There are a few measurements showing that the coats of eland, zebu and European cattle on the equator in Kenya, and the summer coats of the red kangaroo in Australia are 4 mm or less in depth but the lack of knowledge of an elementary coat character is emphasized by the case of a hartebeest that at different periods of its life on the equator had coat depths of 8 mm, 4.3 mm and 3.2 mm. Shallow coats presumably aid temperature regulation in the heat as the summer coats are in general shallower than winter coats.

Coats with a high bulk density have a relatively low penetration by radiation. There are differences between species. The hartebeest mentioned above has a density of 120 mg ml^{-1} in comparison with a mean of 38 mg ml^{-1} for two Hereford and 12 Boran steers measured on the same day.

The summer coats of cattle and of some African bovids such as impala are quite glossy but the importance of this is not known and measurements of gloss have not been made. The deep coats of wool sheep, which can endure hot semi-arid climates, have quite different characteristics from those described above. The value of a deep coat in reducing the heat load from solar radiation in a hot climate has been demonstrated in experiments showing the considerable increase in evaporative water loss when sheep and camels, exposed to the sun, are shorn. This suggests that shallow coats have not evolved to give the most protection against the heat load from solar radiation; other factors such as evaporative cooling, bulk density of the coat, superior sensible heat dissipation at night or in the shade, and behaviour are involved. Sweating may be more effective with a shallow coat because evaporation of sweat from the skin is more effective thermally than from the coat and a deep coat would become wet more readily than a shallow coat.[5]

It is also known that cattle with a shallow summer coat are more heat tolerant in the sun than those which have not shed their winter coat. Many of these very complex, but highly important problems of thermal stresses in cattle have great economic importance as they affect milk-production, hair and wool quality etc. Considerably more research is required in this field as was clearly indicated by Hutchinson.[5]

(b) *Effects of light.* Sunlight has considerable physiological effects on man and the higher mammals. These are fully discussed in Section 5.2. Some of the effects have a periodic nature and are known as photo-periodicity. Light intensity can be expressed in foot candles, lumina or lux. One foot candle (ft-c) is the amount of light from a standard candle reaching a square foot of surface placed one foot from the light source. According to Wurtman,[11] solar radiation reaching the earth's surface provides about 10 000 ft-c of light; in the shade 500 ft-c; in the average laboratory 50–150 ft-c at eye level.

According to Hoffmann[12] the earliest reports of studies on the effects of light intensity on animals are the studies of Rowan[13] on reproduction. He did not believe that light intensity was a factor in the responses of the reproductive system of birds to light once a low threshold was reached. However, when ducks are exposed to rather restricted periods of light, there is a correlation between testicular growth and light intensity.[14] In ferrets, Marshall[15] found that females placed closer to a light source reached their period of sexual activity (oestrous) sooner than those placed farther away. In general in these early studies the intensity of light was varied by placing the animals at different distances from the same light source. When actual measurements of light intensity were reported in some later studies, it was shown that only one footcandle was necessary for normal egg production in poultry and that while a light intensity-response relationship existed for the testicular weight of sparrows exposed to a range of 0.04–10 ft-c, there was no further increase in the response even up to 250 ft-c. The observations on the effect of light intensity on reproduction by Piacsek and Hautzinger[16] and others will be discussed in Section 11.3.

In a number of studies Hoffmann[12] found that the length of the daily photo-period can determine the length of the oestrous cycle in the albino rat, with shorter photo-periods tending to result in more 4-day cycles and longer photo-periods in more 5-day cycles. Similar observations were made in other animals. He pointed out that in most natural settings, animals are exposed to alternations of light and dark and for a given species there will be a concentration of physical activity during either the light or the dark period, so called activity rhythms. When the light-dark signal is removed by placing animals in either constant light or constant dark (or by blinding), animals still maintain a rhythm of activity and rest, but the period, or length of time between the onset of successive activity blocks, is usually no longer exactly 24 h but may be shorter or longer. Hence the active phase will drift with respect to solar time. Through extensive experimentation and observation of the results from many laboratories, Aschoff[17] (1961) suggested the following general rule: 'Animals which are light-active tend to show a shortening of the period between active phases if light intensity (with continuous exposure) increases; dark-active animals tend to show a lengthening of the period with an increase in light intensity'. This rule appears to apply across a wide range of animal species and even to plants.

Some interesting reports have been published on the possible correlation between the amount of light perceived and the age of menarche (beginning of the menstrual cycle) in humans, which may apply also to higher mammals. Zacharias and Wurtman[18] noted that menarche occurred earlier in blind girls than in sighted girls. Magee *et al.*[19] compared girls with minimal or absent vision to those with moderate vision and also found that those with less light perception had earlier menarche. These findings must be interpreted with some caution since a multitude of factors, which were not controlled in these retrospective studies, could be involved.[12]

In this connection the observations of Dumond and others,[20] that women show a maximum light sensitivity during the time of ovulation, and the observations of Zukowski *et al.*[21] in Poland, of Bojlen and Bentzon[22] in Copenhagen, of Valsik[23] in Czechoslovakia, and others, indicating that the first menstrual cycle of girls occurs more frequently in certain months of the year (different months for different countries) may be of interest. They may be related to similar changes in the solar spectrum and its intensity and in light sensitivity during the year;[24] see also Section 5.2.2.

The effects of solar radiation on mammals are due to changes induced in the activity of the epiphysis or pineal gland (located at the posterior lower part of the corpus callosum at the bottom of the cerebrum, see Figs 3.1, 3.2 and 3.5), the hypothalamus (Section 3.3.1) and the anterior part of the pituitary (see Section 5.2.3).

Wurtman *et al.*[11,18] demonstrated that the biochemical activity of the pineal gland depends on the types of visible light under which an animal is maintained (p. 107). Exposure of rats to continuous visible illumination depresses the weight of the pineal gland and obliterates the normal diurnal rhythm in the pineal function, namely in serotonin and melatonin synthesis. Cardinali[25] pointed out that in mammals these effects of light are not exerted directly on the pinealocytes as in lower vertebrates, but are mediated by a complex neuronal pathway which includes the retinas, the optic nerves and chiasm, the inferior accessory optic tracts, the medial forebrain bundles, descending fibres to the upper spinal cords, the preganglionic fibres to the superior cervical ganglia, and postganglionic sympathetic nerves which terminate on or near pineal parenchymal cells.

The administration of melatonin, one of the pineal hormones, is followed by suppression of gonadal development and lengthening of the di-oestrous phase of the reproductive cycle in several species. Conversely light-induced changes in the gonadal activity of seasonal breeders, such as hamsters, ferrets and ducks, are impaired or even prevented if the animals are pinealectomized (Section 11.3).

The cells of the pineal gland are photo-receptors, in lower vertebrates, converting photic energy into neural impulses; they may be photo-endocrine transducers in the avian pineal gland (i.e. they may transform photic energy into a hormonal output) and they are neuro-endocrine transducers in the mammalian pineal gland (i.e. they respond, to the norepinephrine released from pineal sympathetic nerve endings, by synthesizing and secreting the hormone melatonin).

(c) *Effects of altitude.* The physiological effects of high altitude (above 1500 m) on man and on higher mammals have been fully discussed in Section 5.3 and 6.12. Specific problems in relation to domestic animals and wild mammals have been reviewed by Grover,[26] Altland[27] and Morrison:[28] who claim that adaptation to high altitude can be achieved for virtually all species of domestic animals.[26] Abundant proof of this fact is evident on the altiplano of the Andes where one finds large numbers of cattle, sheep, goats, pigs, horses, as well as dogs and cats, in addition to the native populations of llamas and humans, all adapted to living at altitudes above 4000 m. This does not mean that the introduction of domestic animals to high altitude has been without problems. During the first decades of the Spanish occupation of the Andes, reproduction within the human population as well as in the animals they brought with them was very difficult. In the city of Potosi, Bolivia, at 4000 m altitude, there is a monument to the first child of Spanish parents to survive from among all those who were born in the 53 years after the founding of that city. Up to 1950 the mortality rate among lambs born above 3500 m was still 25%. Therefore, it probably takes several generations before the majority of individuals in a population become well adapted. However, once such a population is established, then survival of that species at high altitude is assured.

Altland[27] pointed out that data on vertebrates other than mammals, at altitude, are relatively scarce. Frogs, toads and salamanders have been observed at altitudes up to 3000 m in the Rocky Mountains. In the Himalayan Mountains the presence of 11 species of frogs above 3700 m, five species of lizards and four species of snakes above 3600 m and one species of snake at 5500 m has been recorded. The common European viper has been reported to live as high as 2800 m in the Alps. Hypoxia does not appear to restrict the distribution of amphibians and reptiles, but availability of food and water is a serious limiting factor, as is sunshine. Little is known about the altitudes up to which birds can fly. Some of the earliest records on the capacity of birds to fly at high altitudes were made by

Bert[29] in 1943. He reported that condors and vultures fly at 7000 m without any evidence of distress. Gunacos, swans, turtle doves, sparrows and even humming birds are commonly found at 4000 m in the Andes of Chile. The Bolivian goose or hyallata is found in the Andean highlands at about 3600 m. These geese have blood with a significantly greater affinity for oxygen than members of their class at sea level.

In laboratory experiments conducted at 23 °C sparrows, hawks, pigeons, and gulls became ill at altitudes between 6000 and 9000 m and showed signs of impending death between 10 400 and 11 000 m. If average resistance to decompression is considered, birds of prey appear to be almost as sensitive as other birds with the exception of the chicken. Adult chickens survive acute exposure to 4877 m simulated altitude, but above this level their tolerance is greatly decreased until none survive at 7925 m. Roosters show a greater altitude tolerance than hens, a sex difference contrary to that found in mammals.

The burro (a small donkey) is widely used as a pack animal in mountain climbing expeditions. In the Andes the llama, and in the Himalayas the yak, are used to carry burdens at least as high as 5300 m. In the Pariacaca range in Peru, with peaks around 6096 m, it has been reported that the air is so thin that both man and animals appear to get so out of breath that no spurring is sufficient to budge them.

In various mountainous areas of the world it has been the practice for hundreds of years to graze livestock during the warmer periods of the year at altitudes of 2500–3700 m. At these altitudes many poorly acclimatized cattle have difficulty in breeding and are often afflicted by a disease known as 'brisket' disease which, in its classic form, causes intense dyspnoea, leaves the afflicted animals unable to tolerate even mild exertion, and may lead to cardiac failure. The name is given because of the oedematous swelling in the muscles of the brisket of cattle.

When cattle are taken to high altitude, some individuals prove to be susceptible to pulmonary hypertension while others are resistant. Obviously it is in the susceptible individuals that pulmonary hypertension may progress to such severity that 'brisket' disease develops. Among herds of cattle native to sea level and introduced into high altitude areas, the incidence of 'brisket' disease may be 20–30%. However, among herds established at high altitude, the incidence is only 1–2%.

This brief summary of the major physiological effects of temperature, light, and altitude on cattle and on other higher mammals clearly shows the great economic importance of the study of these problems.

11.1.2. Physiological effects of the weather and climate on sheep

(a) *Thermal effects.* Many studies have been made of thermal effects on sheep, their thermoregulation and the economic consequences of a thermoregulatory dysfunction.[30–34] Some of the major findings can be summarized as follows:

Rawson[33] pointed out that in a hot environment sheep are capable of rapidly reducing the temperature of the brain by panting, while the remainder of the internal body temperature is more slowly reduced. This phenomenon is a protective mechanism in the sheep which enables this animal to reduce the temperature of its brain selectively in a heat stress situation. The cooling of cerebral arterial blood has been suggested to be a means of preventing cerebral overheating and to be partly responsible for the high heat tolerance of the sheep.

In spite of a lowered brain temperature, the sheep continues to pant until the deep body temperature is reduced. From this observation it is apparent that until this is accomplished, a significant heat loss drive still remains; and it is probable that nervous structures outside the brain are sending important temperature information to a central integrator. Experimental cooling of the hypothalamus is capable of inducing a decline in the panting rate even while the animal remains under heat stress. Under these circumstances, the sheep hypothalamus still appears to be operating as a regulator.

These two observations in sheep, that panting selectively reduces brain temperature, and further reduction of brain temperature reduces panting, suggest that hypothalamic temperature receptors and temperature receptors outside the hypothalamus are both involved in body temperature control, including panting, shivering and vasomotor activity.

Thermal receptors outside the hypothalamus, which occur in the skin and spinal cord, are monitors of the internal thermal environment. Since the sheep hypothalamus can be cooled by panting while the deep body temperature remains elevated, one could expect temperature receptors in areas outside the brain and spinal cord.

Several years ago Rawson[33] studied the physiological effects of long term warming of the viscera of sheep. A sudden onset of panting occurred that could not be accounted for by any change in body temperatures that were being measured.

These results indicated that temperature sensors other than those known to exist in the brain, spinal cord or skin were being stimulated. Experiments showed that other thermoreceptors existed that were stimulated in the viscera in the abdomen. Studies by Rawson[33] showed that the response to the intra-abdominal heating of sheep was independent of the vena cava (vein returning blood to the heart) blood temperature.

With the evidence so far available, no specific location for these receptors within the abdomen can be designated. Studies by Wood[35] suggest that the intra-abdominal thermoreceptors probably lie in the wall of the gut. This is an especially attractive hypothesis when one considers that in sheep, fermentation of the ruminal contents can raise its temperature 1.0 °C or more.

A monitoring system close to this source of additional endogenous heat would be most appropriate for the animal. However, recent studies by Riedel *et al.*[36] suggest the possibility that the intra-abdominal thermoreceptors of sheep are situated in the peritoneum (lining membrane) of the abdominal wall. This concept would agree well with the observations made following unilateral splanchnotomy (dissection of the viscera).

Johnson[30] pointed out that apart from the thermoregulatory mechanism described above sheep lose water continuously by evaporation due to respiration and passive diffusion and active expulsion through the skin. Because sheep carry a fleece of continuously changing length, the control and importance of the contributions to total water loss and to heat loss are in many ways different to those in other genera.

The rate of water vapour loss of an animal depends on the water vapour pressure (WVP) gradient from its surface to the adjacent air. The WVP gradient in turn depends on the absolute humidity of the air and on the availability and temperature of aqueous solutions at the interface between the animal and its environment. On the skin where salts accumulate as sweat evaporates, surface fluids may become concentrated; this reduces the WVP and the rate of sweat evaporation.

Sweating and panting lead to loss of body heat only so long as the heat of vaporization of water from body surfaces is drawn from the body and not from the environment. If sweat evaporates at the tips of the wool away from the body surface and draws its heat of evaporation from the air, no cooling of the body occurs.

Unshorn sheep, exposed to a thermoneutral environment which does not increase the rates of metabolism or evaporation, lose approximately 25% of their heat by evaporation, about half being from the respiratory tract and half from the skin. Shorn sheep have thermoneutral temperatures of approximately 25–30°C and at this temperature the proportions of total heat loss attributable to respiratory and cutaneous evaporation and to non-evaporative processes are similar to those of unshorn animals at 10°C.

Below thermoneutrality, both heat production and heat loss increase. According to Johnson[37,38] in shorn and unshorn sheep the increase in heat loss is almost entirely non-evaporative, although exposure to very cold conditions does cause a small increase in pulmonary ventilation and, presumably, in respiratory evaporation. Above thermoneutrality, evaporation accounts for an increasing proportion of heat loss.

In unshorn sheep the increase in evaporation results largely from panting and only slightly from an increased cutaneous water loss. Further studies on water metabolism in sheep were carried out by Macfarlane[32] in Australia and important seasonal changes were observed.

(b) *Effects of light and altitude*. The various physiological mechanisms described under 'cattle' apply to a certain extent to sheep also.

11.1.3. Physiological effects of the weather and climate on pigs

(a) *Thermal effects.* As pointed out by Mount,[39] the pig was one of the earliest animals to be domesticated in primitive human settlements, where it obtained its food by scavenging and foraging in forests. Nowadays in industrialized countries the pig is raised intensively under conditions of confinement, so that for the animal the environment imposed by housing, rather than the out-door environment, has become dominant in many areas. Increasing study of the pig reflects the animal's great economic importance in meat production. The animal is also used increasingly in biological research, and miniature pigs have been bred for this purpose.

The outstanding characteristics of pigs in relation to the thermal environment are that the new-born pig is susceptible to cold and the mature pig is susceptible to heat. Several major adaptations stem from these basic facts. In the case of the new-born, continued exposure of the single animal to cold conditions can lead to death in hypothermia. However, if the young animal is with the other members of the litter and the sow, the behavioural adaptation of huddling means that cold can be withstood more easily and the range of temperature conditions in which the animal can live is considerably expanded.

The poor thermoregulation of new-born pigs in relation to cold, affecting their glycogen and protein metabolism, used to cause the death of 25% of the young animals during the winter in The Netherlands. This resulted in a yearly economic loss of more than 8 million guilders for the farmers. Three to eight weeks old piglets also have difficulties in absorbing enough iron in a cold humid environment which causes their pale skin colour. At present all farmers in The Netherlands keep the new born pigs in sties at 30°C and a relative humidity of 80%. Gradually the temperature is lowered to 17–20°C and the humidity to 60–70%. The floors, walls and doors of the sties are also well insulated to prevent rapid cooling. Practically no young pigs have died since these precautionary measures were taken.

In the case of the large mature animal, the sow or boar, exposure to hot dry conditions can lead to death in hyperthermia, because the pig cannot sweat effectively and its rate of respiratory evaporative heat loss is limited. Large pigs in hot climates must therefore be supplied with sprays or wallow to provide surface evaporative cooling; the pig is basically native to swampy woodland in which it has available this external wetting facility. Pigs of intermediate size, that is the rapidly growing animals which provide pork and bacon, show some of the reactions of the two extremes, but in modern farming the greater concern in respect of environment for these animals lies in the effects on feed utilization and growth, and consequently on energy metabolism, which plays a central role in adaptation to environmental change.

It has already been pointed out that pigs are more affected by summer temperatures than other farm stock because they are non-sweating animals.

Thermoregulation in pigs is achieved by the evaporation of water derived from within the body from the respiratory tract and from the skin surface. Water, which has fallen as rain or has come into contact with the pig's body as a result of wallowing, may also be vaporized by heat from the body; to a limited extent water may evaporate at the tips of the bristle coat where a large part of the latent heat can be drawn from the ambient air. These thermoregulatory processes in the pig have been studied by many workers[39-43] and their findings have been summarized.[1]

Examination of the pig's skin reveals that there are glands associated with each hair follicle and that these glands are relatively large in size. They do not, however, appear to be associated with thermoregulation even though they respond to adrenaline and noradrenaline as do similar glands of other species which do have a thermoregulatory function. Mount[39,41] exposed new-born pigs to a series of environmental temperatures in a calorimeter and measured the total amount of water lost. His studies indicated that the water loss could be accounted for by evaporation from the respiratory tract and it therefore appears that the pig does not sweat, a conclusion also drawn by Brody[44] and by Heitman and Hughes.[45]

Although the pig does not sweat it has the habit of wallowing in mud or water during hot weather and in the absence of water the pig may wallow in its own urine and faeces. The application of mud prolongs the time of water loss by 2 h so that a pig would need to make only infrequent visits to a mud wallow in order to maintain a high rate of evaporative heat loss. Moreover, as the pig is a relatively naked species the mud or water will be in contact with the skin rather than lying on the surface of the hairs. This is important because it means that the latent heat needed to vaporize the water is likely to be drawn in a large part from the pig's skin and so cool the body. Water which is evaporated from the surface of the hair on the other hand may be vaporized by heat mostly drawn from the ambient air and so produce little cooling of the body. In addition the evaporative heat loss achieved as a result of wallowing avoids the problems of water and salt balance which can be encountered in sweating and panting animals.

Under cold conditions after rain the absence of a thick hair coat would of course prove to be a disadvantage, because the cooling of the body by evaporation would be so much greater than in an animal, like a sheep, which is protected by a thick fleece and where the evaporation of water takes place from the tips of the hair rather than on the skin.

On exposure to high ambient temperatures pigs increase their respiratory frequency and decrease their tidal air volume in such a way that the minute volume (ventilation per unit time) increases. At the same time the heat exchange system is by-passed because panting occurs through the open mouth. Unlike most other panting animals (e.g. dog, sheep, cattle and chicken) there is no obvious secondary change involving an increase in tidal volume and decrease in frequency in the pattern of respiration at high body temperatures. The minute volume increases by a factor of about three in pigs with high body temperatures, which is less than that in sheep but similar to that of cattle.

Ingram[46] found that the respiratory frequency depends on both the skin temperature and the internal body temperature so that the respiratory frequency associated with a given hypothalamic and spinal temperature was higher at greater skin temperatures. There is also evidence that there are other regions deep in the body which are sensitive to heat, since even when the hypothalamus, spinal cord and trunk skin are controlled at a constant ambient temperature it was found that the respiratory frequency depended on the rectal temperature. In contrast to other panting animals the pig does not pant in response to the heating of either the hypothalamus or the spinal cord unless the ambient temperature is above the critical temperature. As in some other species heating the scrotum of the pig causes an increase in respiratory frequency while heating a similar region of the skin elsewhere has no effect.

(b) *Effects of light and altitude.* Very little research has been done on the effects of light and altitude on pigs, but it is more than likely that several of the mechanisms described under 'cattle' apply also to pigs.

11.1.4. Physiological effects of the weather and climate on wild ungulates

Ungulates or hoofed animals include many orders not closely related phylogenetically. Cattle, sheep and goats also belong to this large group of animals. The major ungulata orders are the *Perissodactyla* which are odd-toed and the *Artiodactyla* which are even-toed. The distribution of the surviving *Perissodactyla* is limited to Africa, Asia, and central and south America. To this order belong 16 well known species, such as horses, asses, zebras, tapirs and rhinoceroses. The *Artiodactyla* occur on all continents except Australia; they are widely distributed in most natural areas of these continents and have 194 species such as camel, hippopotamus, deer, giraffe and cattle. Because of their wide ecologic distribution in the arctic, desert, grasslands and tropics, ungulates display a variety of physiological patterns of temperature regulation. In general, due to their large size, ungulates cannot avoid climatic stress, as do small mammals, but must actively cope with it.

Thermoregulation studies on ungulates have been mainly oriented toward domestic animals, primarily because of their agricultural and economic value. Since 1961 several good review papers have been written on domestic ungulates,[47–54] e.g. by Bianca, Hafez, Johnson, Lee, Macfarlane, Schmidt-Nielsen and Whittow. However, it has been realized that study of only domestic ungulates might obscure some of the unique physiological processes acquired by wild ungulates living in their natural habitats.

A review of the thermoregulation of wild ungulates was written by Yousef.[55]

He found that rectal temperature (T_{re}) is a simple and an easily measured index both of the body temperature of animals and of the balance between heat production and heat loss. Normal or resting records of T_{re} for 20 species of wild ungulates are summarized in Table 11.1. With the exception of two species, white rhinoceros and hippopotamus, T_{re} ranges between 37 and 40°C. The degree of precision of thermoregulation in these species is reflected in their diurnal variations in T_{re}. The rather pronounced species differences in diurnal variations in T_{re} are illustrated in Table 11.2. While certain species such as the giraffe, impala and hartebeest show rather small diurnal variations in T_{re} (1–2°C), others have a markedly labile T_{re} with wide variations up to 7°C as exemplified by the oryx and eland. Fluctuations in T_{re} seem to be directly related to environmental heat load. In general, T_{re} rises slowly during the day reaching the maximum value in the afternoon and then decreasing gradually to a minimum in the early hours of predawn. This thermal instability represents a useful adaptive mechanism to cope with stressful environments.

Seasonal variations in T_{re} of wild ungulates have not been given attention. Observations on reindeer under Alaska climatic conditions showed no measurable seasonal differences in T_{re} of resting animals. There is no tax-

TABLE 11.1
Rectal temperature (T_{re}) and respiratory frequency (RF) of 20 species of wild ungulates at rest. Compiled by Yousef[55]

Species	Common name	RF (min^{-1})	T_{re} (°C)
I *Even-toed ungulates*			
Tayassu tajacu	Collared peccary	—	37.5–40.0
Giraffa camelopardalis	Giraffe	—	38.8
Ovibos moschatus	Muskox	—	40.0
Oryx beisa	Oryx	17	39.4
Taurotragus oryx	Eland	—	39.8
Gazella thomsonii	Thomson's gazelle	25	39.4
Gazella granti	Grant's gazelle	42	39.8
Cervus elaphus	Red deer	15–70	38.5–39.6
Connochaetes taurinus	Wildebeest	34	39.0
Aepyceros melampus	Impala	45	39.2
Alcelaphus buscelaphus	Hartebeest	28	38.7
Rangifer tarandus stonei	Stone caribou	—	39.0
Rangifer tarandus	Reindeer	50	38.6
Kobus defassa	Waterbuck	25	39.6
Lama guanico	Guanaco	25	38.7
Camelus dromedaris	Camel	12	36.7
Syncerus caffer	Buffalo	—	38.5
Hippopotamus amphibus	Hippopotamus	—	35.4
II *Odd-toed ungulates*			
Ceratotherium	White rhinoceros	—	35.2
Equus assinus	Ass, burro, donkey	15	37.1

TABLE 11.2
Diurnal changes in body temperature (T_b) of hydrated wild ungulates in the field or in a simulated desert environment in the laboratory (22–40°C). Compiled by Yousef[55]

Name	Range in T_b (°C)	Difference in diurnal variation T_b
Collared peccary	37.5–40.0	2.5
Giraffe	37.7–39.0	1.3
Camel	35.2–37.8	2.6
Buffalo	36.9–40.2	3.3
Burro (donkey, ass)	36.5–39.0	2.5
Eland	33.9–41.2	7.3
Oryx	35.7–42.1	6.4
Impala	38.8–40.4	1.6
Hartebeest	38.8–40.4	1.6
Grant's gazelle	39.4–40.4	1.0
Thomson's gazelle	39.0–40.0	1.0

onomic or geographic trend in T_{re} of ungulates. The available data show rather similar T_{re} of arctic, desert and tropical ungulates suggesting that T_{re} is not adaptive to climate. Changes in T_{re} resulting from changes in environmental temperature (T_a) point to the various patterns of thermoregulation. The data for 11 different species show that:

(a) the extent of changes in T_{re} over a wide range of T_a are not related to ecologic distribution or to phylogeny;

(b) at T_a above T_{re}, i.e. 45–50°C, all wild ungulates studied to date are able to thermoregulate effectively as shown by a stable high T_{re} that is well below T_a.

The highest T_{re} recorded at 50°C was 41.5°C in the impala and hartebeest. The ability to tolerate varying degrees of hyperthermia is noteworthy since it is exhibited by both arctic and desert species, large and small.

In recent experiments by Yousef on burros[55] he found that at a T_a of 46°C they were able to maintain a T_{re} of about 38.5°C for a period of 24 h without any apparent distress (one of the burros did not even decrease its food consumption). In addition, various species of antelope were able to maintain a remarkable degree of regulated hyperthermia for as long as 12 h.

The heat loss of ungulates is the same as in all mammals and includes non-evaporative and evaporative pathways as described in previous sections. The evaporative cooling involves both respiratory and cutaneous cooling. As in the case of sheep, the coat of ungulates has a great insulative value. This phenomenon was studied particularly by Irving,[56] Scholander[57] and Whittow.[58] Reindeer under winter conditions in Alaska are able to decrease the skin temperature of their extremities to near freezing, thus conserving energy and preventing hypothermia in the body.

The pattern and the control of the sweat glands in wild ungulates have been studied by many workers.[59–63] In all the species studied, thermal sweating is under adrenergic neuro control. Sweat glands respond to adrenergic drugs but not to cholinergic drugs. The sweat glands of the various species studied release their secretion in three different patterns: cyclical discharge, a continuous but stepwise rise in discharge and a continuous but immediate rise in discharge. No correlation was found between the patterns of sweating and the phylogenetic position of the species. Bligh[59] assumes that periodic or cyclic discharge by the sweat glands is caused by the contraction of the myo-epithelium surrounding the gland while the continuous discharge of sweat could be caused by a sustained overflow of fluid onto the skin surface. Relatively little work has been done on the thermoregulation of arctic ungulates. The one species that has been given attention is the feral reindeer. Studies by Scholander[57] have revealed a good correlation between the thickness of fur and insulation, but no correlation was found between insulation and body size among large mammals i.e. large mammals have about the same insulation per unit of surface area. Also it was shown that large arctic animals have a very wide thermoneutral zone (from +30 to −40°C). This means that total insulation would reach a maximum value only at a T_a as low as −40°C.

The main problem for animals in the desert is not how to stay warm but how to keep cool with and without a source of water. Wild ungulates in the desert utilize several mechanisms to cope with heat and aridity:

(a) *Labile homeothermy (i.e. variable T_{re})* Schmidt-Nielsen[53] was the first to demonstrate that variable diurnal T_{re} of the camel is an important mechanism for survival in the desert. Hyperthermia during the hottest part of the day results in both conservation of water, as a result of heat storage in the body, and a reduction of the amount of heat gain from the environment, as a result of reducing the gradient between T_{re} and T_a. This mechanism has been found to be used to varying degrees by several other species even when hydrated and resting.

(b) *Reduced heat production*. The recognition of labile homeothermy raises the question of how the brain and other heat sensitive areas of the body can tolerate this degree of hyperthermia.[64–66] In some ungulates the internal carotid artery is absent and blood is supplied to the brain primarily by the external carotid artery. At the base of the brain, the carotid artery forms a network (rete) of small blood vessels consisting of hundreds of small parallel arteries which reform to supply the brain with blood. This rete lies in a venous pool, the cavernous sinus. Venous blood returning from the nasal passages is cooled by evaporation as the animal breathes and drains into the cavernous sinus. This cool venous blood exchanges heat with the arterial blood thus cooling the blood supplying the brain. This counter-current cooling of the brain is of importance for the tolerance of extremely high T_{re}.

(c) *Increased dehydration tolerance*. The maximum ability for tolerance to dehydration is not known but some ungulates are known to tolerate a loss of water corresponding to 30% of their original body weight even at T_a of

40 °C. During dehydration, most of the ungulates which have been studied have shown various degrees of decreasing heat production and evaporative water loss. In some species, such as Grant's and Thomson's gazelles, dehydrated animals abandoned evaporative cooling and instead allowed T_{re} to increase to a point where T_{re} and T_a were about the same. Dehydrated gazelles began to pant when T_{re} exceeded 40 °C.

During dehydration, ungulates decrease their urine output and water loss in faeces. Associated with the reduced urine volume some ungulates excrete concentrated urine whereas others such as donkeys appear unable to do this.

The mechanisms involved in increased tolerance to dehydration seem to be related to the fact that some ungulates are able to maintain their plasma volume during dehydration as has been reported on the camel and the burro. However, in other ungulates such as the guanaco and reindeer, plasma volume decreases significantly with dehydration. It seems therefore, that there is no one common mechanism to explain tolerance to dehydration and further studies are needed to reveal all the mechanisms involved.

The rapidity of rehydration is remarkable in wild ungulates. In about five minutes or less, the burro can drink enough water to restore its body water deficit. The nature of the signal that stops drinking when water deficit is restored is a mystery. It is suggested[55] that a hypothalamic inhibiting receptor responds to gastric distension with a turning-off signal. The ability to fill up with water rapidly, to escape predatory enemies lurking at water holes, has been suggested as having an important survival value.

Several species of antelope continue to inhabit inhospitable areas of the desert even during prolonged droughts. However, when the minimum water requirements of these same animals were measured, Maloiy and Hopcraft[67] found that they required between 2–3 l/100 kg of body weight per day. How then can these animals survive without drinking? Taylor studied this problem and found that the eland which seeks shade during midday beneath Acacia trees and eats their leaves, can obtain a sufficient amount of water from these leaves and probably can live indefinitely without drinking. The oryx and the gazelle, however, neither seek shade nor eat moist leaves. These animals eat dry grasses and shrubs. Taylor[68-70] confirmed the early observation by Buxton[71] that some dried grasses collect moisture from the desert air at night when relative humidity is highest, i.e. these plants are hygroscopic. It is then possible to see that by grazing at night, these wild antelopes could meet their water requirements without drinking.

The various mechanisms described above partly explain why wild ungulates are able to live in the most inhospitable arctic and desert environments. However, many more physiological studies are required before all the meteorotropic problems of this large group of animals can be solved.

11.1.5. Physiological effects of the weather and climate on ruminants

The term 'ruminant' is used for a large group of herbivores that ruminate and have a highly specialized and complex stomach. They include the camels, the Bovidae, Cervidae, Giraffidae etc. They are distributed over a variety of climatic and vegetational zones. Of the ruminant species so far studied evaporative heat loss mechanisms are very similar; under hot conditions they all pant and sweat, the amount of heat lost by either of these two mechanisms varying with the species. The physiological mechanisms involved have been fully described.[1,72-77]

The mechanism of panting was described by Robertshaw[62] as follows: 'panting is initiated by a specific thermoreceptor input; after initiation, the typical rapid shallow pattern of respiration is the net result of a specific thermoreceptor drive and possibly of heat stimulation of the medulla oblongata to increase total respiratory ventilation.'

Robertshaw[75] pointed out that all ruminants are exposed to high potassium intakes since normally the plants they eat are made up of cells containing potassium but with little sodium content. Tropical pastures are even lower in sodium than temperate zone grasses (by a factor of 10 to 20) since sodium is leached from tropical (and mountain) soils by the high rainfall. Potassium concentration also is usually greater in the temperate zone grasses than in tropical pasture. Potassium acts as a diuretic and it moves rapidly through the kidneys, partly by direct passage and partly by secretion from the distal tubule.

In mountains and tropics where there is little sodium, aldosterone is secreted in order to retain renal sodium and thus maintain the extra-cellular volume. In the desert, however, there are many plants which accumulate sodium and some of them also store potassium (e.g. *Kochia*). Leaves of the saltbush family (*Atriplex*) may contain over 15% of their dry weight as sodium salts (3 mol kg^{-1}) with potassium at 1 mol kg^{-1}. When such food is eaten, ruminants compensate by drinking extra water proportionally to the salt load, which is thus excreted.

One of the most interesting groups of ruminants is the camels. They were particularly studied by Schmidt-Nielsen[76] and associates, by Macfarlane,[72] Maloiy[74] and Siebert.[78] Some of their most characteristic mechanisms in relation to thermal stresses were summarized by Maloiy.[74]

The ability of the camel to cross hot waterless areas, for days or even weeks without drinking water has always puzzled environmental physiologists. It is only recently that physiological and behavioural mechanisms which enable the camel to survive in these hostile environments have been studied by biologists.

The camel weighs over 500 kg and it cannot escape the heat of the day as can for example a small desert rodent, the Kangaroo rat, which escapes by retiring to its underground burrow. However, by appropriate behaviour the camel does reduce the amount of heat it receives from its surroundings. In addition the camel keeps cool by sweating following heat exposure. In the desert, therefore, the camel uses water for evaporative cooling. This explains why in the Sahara Bedouin still require to water their herds of camels regularly.

The water metabolism of the one-humped camel (*Camelus dromedarius*) has been studied in the field in the Sahara and central Australia and in Kenya under controlled conditions inside a climatic chamber.

Schmidt-Nielsen[76] made studies of the body temperature of camels in relation to the water economy and metabolic rate. When exposed to high air temperatures (40–45 °C) and solar radiation, the camel dissipates excess heat by increasing cutaneous evaporation of water (sweating). At the same time, at these high temperatures and especially when dehydrated, the camel allows its body temperature to increase by as much as 3–5 °C, from below 35 °C in the early morning to 40 °C after eight or more hours of exposure to a heat load and solar radiation in the desert.

Following dehydration in the desert, the camel reduces its faecal water loss. Thus a camel exposed to the desert heat and deprived of water for long periods will excrete small dry faecal pellets. There is greater water and salt absorption from the large intestine following severe dehydration in the camel, than in other large herbivores. Another feature of a camel's water metabolism is the remarkable extent of its ability to reduce urinary water loss following dehydration.

Due to these various mechanisms the camel can tolerate extreme water depletion. Camel, donkey and sheep tolerate without ill effects a loss of water corresponding to 30–31% of the initial body water. Man is only able to withstand a loss of water corresponding to 10% of the body weight and he will collapse and die after a loss which exceeds 12%.

Coupled with this high degree of tolerance to water loss, the camel like other ruminants, rehydrates itself quickly following dehydration. A dehydrated camel can drink 40–60 l of water or 60–80% of its loss of weight in 10–15 min to restore the water deficit. The rate of water absorption across the rumen wall (first stomach of ruminants) seems to be slow in camels. During the period of rehydration the red blood cells do not undergo haemolysis as has been observed in other animals such as cattle. This helps the camel to ingest large quantities of water without any threat of red blood cell rupture.

11.1.6. Physiological effects of the weather and climate on birds in general and on domestic chickens in particular

The weather and climate influence birds indirectly, through food, and directly, to the extent that certain meteorological factors (particularly temperature and solar radiation) affect the production of sexual hormones, the sexual ratio, the laying rhythm of fowls, etc. According to some authors, weather and climate affect the colour and appearance of the feathers. Also, the direction-finding capacity of migrating birds has been ascribed by several research workers[79] to the capacity of various animals to locate themselves by means of the rays of the sun, their direction, intensity and degree of polarization.

In the case of birds a sharp distinction must be made between the effects of the thermal environment and the effects of light, the so-called photo-periodic effects.

(a) *Thermal effects.* Many authors have studied the thermoregulation, metabolism and endocrinal functions of birds in relation to the meteorological environment. Excellent review papers can be found in Animal Biometeorology.[1,25,80-86] The following section summarizes some of the major findings.

In contrast to poikilothermic reptiles birds, like mammals, are homeothermic animals which are small and consequently have relatively large surface areas from which heat is lost; they have high normal body temperatures (commonly higher than 40°C, i.e. higher than those of mammals). Therefore the problems of heat loss are potentially acute for them.

(i) *Effects of cold*: Rising[81] pointed out that apart from an increase in the metabolic processes, certain non-physiological responses to cold (such as the development of highly effective insulating layers of feathers and fat), modifications of the circulatory system to reduce the rate of transfer of deep body heat to superficial areas, and communal nocturnal roosting have received much emphasis. Modifications of the insulating coat play a key role in the annual adjustments of a bird in response to hibernal cold; cold-adapted species possess better insulation than do warm-adapted ones.

With exceptions among certain avian orders, body temperatures of birds in an inactive state in moderate temperature environments range from 39–44°C during the daily waking stage.[82] This range contrasts with the range of 36–40°C under similar conditions for higher mammals. Although many factors may influence body temperature (age, sex, breed, size, activity, diurnal variations and the insulating value of the outer integument), the above comparison indicates that in periods of low environmental temperature birds may require a higher metabolic intensity to maintain relatively constant deep body temperatures by endothermic means.

According to Siegel[82] birds living in polar regions are remarkably tolerant to cold. Although larger species may have sufficient feather insulation to live with no special metabolic effort, smaller species are not so well protected. For example redpolls, chickadees and gray jays, ranging in weight from 12–60 g, live at temperatures of −20 to −50°C although their insulation appears to be inadequate. The daytime critical temperature of redpolls in Sweden is about −20°C their metabolic rate is some 300% greater than their resting rate. Similarly, pigeons (*Columbia livia*) with critical temperatures of −20°C have metabolic rates about 220% above resting rates at −30°C. The increases in food intake in subsequent days indicate that acclimatization is related to the ability to sustain elevated heat production by metabolic means and that thick feather coats alone cannot protect birds against severe cold.

Several studies have shown that heat production of birds without shivering (so-called non-shivering thermogenesis) is a relatively unimportant source of heat in birds. It was found that the shivering reflex is already present in newly hatched chicks.

(ii) *Effects of heat*: the large numbers and the diversity of species that live in hot climates are indicative of the adaptability of birds to high temperature environments, moreover, because most birds are diurnal and generally do not utilize underground burrows for shelter, their chances to avoid the midday heat are further reduced.[82]

In ambient temperatures below the upper limit of thermoneutrality, the bird loses heat to the environment by physical means (radiation, conduction, convection and evaporation). But as temperatures approach the upper critical limit, sensible heat loss becomes negligible and insensible heat loss (i.e. evaporative) becomes the only method for heat dissipation.

In recent years evaporative water loss in birds has become a subject of extensive study. It has been commonly assumed that most birds are unable to dissipate more than approximately half of their metabolic heat production by panting and that evaporation from the skin is a negligible part of the total water loss of the bird because the bird lacks sweat glands.[83] However, Lasiewski *et al.*[87] found that a wide variety of birds are capable of maintaining their body temperature below environmental temperatures of 44 to 46°C for periods of from one to four hours by some evaporative cooling, provided that the water vapour tension is low. This observation was confirmed in 1969 by Schmidt-Nielsen *et al.*[88]

An economically important group of birds is the domestic chickens. It is generally agreed that the modern chicken arose from the jungle fowl of Southeast Asia. Probably as a result of more than 3000 years of domestication, the chicken has become acclimatized to a much broader range of environmental conditions than exist in a tropical jungle. Consequently, the chicken has adapted reasonably well to most areas of the Temperate Zones of the world. Man, of course, has assisted in the adaptation process by providing varying degrees of protection against meteorological stresses.[80]

The thermoneutral range, which is defined by Sturkie[89] as being the environmental temperature range within which there is little or no change in heat production by the bird, is an indicator of the optimum environmental temperature for the bird. According to Reece and Deaton[80] the range for the young chicken is very narrow; for the mature chicken it is also rather narrow when it is compared with that for smaller birds that are native to the Temperate Zones. The optimum environmental temperature for chickens starts at 35 ± 1°C during the first week and decreases to

33.5 ± 1.5 °C by 3–4 weeks of age. At maturity the optimum temperature is 21 ± 3 °C.

The optimum temperature for maximum growth of 1-day-old chicks seems to be the thermoneutral range; but at 5 weeks, the optimum temperature is lower than the thermoneutral range. Because of the rather rapid change in temperature required during the early stages of chicken growth, the domestic growing of chickens, after they hatch, is usually divided into two phases, the 'brooding' phase and the 'grow-out' phase.

The brooding phase in commercial production covers the period from hatching to three or four weeks old, temperature control is usually accomplished by providing a heater that directs heat downward over an area about 2 m to 4 m across, where from 500 to 1000 chicks are placed shortly after they hatch. The heater is equipped with controls to permit the temperature to be regulated rather precisely at about 35 °C.[80]

The grow-out phase covers the period from after four weeks of age (end of brooding phase) to the age at which the chickens have reached the degree of maturity for the intended use. For meat-type chickens the age at slaughter may range from about 6½ weeks for specialty use to about 10 or 12 weeks for general use, in areas where a heavy dressed chicken is in demand.

Chickens can be grown over a much wider range of temperatures during the grow-out phase than they can during the brooding phase. However, the temperature for optimum performance is 18.3 °C, which is the thermoneutral range for mature chickens. In the temperate zones of the world, on a seasonal basis, the climatic temperatures differ widely from the optimum temperatures required for chicken growth. Therefore for growing chickens commercially, shelter against thermal stresses must be provided by appropriate housing.

During the winter season, poultry house design, insulation, ventilation, stocking rate, and supplementary heat are balanced so that the temperature in the house can be maintained within the range from about 5 to 21 °C during the grow-out phase. As rearing temperatures drop below 18.3 °C, feed consumption increases. This increase in feed consumption can be offset by maintaining the house temperature above 18.3 °C by improving the housing or heating the house, or a combination of both. The actual temperature to be maintained will be based on an economic analysis that involves housing costs, the cost of fuel to heat the house and the cost of feed.[80]

During the summer season, poultry reared in the temperate zones of the world may be exposed to temperatures ranging from about 15 to 44 °C. The diurnal temperature range is quite variable, depending largely on the geographic location.

High temperatures during summer, produce two major effects, depressed growth and decreased survival rate. The excess body heat of homeiotherms in temperatures above the thermoneutral range must be dissipated in two forms: the so-called sensible heat, which is lost by conduction, convection, and radiation to the surrounding environment; and the latent heat, which is lost through the evaporation of water from the skin surface and the respiratory tract. In the avian species, practically all of the latent heat loss is from the respiratory tract. Therefore, the relative humidity of the ambient air has a pronounced effect on poultry growth and survival at high temperatures.

(b) *Effects of light.* On pp. 267–299 the various effects of light on animals in general and on cattle in particular were discussed. The effects on birds have been studied, particularly by Benoit[90] and Assenmacher[91] in France since 1953. Their findings and the studies of Miline[92,93] and Hollwich[94] on light effects have been discussed extensively in Section 5.2.3.

Benoit[90] and Assenmacher[91] observed an increased secretion of gonadotrophic hormones after light stimulation. Cardinali[25] pointed out that the relative effectiveness of different wavelengths of light to induce gonadal changes in birds has been tested in different species. Light from the red end of the visible spectrum has been shown to stimulate gonadal development in starlings,[95,96] turkeys,[97] sparrows,[98] ducks[99] and quails;[100] in contrast, blue or green light have little or no effect on the activity of the gonads in these avian species. White leghorn pullets exposed to short wavelengths attain sexual maturity earlier than pullets exposed to red or white light.[101,102] Since in this experiment the photoperiod length was decreased during the rearing period the observations were interpreted as the inhibitory effect that results from reducing daily hours of a gonadal activating radiation, i.e. red or white light, rather than as a stimulation caused by green or blue light.

Photoperiodic testicular responses in birds involve retinal and extra-retinal photoreceptors. Increases in gonadal function by light can be induced or maintained in several avian species after removal of the eyes. Though the exact neuroanatomical location of extra-retinal photoreceptors is uncertain, there is evidence that they exist in the brain, in particular in the hypothalamus.[103] Moreover extra-retinal photoreceptors are more photosensitive than retinal photoreceptors in mediating gonadal changes in birds. In the duck, where retinal receptors respond only to orange-red light, extra-retinal receptors respond to light throughout the visible portion of the spectrum.[90]

From a practical point of view the most important studies have been carried out in relation to season and the laying rhythm of hens. An interesting summary was published by Van Albada in 1958.[104] He pointed out that, whereas in wild birds the laying of eggs is limited to a few months in spring time, prolonged domestication and selection have extended the period of egg production in the domestic fowl to the entire year. Still, many domestic fowls are sensitive to seasonal influences. Thus the laying results, in temperate zones of the northern and southern hemispheres, indicate a maximum production in spring, whereas in the tropics the production is equal throughout the year.

Investigations have shown that the change of hours of daylight is a dominating factor in the seasonal effect on laying rhythm. Light stimulates the pituitary. It secretes follicle stimulating hormone. Hen's ovaries are stimulated to activity and the egg follicles develop and mature. The almost mature follicle sends a, probably hormonal, stimulus to the hypophysis, under the influence of which the latter causes the follicle to complete maturity and ovulation. Between the secretion of the ovulation inducing hormone and ovulation itself a fairly constant period of about 8 h elapses.

It is not only the egg production which is affected by season, but the age at which sexual maturity is reached as well. In the northern latitudes it is well known that 'early' pullets, hatched between January and March, will reach sexual maturity earlier than 'late' pullets, hatched between March and May.

Particularly interesting are the effects of light on the sexual ratio of birds (see Section 8.4).

According to studies by Riddle and Fisher in 1925,[105] using three species of pigeons kept on the same diet throughout the year, in early spring with increased thyrotrophin production and thyroid activity there is an excess of male offspring, whereas, during periods of less activity and smaller sized thyroid (in summer), there is an excess of female pigeons.

11.1.7. Physiological effects of the weather and climate on fish

Although the phenomena of the hydrosphere* do not belong to biometeorology a few words must be said on the thermal conditions of fish. The great majority of fishes are poikilotherms, expending no energy on maintaining differential body temperature by specialized metabolic or behavioural means. They are free, however, to use temperature preference within the environmental circumstance available, to select a suitable body temperature. As such, they are thermal conformers; every response and every process proceeds within a temperature range dictated by the immediate environment. Consequently growth in temperate climates is strongly seasonal, almost stopping in winter months for some fish. Yet fish are particularly plentiful in sub-arctic regions of the ocean, where nutrient rich water provides abundant food; they may even grow to a fair size while inhabiting ice tunnels under the polar cap. Adaptive or compensative mechanisms have evolved to meet the wide range of environmental temperatures. In general, tropical fishes grow fast and mature early; the reverse applies to polar species. Studies on the temperature regulation of fish were particularly carried out by Brett,[106] Ferguson,[107] Fry,[108] Sullivan[109] and others. In Table 11.3 the body temperatures of a number of fishes are given

*Although there is an interrelationship between the hydrosphere and the atmosphere, biometeorology, by definition, is the study of only atmospheric effects.

according to Fry[108] based on the work of Bennett,[110] Ferguson[107] and Spigarelli and Thommes.[111]

With the exception of tunas and sharks, fish can attain a certain constancy of temperature by behavioural regulation; i.e. they are likely to be found in waters of congenial temperature and live in a range of temperatures which is characteristic for the species.

There is evidence that both peripheral and central components of the nervous system are involved in behavioural temperature regulation in fish as in other vertebrates. While the vast majority of fish achieve a certain constancy of body temperature, which they attain by thermokinetic behaviour, there are two groups of fish, the lamnid sharks and the tunas, which have a limited physiological regulation of their body temperature, chiefly of the muscles. In these fish, which are open water species and swim continuously, the heat produced by the activity of the red swimming muscles is conserved by a special arrangement of blood vessels to give a counter current heat exchanger.[107–109, 112]

In all fish the blood temperature is brought almost exactly to water temperature as it passes through the gills and all blood passes through the gills after each passage through the heart before being returned to the body. Fish are drained of their metabolic heat after each circuit of the blood except of the minor heat exchange between the ordinary parallel venules and arterioles, which in large fish can give an excess temperature of the deep muscles of a degree or so Centigrade. In the lamnids and the tunas, however, the red muscle which supports continuous activity is segregated into a discrete muscle running along the length of the fish and imbedded in the white muscle which supports the burst activity.[113,114] In these fish, instead of being fed from the dorsal aorta, the red muscle receives its major supply from branches of the greatly developed lateral artery and is drained by tributaries to the lateral vein. The cold entering arterial blood is warmed by the leaving venous blood and the metabolic heat is thus conserved in the muscle.

TABLE 11.3
Body temperatures of various fish captured in warm effluents, in the vicinity of a thermal power plant, in relation to their temperature preference as found in field observations or laboratory experiments. From Fry[108]

Name	Species	Lake temp. °C	Body temp. °C
Lake trout	Salvelinus namaycush	8.2	11.9
Brook trout	Salvelinus fontinalis	10.2	15.4
Rainbow trout	Salmo gairdneri	9.7	14.4
Brown trout	Salmo fario	11.3	16.1
White sucker	Catostomus commersomi	13.9	18.7
Carp	Cyprinus carpio	15.9	23.1
Yellow perch	Perca flavescens	17.2	22.5

In very large tunas the deep muscle temperature can be very constant over a wide range of water temperatures but in fish of moderate size such constancy is limited.

For further details see *Progress in Animal Biometeorology.*[1]

11.1.8. Physiological effects of the weather and climate on insects

The class Insecta belongs to the Phylum *Arthropoda* which comprises about 800 000 species. Of these, about 75% of all species are insects, each having their own ecological relationships. Their function in the biosphere can be both favourable and unfavourable. They may cause or transmit dangerous plant, animal and human diseases.

The *Arthropoda* consist of terrestrial, marine and flying organisms, the latter group has a very wide distribution due to aerial transport.

The distribution of terrestrial arthropods depends primarily upon meteorological factors. The *Arthropoda* are poikilothermic, i.e. their body temperature varies with that of their environment.

The influence of weather and climate on insects was reviewed by Haufe[115] who pointed out that the response in insects to the physical environment differs in many respects from that in man and in other higher animals. Mammals, including man, cannot discern air humidity as such; only under conditions of high water vapour pressures and high environmental temperatures, which reduce dissipation of heat by evaporation, they experience a great discomfort. Insects, on the other hand, react rapidly to changes in humidity through extra sensory organs. For example tsetse flies have humidity perceptors in their respiratory spiracles which enable them to control respiratory water loss. Insects are equally sensitive to air temperature and avoid unfavourable thermal conditions. They generally exhibit a higher sensitivity to factors in the physical environment than do higher animals and this is manifest in a highly developed degree of orientation to changes in their surroundings.

Insect activity depends greatly on the following six environmental factors:

(a) *Temperature* is a dominant factor in the development, behaviour, reproduction and survival of insects. Temperature-rate relationships for development and growth in insects are not considered to be linear; comparison of the effects of exposure to variable temperatures with those to constant temperatures should be considered with some reserve. For example at lower temperatures the development is more rapid for fluctuating temperatures than for constant temperatures; while at higher temperatures growth is retarded.

(b) *Humidity* conditions in insect habitats have an important influence on activity, distribution and survival. Transpiration affects water balance, depresses metabolism and retards development in some species. Most observations on insects in the field have been correlated with mean humidity levels and very little is known about the importance or influence of continuously varying humidities in nature. Low variable humidities stimulate egg production in some insects whereas females of other species oviposit more freely when humidity is high. Humidity is highly correlated with flight and dispersal and is a limiting factor in the duration of activity in the cycling of flight.

(c) *Light stimuli* modify both temperature and humidity effects. Light influences the behaviour of some insects, and diurnal responses to light may have adaptive advantages for many species. Alternating light and darkness provide an essential clue to the oviposition cycle, for example in *Aedes aegypti*, since cyclic behaviour breaks down in both constant light and constant darkness, and since aperiodic behaviour occurs in constant light irrespective of the light conditions during larval upbringing. Photoperiodism influences the development of insects; with thermal stimuli it is a factor in the induction of diapause, i.e. a dormant state which is under hormonal control and which facilitates survival of the species in certain stages of metamorphosis during unfavourable seasons. The abundance of some nocturnally active biting flies such as tsetse flies and mosquitoes is also related to lunar cycles within the fly season.

(d) *Atmospheric pressure*: changes in atmospheric pressure influence the activity of mosquitoes considerably, either as a direct mechanical effect or indirectly through alteration of atmospheric constituents whereas in man small changes in atmospheric pressure have little or no effect. Winds have a considerable influence on the activity and orientation of flying insects, especially in their dispersal from breeding grounds.

(e) *Rainfall* is directly related to the abundance of many insects such as mosquitoes and tsetse flies.

(f) *The effects of electric fields* (electrostatic and electromagnetic fields and air ions) on insects have been much studied in recent years. Contrary to the case with humans several observations suggest considerable effects of electric fields and air ions on insects.

(i) The possible effect of electrostatic fields (potential gradients in the atmosphere) on insects has been discussed by many authors. Fabre,[116] Uvarov[117] and Wellington[118] described the effects on insect behaviour in particular during thunderstorms. Schua[119] showed that worker bees increased the number of their foraging flights during periods of electrical activity in the atmosphere. Guard bees became hostile when alternating potentials were placed at either side of the hive entrance. Flying insects are repulsed when they approach a positively charged screen.

(ii) Levengood and Shinkle[120,121] observed in *Drosophila melanogaster Meig.* an increase in the average progeny yields of 34.4% as compared with the controls if the insects were subjected to electric fields. They believed that the actual cause of the effects is due to changes in natural ion levels.

(iii) Field observations by Maw[122] and others have shown that there is an apparent relationship between the location of male swarms of *Aedus trichurus Dyar* and the natural electrical gradients that exist around trees and shrubs in a forest.

(iv) Studies by Haine and König[123] have shown that moulting of the aphid *Myzus persicae Sulz* was reduced when electromagnetic waves of 9 Hz and high intensities were applied. Moulting increased with medium intensities. Low electric intensities were followed by a delayed increase.

(v) The influence of an excess of positive or negative ions on insects was studied by Haine.[123] She reported effects on the aphid *Myzus persicae Sulz*. It was found that under constant temperature, humidity and light conditions moulting increased in negatively ionized air after 10–20 min treatments.

(vi) Maw[122] observed that the emergence of the pupae of *Neodiprion lecontei Fitch* was considerably retarded if they were previously subjected to negatively ionized air. The insect *Calliphora vicina R.P.D.* showed an increase in flight activity were previously treated with positively ionized air. The peak occurred about three quarters of an hour after the first ionic exposure.

(vii) According to Maw[122] the blow fly *Phaenicia serocata Meig* shows greater flight speeds if treated with positive ions, while negative ions resulted in a more moderate flight, with fewer bursts of speed. Exposure to ions of alternating polarity after a period of exposure to positive ions produced an increase in flight speed.

Several of the factors mentioned above, particularly temperature and humidity, interact to a high degree in affecting insect activity, development, and survival. Reproduction and development in some species are influenced differently by temperature and humidity together than by either factor alone. Humidity affects the temperature preference of some species while temperature modifies humidity preferences in other species. Research workers are not in complete agreement on the parameter that best accommodates the combined effects of these two factors on an insect. Wellington[118] found that evaporation was a better indication of controlled variations in temperature and humidity on insect behaviour than either of the controlled variables alone. The evaporative power of air is influenced by wind velocity as well as by temperature and relative humidity. A saturation deficit of the air is closely related to the transpiration of moisture from some insects in controlled-humidity environments in the laboratory.

As many communicable diseases, such as malaria, trypanosomiasis, yellow fever, and filariasis, depend on an insect vector for transmission either among humans or between humans and certain other animals, that act as reservoir for the parasitic disease organism, it is evident that the problem of the influence of weather on insects and diseases is an extremely complicated one.

Ross,[124] in 1899, was the first to realize that the universal disease malaria, caused by the haemosporozoan *Plasmodium* is transmitted by mosquitoes. It exists where certain *anopheline* mosquitoes maintain themselves in sufficient numbers to ensure transmission. Of the more than 25 identified *anopheline* mosquitoes the disease is specifically attributed to *Anopheles gambiae Giles* which acts as the main vector. This species is able to maintain sufficient numbers to ensure transmission of the disease, feeding on the blood of man and other animals. The annual rainy season in tropical countries is usually followed by a dry season during which water pools, suitable for the life-cycle of *A. gambiae*, dry out. Thus, the numbers of mosquitoes diminish, and those left are believed to undergo specialized physiological changes to maintain themselves until the next rainy season.[125]

Malaria mosquitoes are very sensitive to differences in smell of the human body. Studies by Muirhead-Thompson[126] in Jamaica and Trinidad revealed that babies are bitten more than adults, males more than females, and Thomas[127] observed the same in Sierra-Leone (Africa).

Similar relationships were observed in the case of yellow fever. It is a disease that has been known for a long time. It was not until the beginning of this century, however, that the causative agent was isolated by Reed and his colleagues in America. They found that the blood of patients during the first three days of the disease contained a filter passing virus which was highly infective. They also realized that the disease is transmitted by the bite of the mosquito *Aedes aegypti Linnaeus* which is predominantly a domestic species. It is diurnal in habit and its activity and development are greatly affected by temperature, though no natural environmental temperature has proved fatal to the species in the field. At temperatures above 23 °C the biting cycle of the mosquito starts in the morning and continues until late afternoon.[125]

Until recently, yellow fever was endemic throughout the Panama zone, Gulf of Mexico, West Indies and the Brazilian Coast. In Africa the fever was thought to be confined to a narrow zone on the West Coast. Smart and Brown observed that among persons of the same race and skin colour, *Aedes aegypti L.* is most attracted by the person with the highest skin temperature. A heated or cooled hand of the same person is less attractive to mosquitoes than the normal hand. Dark skins or clothes attract mosquitoes most (according to Gjullin[128] the attraction decreases from black through red, brown, yellow to white).

Brouwer[129] was able to eliminate various possible causes of the above observations by carefully planned laboratory studies, using an air-current olfactometer. The differences in attraction are not due to differences in radiation or movement of the subject. Nor are differences in temperature and humidity primarily responsible. It was found that the main cause is the difference in smell, which is due to different concentrations of propionic acid, isobutyric acid and other fatty acids in the sweat of the skin.

It is evident that all meteorological factors affecting the composition and the amount of sweat on the skin, its evaporation and spread of the smell in the air are likely to play an important part in the spread of malaria, yellow fever and other similar diseases. It is also due to the local microclimate near houses that, in the tropics, the majority of mosquitoes are found on the lee side of the huts in which the native population is living.

Another important group of insects responsible for the transmission of serious diseases are the sandflies (*Phlebotomus*). They are widespread in arid and savanna areas in north Africa and are responsible for transmission of Phlebotomus Fever, Cutaneous Leishmaniasis or Oriental Sore and, most important of all for Visceral leishmaniasis, known more commonly as Kala-azar. Sandflies are also known to cause Harara, a skin condition resulting in urticaria as a reaction to their bite. This group of insects consisting of 36 species was particularly studied in the Sudan by Abushama and his colleagues.[125] Kala-azar disease is caused by a protozoan parasite of the genus *Leishmania*, which infects the liver, spleen and other visceral organs as well as the bone marrow of man.

According to Abushama[125] animal burrows and cracks in the ground are important for the development of sandflies. Long-lasting fissures in cracking clay during the hot dry seasons which are saturated during a short rainy season preceding several months of dryness, offer ideal breeding areas for the flies. The optimum physical conditions for all species of sandflies are: still air, darkness, a temperature of approximately 28 °C and a high relative humidity. When conditions are not optimal the females of some species refuse to feed and die of starvation.

Sandflies show marked seasonal variations in numbers depending on the prevalent physical factors and meteorological conditions. They disappear during heavy rain to reappear during the following dry period just before the next rainy season. They show diurnal periodicity and are mainly nocturnal. During the daytime they rest in dark damp corners in houses, behind cupboards, in crevices, trees etc.

Another important group of flies are the house flies which are well known as an actual menace to human health.[125] They habitually enter the houses, and come in contact with human food or drink, after breeding or feeding on excrement or dead animals. These flies are capable of transmitting pathogenic organisms that cause serious human disease, such as typhoid fever, dysentry, cholera and tuberculosis.

The house fly develops in almost any fermenting organic matter. It has been bred from pig, chicken and cow manure, decaying vegetation etc. Warm dry weather prevails for most of the year in the greater part of north Africa and small amounts of refuse usually dry quickly when exposed to the sun. In the Sudan flies increase in February and March before the hot weather of April. A similar increase in the middle of the dry season sometimes occurs on the Red Sea coast where flies increase during the rainy season which occurs there in winter.

The house fly transmits pathogenic organisms largely because of its method of feeding. From human excrement, open sores and putrifying matter, flies may visit man's food or drink, healthy mucous membranes or uncontaminated wounds and directly transmit infective organisms. Therefore house flies play an important role in the transmission of many human and animal diseases. They have been found to disseminate typhoid fever, infantile diarrhoea, cholera, bacillary and amoebic dysentery, anthrax, leprosy, various types of opthalmia, trachoma and tuberculosis. They also ingest and transmit the cysts of intestinal protozoa and the eggs of certain Helminth worms.

Trachoma, a very serious viral disease of the eye, which is very common in north Africa, the Middle East and India (see Section 6.6), is transmitted by the house fly within 24 h after feeding on an infected eye. It is one of the main causes of the very high incidence of blind people in these areas.

Finally a few words must be said about the tsetse fly, whose principal characteristics were also summarized by Abushama[125] as follows. The family *Muscidae* of the order *Diptera* contains 20 species of *Glossina*, flies collectively known as the tsetse flies. Both males and females of these flies are blood-sucking insects and some of them transmit blood parasites. Tsetse flies are peculiar to Africa south of the Sahara and to the Aden peninsula. On the whole, they occupy an area of over four million square miles. Their economic importance in Africa is very great since they transmit protozoan trypanosomes pathogenic to man and his domestic animals. The tsetse fly *Glossina palpalis Newstead* transmits *Trypanosoma gambiense* the causative agent of human sleeping sickness, while *Glossina morsitans Westwood* transmits *T. rhodesiense* which causes the more localized Rhodesian form of the same disease. *T. brucei*, transmitted by *G. morsitans*, is responsible for the disease Nagana which is fatal to domestic animals.

The tsetse flies feed on the blood of reptiles, birds and large ungulates fairly indiscriminately. They find their hosts primarily by sight and secondly by smell. They are attracted by the movement rather than by the form of the body. In general ungulates seem to be favoured as hosts and when these disappear from one district the numbers of the flies either decrease sharply or even disappear.

The tsetse fly can breed throughout the year except at the height of the rains when the excessive humidity and low evaporative powers of the air result in a drop in the reproduction rate for about a month. The flies increase in number mainly during the early rains when the temperature, saturation deficiency and availability of food are more favourable. This also happens when the sites for larviposition are more numerous. In the dry season, the fly population is at a minimum and becomes concentrated in certain limited areas where the vegetation provides a suitable microclimate.

In Africa the tsetse fly is restricted to a longitudinal belt from 14°N to 29°S from coast to coast. Even within these limits, the tsetse fly is still not universally found, but occurs in belts or isolated pockets. Climatic conditions are the fundamental reason for the northern and southern limits of tsetse fly occupation.

The northern limit seems to be the 30-in isohyet, at least in the north of Nigeria. Above this line the rainfall is insufficient to support shrub thick enough to produce the critical microclimate. Individual species of *Glossina* are limited by different climatological features, which explains their discontinuous distribution. *G. morsitans*, for example, is limited in Rhodesia to a mean annual temperature of 20 °C. The tsetse country of the Sudan is that part which is both far enough south to have sufficient rain and high enough to be above the grass swamp.

The second large group of meteorotropic insects is the desert locust.[130,131] A locust passes through three important stages: eggs, wingless nymphs or 'hoppers' and winged adults. The latter eat daily the equivalent of their weight from the crops they encounter, i.e. 0.04–0.09 of an ounce.

During one of the serious locust invasions in Ethiopia in 1958 almost 167 000 tons of grain were destroyed, enough to feed more than one million Ethiopians for a year.

One of the first great research workers on locust was Sir Boris Uvarov in England.[132] He demonstrated that after heavy rains the locust eggs in the desert sands began to hatch. The hoppers touched each other, triggering a change in behaviour and colour, the colour changing from yellow to black and red. In the Anti-Locust Research Centre in London these different changes have been also studied experimentally.

In 1953, at Tananarive in Madagascar, representatives of the meteorological services of 29 countries of Africa stated that the further development of the application of synoptic meteorology to the forecasting of locust movements was economically one of the most important problems of applied meteorology in Africa. With the support, professional and financial, of WMO and FAO since 1961, further work has made it possible to utilize current daily synoptic charts of the whole area which is subject to invasion by the desert locust. The first step towards an understanding of the relationship between locust migration and synoptic meteorology was provided by the systematic mapping and analysis of current locust reports from all infested countries since 1928, at what subsequently became the Anti-Locust Research Centre in London. This centre soon demonstrated that areas and seasons of breeding by the desert locust coincide with areas and seasons of rainfall, and it subsequently became clear that the migration of the swarms in general takes them out of areas in which the seasonal rains are almost finished, and into areas in which the seasonal rains are beginning.

Since 1961 several reports have been published by the Food and Agriculture Organization in Rome, and by the W.M.O. in Geneva (e.g. Technical note 68, 1965) on the meteorological effects on locust development. The London Anti-Locust Centre sends its warnings for invasion to the F.A.O. in Rome, which acts as a strategic command for locust control and sends teams of experts to whatever countries need them.

Both soil moisture and temperature are important micro-meteorological factors for the development of locust eggs. The egg must absorb its own initial weight of water in order to develop. This means in the field a 20 mm rainfall at temperatures of 20–33 °C before development takes place. The further development of the hoppers requires temperatures between 20 and 46 °C. The period of development varies between 62 days (at 27 °C) and 20 days (36–40 °C). For further details see references mentioned above.

Rainey[130] pointed out that some other major insect pests of tropical agriculture whose effective control like that of locusts can depend on adequate meteorological forecasts, exhibit patterns of long-range migration which appear to be dominated by winds and convergence systems in a manner similar in many respects to those of locust migration. For example, the breeding season of the African Armyworm (*Spodoptera exempta Wek*), varies markedly and consistently in different parts of eastern Africa, with outbreaks of larvae in Kenya, Tanzania and Uganda confined to November–July (mainly December–May), and alternating with outbreaks in Ethiopia which are very largely confined to April–August.

Extensive work in China on the Oriental Armyworm (*Mythimna separata Wek*) has included recaptures of marked moths at distances of up to 1400 km from the point of release. There is evidence that overwintering does not occur north of 36 °N, while north of 39 °N there is a well marked moth invasion of the north-eastern provinces in May–June, followed by one or more damaging generations until August.

The first arrivals in 1937 of the Cotton Leaf worm (*Alabama argillacea Hb*), a kind of moth, as far north as New York State coincided with a spell of warm southerlies associated with the passage of a deepening westerly depression. In Brazil, in 1966, the initial invasion of the moths into the Fazenda Sao Miguel cotton-growing area of the Rio Grande del Norte coincided closely with the beginning of the seasonal rains, on 6 February 1966.

These few examples demonstrate clearly the great importance of the studies of Entomological Biometeorology both for agriculture, and for veterinary and medical sciences.

11.1.9. Physiological effects of the weather and climate on arthropods in general

The distribution of many arthropod species is dependent primarily upon meteorological factors, in particular light and temperature.

(a) *Effects of light* It is difficult to separate the effect of visible light from that of thermal effects in the field, but the general biological significance cannot be over emphasized. In addition to the importance of visible light in stimulating the activities of many diurnal insects and arachnids, it has wide effects upon the speed of development, vitamin synthesis, pigmentation colour change and so on. The secondary effects of light upon behaviour, however, are often more obvious. Many insects, arachnids, myriapods and woodlice avoid the light and can find their food in darkness; at the same time they escape the attention of their predators and other enemies. Conversely many diurnal forms require the light in order to find their food.

(b) *Effects of temperature*. Temperature acts upon fertility, development, activity and longevity. There are four temperature barriers in the distribution of arthropods.

(i) The lower lethal temperature which is an absolute barrier to the distribution of many insect pests.

(ii) The lower mean temperature below which activity ceases and hibernation takes place, which is the chief barrier to the dispersal of many tropical forms such as tsetse flies.

(iii) Upper mean temperature above which dormancy or 'aestivation' may set in.

(iv) Upper lethal temperature which is fatal to the organism.

Certain arthropods, such as the inhabitants of hot springs, can tolerate very high temperatures; the larva of a Nigerian arthropod *Polypedilium*, which can survive the loss of most of its water content for several years has, in the dehydrated state, an upper lethal temperature of nearly 70°C. It can withstand temperatures well above 100°C for short periods.

Many *Arthropoda* withstand unfavourable climatic conditions in the diapause (see Section 11.1.8.(c)), which is usually characterized by reduced metabolism and enhanced resistance to meteorological factors such as heat, cold or draught. In the case of plant-eating insects and mites, the onset of diapause frequently coincides with some distinctive phase in the growth cycle of the host plant. Diapause is usually triggered by changes in day length but is independent of light intensity provided that this exceeds a threshold value greater than that of moonlight.

A certain degree of temperature control is achieved by behavioural means. Animals of small size have a surface area so large in proportion to their weight that the maintenance of a constant body temperature is not possible. This difficulty has been overcome by social insects such as honeybees which, unable individually to control their body temperature, are nevertheless to some extent able to keep their hive cool in summer and warm in winter. A fall in the outside temperature induces the bees to form a dense cluster but, below about 13°C, they become restless and begin to move about until the temperature rises to about 25–39°C when the movements cease. The temperature in the cluster then begins to sink until movement is again engendered. In summer, however, when the air temperature goes beyond 35°C, the nest is cooled by fanning movements of the wings.

(c) *Effects of humidity*. Environmental humidity affects the rate of development of arthropods and their behaviour. Certain ant and termite species have square heads which they use to close the entrance to the nest against drought and other unfavourable climatic conditions.

(d) *Effects of wind*. Wind currents play a great part in the distribution of smaller organisms including both winged and wingless types such as spiders, mites etc.

11.2. BIOMETEOROLOGICAL EFFECTS ON MEAT QUALITY; MILK, WOOL AND EGG PRODUCTION

11.2.1. Effects of the weather and climate on meat quality

Several studies have been made on beef quality, particularly by Berg,[134] Butterfield,[135] Dowling[136,137] and others. In the past, cattle in the tropics, especially those under stress due to lack of animal care and management, were reported to be of poor quality and not suitable for beef. Usually beef produced under such conditions is tough.

High temperature and humidity are the principal meteorological factors which determine beef production and quality in the tropics. Great atmospheric heat, especially if accompanied by high humidity, tends to affect the appetite and growth of cattle. However, normally hot atmospheric temperatures *per se* need not suppress production. Provided ample water and fresh nutritious green forage are available in a warm environment, with low humidity and with air movement which enhances the effective vaporisation of the sweat, cattle need not be unduly stressed. Webster[138] demonstrated that cattle can also adapt to intensely cold conditions of −11 to −22°C. The adapted cattle will grow steadily throughout the year. Consequently an overall greater yield is possible. Similarly yields of cereal crops required by cattle are low in the tropics compared with yields in temperate areas, but in the tropics more than one crop can be grown per year for an overall higher production. Certainly grass forage yields are also higher but until recently appeared to be less digestible and of lower value.

Modern technology can overcome climatic factors at both extremes of cold and heat. Therefore high intensity systems of animal production can also work in hot areas. Of course intensification requires sufficient capital and equipment, and the application of knowledge, to obtain economical and beneficial results which can only accrue with trained personnel. Confinement of animals in most instances has improved both quantity and quality considerably. At the same time confinement and closer animal contact increases the possibility of disease and stress problems.

Biometeorological effects on meat quality are not easy to ascertain. Meteorological factors need not directly affect the quality of meat of animals under proper nutrition, sound management, with suitable animals under a positive animal health and production regime. Meteorological stress undoubtedly affects the tenderness, flavour and juiceness of meat. Even fat colour is affected by pre-slaughter stress and post-slaughter conditions. For example, high humidity and slow chilling in still air make the fat more yellow. Advanced technological care of animals is required to ensure maximum results.

Dowling pointed out that with the control of disease and using the *Bos indicus* type of cattle, beef production under hot conditions is highly successful. Similarly, *Bos bubulos*, the water buffalo produces tasty meat that can be produced under still more stressful tropical conditions. Hence it would appear that the meteorological effects on meat quality can be overcome anywhere under intensified conditions if costs are not a limiting factor. Thus the world's greatest potential for meat production lies in the vast areas of the humid tropics with year-round warm weather, high rainfall and deep porous soils.

11.2.2. Effects of the weather and climate on milk production

Many studies have been made on the influence of meteorological factors on the milk production of cattle.[139] Bianca[140] and H. D. Johnson[141,142] in particular summarized the various observations made in developing countries. According to Johnson[141,142] the principal meteorological factors which may affect milk quantity and quality are temperature, humidity, wind velocity, radiation and altitude. An adverse meteorological factor can influence milk production by alteration of the animal's thermal balance, water balance, energy balance and behavioural activity. These functions are interrelated and involve the neuro-endocrine system. Indirectly, these meteorological factors express their effect on the animal's milk production by altering the quantity and quality of feed and water intake and also may determine whether diseases or parasites are present in the environment. In the tropics, meteorological factors of high temperature, relative humidity, and the indirect factors of inadequate nutrition and parasites are reported to severely limit milk production. Environmental pollution of air, water, and soil by animals themselves, or by industry in more urban areas, provide chemical modification of the environment which may influence production.

Some dairy breeds acclimatize more readily to various climatic environments and this results in less impairment of the lactation functions. Highly productive *Bos taurus* cattle are often overbred with more heat tolerant and disease resistant *Bos indicus* cattle in order to increase milk production and still to retain heat and disease tolerance under tropical management conditions. In Tanganyika, Mahadevan and Hutchison[143] found that half-breeds were most productive, but in Jamaica, according to Lecky,[144] the three-quarter Jersey animals were most productive. There are wide differences between *Bos taurus* and *Bos indicus* in their ability to produce milk under tropical or temperate conditions. The east African zebu cows have only about half the potential for milk production of Indian sahiwal cows emphasizing the genetic variation that exists within the group of zebu-type cattle.

(a) *Influence of temperature*: Bianca[140] and Johnson[141,142] have shown that in general the yield and composition of milk are much the same within the temperature range of 10–20°C. From 20 to 27°C, the yield decreases slowly and the fat percentage is reduced, but beyond 27°C, the decline in yield is much more marked whereas the fat percentage increases and the content of non-fat solids is usually decreased.

Popoff[145] found that cows produced less milk at a temperature of −10 to 4 °C, than at a temperature of 5–12 °C. Kelly and Rupel[146] observed that changing cows from a warm barn to an open lot (average temperature of −8 °C) resulted in a daily decrease of 0.38 kg/cow. However, extra feeding minimized this effect. Dice[147-149] reported similar results on cows and heifers.

Regarding high temperatures, Regan and Richardson[150] observed, in Holstein, Jersey and Guernsey cows, a 40% drop in milk production when the temperature was increased from 4 to 35 °C. Early studies in the Missouri Climatic Laboratory indicated a decline in milk production at 24–27 °C for Holstein, 29 °C for Jersey cows and 29 °C for Brown Swiss cattle. The breathing of air cooler than the environmental temperature has been found to be significant in preventing a decline in milk production.[151,152]

Summarizing it can be said that most studies point to the depressing effect of hot summer environments for milk production in most temperature zones. Not only is the total milk production reduced with rising temperatures but also the solid-not-fat content of the milk decreases. For further details see Johnson.[153]

(b) *Influence of humidity*: according to Johnson[141,142,153] milk yields are related to humidity at temperatures above 27 °C. To quantify these relationships a temperature-humidity discomfort index was established for milk production. An environmental index that quantifies the contributing effects of radiation, humidity, wind, and diurnal cycles on lactation is of course more meaningful since dry-bulb temperature is always accompanied or qualified by the other variables. Such an environmental 'equivalence index' should be established. An index of this nature would relate more directly to the heat balance and comfort of the animal than dry-bulb temperature. The application of the temperature-humidity index to milk production was a first effort in this direction.

(c) *Wind or ventilation* increases convective and evaporative exchange of heat between the animal and the air about it. Data from the Missouri Laboratory show that increased wind at temperatures below the comfort zone may become detrimental to milk production.

(d) *Radiation*: when the air is hot, radiation will augment the decline in milk production. The difference in the animal's response to shade at different locations points out the importance of radiation in the microclimate and the animal's energy budget. Shade does not effectively change either the air temperature or humidity at a particular location, but it does change the radiation items in the energy balance of an animal.[154]

(e) *Altitude*: according to Montero,[155] at the higher altitudes in central and south America with moderate temperatures and improved nutrition, higher producing European breeds of cattle are reasonably well adapted; however, quantification of the effect of altitude on milk production is still missing. The higher temperatures and associated humidity and radiation, poorer natural forage, and diseases in the lower altitudes or coastal regions forces farmers to use hybrid milk cattle or breeds indigenous to the tropics.

11.2.3. Effects of the weather and climate on wool production

Meteorological effects on wool production have been studied in Australia and New Zealand.[10,156-162] Wool production comprises the following three aspects.

(a) *Wool growth.* Brown[156] reported from Australia that productivity per head decreases from south to north. In the coastal districts the locality is confused with breed, but in the semi-arid zone the sheep are Merinos with, however, a concentration of the high producing south Australian strong wool strain in the north of south Australia and parts of New South Wales. Brown[156] thinks that the gradients of productivity might be the consequence of local variation in the availability of feed rather than the direct result of climate.

Thwaites[10] has shown, in laboratory experiments, that a hot environment can decrease wool growth in Merino sheep but that the decrease can be accounted for by a decrease in feed consumption. However, it maybe that heat during gestation may affect follicular development in the foetus so that wool production in the adult is reduced. Cold does not appear to affect wool growth provided that the site is protected from cold. If the site is not protected, wool growth is decreased and the decrease may be large.

Hutchinson[159] pointed out that on a constant food ration there is an annual rhythm of wool growth. In all the breeds examined, except the Bikaneri, maximum growth is in the summer and minimum in the winter. In Bikaneri ewes a reversed rhythm has been observed. The rhythm is small in most Merino strains and larger in other breeds examined. Several investigations have shown that it can be entrained by the photo-periodic rhythm. The rhythm has strong endogenous properties in that it can continue without loss of amplitude or timing for at least 2½ years in an equatorial day length with exaggerated reversed seasons of temperature and humidity. It has been suggested that the circadian rhythm of wool growth may be a modification of an archaic pattern of shedding and regrowth in which the shedding phase has been lost. The rhythm of shedding and regrowth of the kemp fibres of the Limousine sheep is entrained by day-length. According to Slee[160] in Wiltshire Horn lambs, born and reared in continuous dim light, the first season of shedding is protracted and incomplete, as if photoperiodic change or brighter illumination were required to initiate the annual rhythm of shedding and regrowth. Symington[162] found that Persian Blackhead rams did not shed their coats when they were housed for eight months.

In the field, the seasonal variations of wool production are different from those on a constant food ration. In Merinos the effect of nutrition overrides that of the weak circadian rhythm and wool growth depends mainly on the green feed available.[156] Consequently, in a mediterranean climate the maximum production is in winter and spring and the minimum in summer. On the other hand in the Cheviot and Scottish Blackface and perhaps some other breeds, nutrition in winter has little effect on wool growth; at this season production appears to be controlled by the circadian rhythm. In summer wool growth is sensitive to nutrition.

(b) *Wool quality.* According to Hutchinson[159] ultraviolet light degrades the tips of the wool fibres so that they may break off during processing. This decreases the value of the wool. Vegetable matter and dirt get into the wool in quantities depending on the locality and therefore on the climate. Wool with more than about 6% of vegetable matter and dirt is not of high quality. In some districts of Australia, which in other respects are suitable for fine wool production, 12–15% of vegetable matter is not uncommon.

Fleece-rot, a bacterial dermatitis with exudation, matts the wool fibres and is usually accompanied by staining of the wool. The cause is the presence of moisture on the skin over several weeks[158] and it has been associated with particular patterns of rainfall. In India 'Canary' colouration of the wool, in this case without fleece-rot, is a serious economic problem. In some districts sheep have to be shorn twice a year to separate wool grown in the wet season from that grown in the dry season. Consequently, there have been many recent investigations in India on the influence of climate and fleece type on the incidence of staining.

Fly-strike is caused by the development of blow-fly maggots under the skin and outbreaks can be severe after wet weather. It is most common in the neighbourhood of the tail, anus and vulva, but if fleece-rot is present there may be fly-strike under the fleece which reduces the value of the wool.

Mycotic dermatitis, an infection of the wool follicles by *Actinomyces dermatonomus*, is associated with high rainfall, like fly-strike, and is sometimes responsible for economic loss.

(c) *Effects of shearing on the thermoregulation of sheep.* Shearing of sheep disturbs their thermoregulation. The fleece protects the animals from the heat of solar radiation as well as from cold. After shearing sheep often die from cold but not from heat. The loss in Australia averages less than 1% but can occasionally be catastrophic in localized areas in both Australia and New Zealand. The heavy losses occur only in a combination of heavy rain and strong wind. Cheap polyethylene coats to protect shorn sheep have been developed recently.

11.2.4. Effects of the weather and climate on egg production

The physiological effects of thermal stress and light on birds and domestic chicken in particular, have been reviewed on p. 281. Later on p. 320, the general influence of meteorological stimuli on the poultry industry and in egg production will be described. Some more specific relationships between the weather and climate and egg-shell quality and egg production, summarized by Wilson,[85] are:

(a) *Factors affecting egg-shell quality.* According to Walford[163] and Tanaka[164] high temperatures in particular have a great effect on the egg-shell quality. Shell thickness may also be experimentally affected by a high carbon dioxide concentration, while ammonia, at levels below 105 ppm, is without effect.

Dirty egg-shells result when the litter moisture is high. Winn and Godfrey[165] found that above 80% relative humidity problems arose with wet litter. According to Deaton *et al.*,[166] space allotment per bird is as important as is relative humidity. Heat and high humidity significantly decreased the thickness of the whole shell and of all its layers except the membranes.[167] Structure as well as ultra-structure is unfavourably affected by climatic stress. The cuticle degenerates, the mammillae are rounded at their bases, the calcareous deposits are loose, irregular, lack normal columnar orientation and develop cavities. The organic spongiosa matrix is of normal density but shows larger vesicles, some cavities and fewer and shorter fibrils which are not so well oriented.[167]

According to Lacassagne and Jacquet[168] the number of double-yolked, thin shelled eggs, and of eggs without shell membranes, was related to early sexual maturity. The lighting regimen determines the age of sexual maturity. There is also an association between ambient temperature and the number of blood spots in eggs.[85] Hens kept at a temperature of 21 °C laid more eggs containing blood spots than hens kept at a temperature of 32 °C.

According to Payne,[169] egg size tends to be increased if a decrease in temperature takes place during the last three hours of the light phase.

(b) *Factors affecting egg production.* Ambient temperature, humidity, air velocity and light all affect the body temperature and comfort of chicken and therefore their egg production. Hyperthermia in a serious threat for the domestic chicken since biologically the chicken is poorly adapted for heat dissipation. Within the laying house an ideal temperature is between 13 and 30 °C. The day-old chick is sensitive to both hypothermia and hyperthermia if subjected to extremes; some will die and the growth rate of survivors will be retarded.

The velocity of air is a factor in convective heat control. For chickens subjected to environmental temperatures below body temperature and an air velocity of 150 m min^{-1}, body temperature control is less dependent on respiratory evaporation than with a velocity of 15 m min^{-1}; however, at temperatures above body temperature high air velocity appears detrimental and survival times are reduced.[82]

The influence of light on reproduction and growth has probably received more research attention in recent years than any other parameter in poultry husbandry. Natural light springtime photo-periods enhance egg production. Light penetrates the retina of the eye and ultimately causes the secretion of the gonadotropins necessary for gonadal function. The three types of light regimens that induce gonadal growth in photo-sensitive birds have been listed by Wilson *et al.*[170]: long photo-periods, frequent photo-periods and conditioned cycles. The age of onset of semen production is inversely related to the photo-period length during the growing phase but does not alter the fertilizing capacity of the semen.

For further details see *Progress in Animal Biometeorology.*[2]

11.3. EFFECTS OF THE WEATHER AND CLIMATE ON FERTILITY AND REPRODUCTION

An excellent review of the general problem was prepared by Smidt,[171] who concluded that meteorological influences on reproduction and fertility can act both directly and indirectly. In females, sexual development, manifestation of sexual rhythms, sexual behaviour and reproductive processes from the time of ovulation to parturition of offspring are modified by meteorological factors. In males, sexual maturity, reproductive behaviour, spermatogenesis and fertilizing capacity of sperm are similarly influenced. The more important direct meteorological factors affecting fertility are light, temperature and humidity. Other factors include altitude, air movement and radiation.

11.3.1. Effect of light

Light is an important timing factor for reproductive rhythms, as was discussed in Sections 5.2, 8.9, 11.1.1 and 11.1.6. In wild mammals and birds sexual rhythms will be activated by light in such a way that the birth of the offspring will take place during a favourable climatic period. Through domestication an alteration of the correlation between light and reproductive rhythms has taken place. This correlation is still pronounced in domestic animals with seasonal fertility, e.g. sheep, goats and horses. The fertile period in sheep is initiated by increasing daylight length.

There is a notable variation in the duration of the fertile season within species. Since the length of the fertile period is partly responsible for reproductive efficiency, light is an important criterion for productivity in the respective species in terms of animal production. Species which are polyoestrous throughout the year, such as cattle and swine, exhibit only slight seasonal variations in reproductive traits. In birds laying periods and brooding time are controlled by light. This is especially true for the control of the ovulation sequence by the diurnal light/dark change.

11.3.2. Effect of temperature

According to Smidt[171] the influence of temperature on reproductive processes must be considered in relation with thermoregulation. Temperatures within the range of thermal neutrality exhibit pronounced influences on reproductive functions. But those temperatures either above or below the critical temperatures put heat stress or cold stress upon the animal. In this respect the reproductive organs seem to be a *locus minoris resistentiae.*

Temperatures below the freezing point exhibit specific influences on reproductive efficiency. In animals not adapted to cold, low temperatures may negatively affect libido and reproductive efficiency. Local influences of deep cold on the testes may cause disturbance of spermatogenesis and fertilizing capacity. Also, embryonic development may undergo negative influences. Low temperatures are highly critical for newborn animals because the thermoregulatory mechanisms involved are not yet well established. Natural or artificial shelters against cold diminish the relevance of low ambient temperatures on reproductive efficiency.

More important than low temperatures are high ambient temperatures. Generally, extremely high temperatures have a negative effect on reproduction in non-adapted animals. The onset of sexual maturity is often delayed in both sexes under the influence of heat. However, this could partly be channelled via retarded growth. Also, the female sexual cycle can be prolonged or even totally suppressed by heat stress, this is especially common in cattle.

If copulation or artificial insemination takes place with high ambient temperatures, a reduction in the fertilizing rate could occur as well as a risk of endangering embryonic development. Direct temperature effects will be established both through the maternal metabolism and through a direct influence of heat on the first development states of the embryo. Embryos which overcome this critical period have a delayed foetal growth due to high ambient temperatures, perhaps resulting in low birth weights. Thermoregulatory problems deriving from lactation, if prevailing synchronously with cyclic activity and conception, may cause additional negative effects on reproduction.[171]

Very high temperatures will result also in a decreased sex drive in male mammals. The well known negative effect on spermatogenesis must be interpreted as a local influence of high ambient temperatures on the gonadal tissue and on metabolism. Extremely high ambient temperatures may cause complete cessation of spermatogenesis. Most mammals, however, have local regulatory mechanisms to keep the temperature inside the gonads at an optimal level, that is 3–5 °C below the body temperature. In artificial insemination the semen may be exposed to meteorological influences. If semen is exposed to high temperatures in the genital tract of females suffering from heat stress, it may retain its fertilizing capacity but cause embryonic mortality.

As stated before in poultry high ambient temperatures result in decreased egg production and egg quality. The hatching of eggs will not only be negatively influenced by high temperatures prevailing during semen deposition and fertilization but also by temperatures above the optimal range during the time of storage of eggs until the onset of hatching. The effect of the size and motility of spermatozoa and their sulfhydryl content on fertility were discussed, in connection with the sex ratio of animals, in Section 8.4.

11.3.3. Effect of other stimuli

Apart from light and temperature, humidity and air movement are also important factors during periods of high ambient temperatures, as they seriously affect the internal thermal balance of the animal. The same is true for radiation. Most likely high altitude also affects the fertility of animals.

11.3.4. Indirect meteorological influences

In addition to the above direct meteorological influences, indirect influences on fertility and reproduction have also been reported. The most important of these are soil quality and vegetation, both of which are very important factors for the fertility and reproduction of animals. Certain meteorological conditions, such as long dry periods, may cause food shortage and decreased food quality. Also specific soil conditions, which depend on meteorological factors, influence the quantity and quality of foodstuffs of plant origin.

In view of the relationship between nutrition and fertility meteorological factors could thus influence reproduction indirectly. Disturbed reproduction results from starvation and from an insufficient supply of protein, energy, minerals, trace elements and vitamins as far as they are essential to, or important for, reproductive functions. The supply of drinking water is also of importance.

11.3.5. The principal physiological mechanisms involved in meteorological effects on reproduction

They have been summarized by Smidt[171] as follows:

'Meteorological influences on reproductive functions are mediated by a very complex control system. Receptors for light, temperature etc. receive the environmental stimuli which, by neural pathways, are led to the respective central nervous structures. The influences on reproduction are then channelled by neurohormones which are responsible for the gonadotropic activity of the pituitary gland. The gonadotropic hormones influence the functions of the gonads and their target organs.

Meteorological influences may also be mediated through extragenital endocrine systems which control reproduction via the general metabolism. Since the autonomic nervous system not only reacts to meteorological stimuli but also plays a part in reproductive control, it is also of considerable importance in this context.

Some of the temperature effects act locally in the genital organs themselves. The temperature inside the female genital tract will be affected by heat stress conditions of the animals. Male mammals have special thermoregulatory mechanisms to prevent negative local effects of heat stress, such as extracorporal location of gonads, anatomy and function of blood vessels, sweat glands in the scrotal skin, cremaster muscles (internal muscles of the abdomen, which elevate the testes) etc.'

11.4. EFFECTS OF THE WEATHER AND CLIMATE ON THE CAUSATION AND DEVELOPMENT OF ANIMAL DISEASES

Review papers on meteorological effects on animal diseases were prepared by Crawford,[172] Smith,[173] Ferguson and Branagan[174] and others. In particular the effects of meteorological factors upon parasites and nematodes causing both animal and human diseases, were compiled in the *Proceedings of the Twelfth Symposium of the British Society for Parasitology* in London by Angela E. R. Taylor and R. Muller.[175] It would be beyond the scope of the present book to review all the reported meteorotropic animal diseases, but some of the major phenomena reported are summarized in Table 11.4.

Crawford[172] pointed out that weather conditions have a very marked effect, both direct and indirect, on the well-being of farm animals. In fact they determine to a considerable extent the type of animal kept and the kind of husbandry adopted in any particular area. This is particularly true of grazing animals which spend the greater part of their lives in the open.

Weather conditions affect livestock indirectly by determining the types of fodder plants which can be grown in any particular area and this indirect effect on the nutrition of the animals influences the incidence of the various metabolic disorders and deficiency diseases to which they are subject.

The direct effects of weather are perhaps more apparent in relation to the infective and parasitic diseases. The incidence of these diseases can be influenced by weather conditions acting upon the infective agent, the vector, the intermediate host or the host animal.

Finally weather conditions play a large part in determining the disease picture in any particular area. According to Crawford[172] if a disease problem is being investigated one of the first questions the veterinarian is likely to ask is 'Does the incidence vary with the season of the year?' Information on this aspect of a disease problem can be very helpful in elucidating its nature and causation.

11.4.1. Indirect influences

Weather, climate, season and soil determine not only the type of plants growing in a particular area but also their chemical composition (particularly trace element and vitamin contents). As animals, more than man, live largely on the food grown in the immediate surroundings of their living quarters, local changes in weather and climate may affect the nutrition of animals and are therefore responsible for various metabolic disorders and deficiency diseases observed in farm animals.

As pointed out before, poisonous plants and toxic fungi in meadows may be responsible for animal diseases. The abundance or scarcity of such plants and the severity of the poison often depend on specific weather conditions. One of the many examples of this type of meteorotropic disease is known as *Facial Eczema* in New Zealand, a liver disease common among sheep, but also liable to attack cattle and horses. The liver damage is caused by a photodynamic substance (see Section 5.2.4), phylloerythrin, a product of the digestion of chlorophyll in ruminants, which normally is eliminated in the bile. However, when it damages the liver, it reaches the peripheral arteries and upon exposure to sunlight, severe skin lesions occur. Recently it was found that the liver injury is due to a toxic fungus in certain rye grass pastures in New Zealand. The development of the fungus (and of the disease) is clearly seasonal. Outbreaks occur in those autumns which follow a hot dry summer, but seldom after a cool wet summer or during a late cold autumn. Outbreaks occurred particularly after rain (only half an inch is required) which stimulates the new growth of dried-up pasture. Supported by present knowledge, Filmer, Mitchell, Walshe and Robertson[173] were able to develop a system of meteoropathological forecasting for this particular animal disease.

Another example is *Grass Tetany*, a disease characterized by tetanic convulsions frequently causing death. It is due to a shortage of magnesium in the serum of grazing cattle. It is particularly observed in cattle and sheep in areas where the magnesium content of grass falls below a critical level. The disease is being extensively studied in England by Allcroft[176,177] and in The Netherlands by t'Hart and Kemp.[178] The specific meteorological conditions which seem to activate the magnesium deficiency are those which, during the preceding period, have caused a strong cooling effect. As a result low magnesium values are found in cattle, specifically in those periods when the mean minimum temperature is lowest for the year, but also in all periods when the temperature falls below 5.5 °C and sunshine is scarce, rainfall preceding the period is high and of frequent distribution, winds are strong and frequent, and precipitation is accompanied by hail or snow. In other words, many periods preceded by active cold fronts and the influx of cold polar air are particularly favourable to the creation of magnesium deficiency and its clinical consequences.

11.4.2. Direct influences

Two main groups of animal diseases should be distinguished, diseases of domestic animals (cattle, horses, sheep and pigs) and diseases of poultry (see Table 11.4). The various diseases have been classified according to different principles. Crawford[172] distinguishes between thermoregulatory (due to non-sweating) diseases, diseases due to metabolic disorders (including nutritional diseases), bacterial diseases, virus diseases, diseases transmitted by arthropod vectors (insects, e.g. mosquitoes and flies; ticks etc.) and parasitic diseases due to helminths (flukes or tapeworms) and the roundworms or nematods.

Smith[173] also distinguishes windborn diseases (diseases caused by bacteria and viruses which are transported by air currents) and fungal diseases. This latter group belongs actually to the nutritional diseases.

Ferguson and Branagan[174] distinguished only between infectious and non-infectious animal diseases.

TABLE 11.4
Important animal diseases[a] and their meteorotropic relationships

Name of Disease	Clinical Symptoms	Causative Agents	Animals Involved	Meteorotropic Relationships
	Domestic Animals (Cattle, Horses, Sheep, Goats and Pigs)			
Thermo-regulatory diseases	*Anhydrosis* Progressive loss of ability to sweat even when exercised in very hot weather (after a race the horse skin is dry) upsetting heat regulation. During the race temp. rises to 45° C and the horse may die during the race	Seriously disturbed heat regulation	Horses	Only occurring in areas with high air temp. and high RH, not in very hot arid countries.
High altitude diseases	*Brisket* diseases See p. 270	Serious dyspnoea due to high altitude above about 2500m	Cattle	Strong reduction in partial oxygen pressure in mountainous areas above 1500m.
Bacterial diseases	*Braxy* Very acute fatal disease caused by infection of stomach and intestinal wall	*Clostridium septicum* (widely distributed in soil)	Sheep	Particularly occurring with onset of severe frost in late autumn or early winter (after eating very cold grass stomach is chilled, permitting invasion of bacteria causing sudden death the morning after severe frost).
	Haemorrhagic septicaemia	*Pasteurella septica*	Cattle (particularly in India and often parts of Asia)	Outbreaks begin soon after onset of monsoon rains. Perhaps sudden strong growth of young grass upsets digestive system.
Virus diseases	*Rift Valley Fever* High fever	Transmitted by certain mosquito species	Mainly Sheep, Goats, Cattle (particularly in Africa)	Usually appearing in summer and autumn in wet periods in valleys and low-lying areas.

[a] See also various animal diseases transmitted by mosquitoes and flies discussed on p. 293. For more details concerning meteorotropic animal diseases see *Progress in Human Biometeorology*, Vol. 1, part III (1972) pp. 93–108 and 146–148 and references in Smith, pp. 43–49.

TABLE 11.4—contd

Name of Disease	Clinical Symptoms	Causative Agents	Animals Involved	Meteorotropic Relationships
Virus diseases (contd.)				Tendency of 5 years of quiet periods between epidemics.
	Foot-and-mouth disease (*Aphtae epizooticae*) Acute, highly infectious disease. Onset abrupt with high fever, followed by vesicles in the mouth. Dripping of mucous on the legs causes similar infectious vesicles on the feet which become swollen and painful. Mortality low, usually about 5%, but serious health deterioration	Picorna virus	Cattle and Swine most susceptible Sheep and Goat less	Virus does not easily survive excessive sunlight and high temp.; damp cold weather favours survival and spreading. According to Smith: absence of UV, high RH and strong downwinds (i.e. rainfall at night) favours development of epidemics.
Tickborne diseases	*Babesiosis* High fever	Transmitted by a tick *Boophilus microplus*	Cattle (particularly in Queensland, Australia)	Size of transmitting tick population depends on survival time of female tick to lay eggs etc. which is strongly affected by atmospheric humidity. Periods of drought and lack of morning dew kills the tick population. During wet periods pastures must be destocked.
Helminth diseases		Different types of infections by roundworms, tape worms and flukes during free living stages outside animal body	Particularly in herbivorous grazing animals	Indirect effects because of seasonal changes in herbage on which host is feeding. Direct effects of temp. and RH on hatching of eggs and development of larvae outside host; i.e. microclimate of herbage cover is critical for development of worm parasites.

TABLE 11.4—contd

Name of Disease	Clinical Symptoms	Causative Agents	Animals Involved	Meteorotropic relationships
Helminth diseases (contd)	*Fascioliasis* or Liver Fluke Disease Damaged liver	Caused by *Fasciola hepatica* transmitted by a trematode. Parasite has has complicated life cycle in the snail *Lymnaea truncatula*	Sheep	Specific humidity and temp. conditions required for a successful life cycle of parasites.
Diseases due to metabolic disorders usually indirect meteorotropic diseases	*Grass tetany* p. 307 Suddenly developed tetanus convulsions	Hypo-magnesaemia	Grazing cattle and Sheep	Hypo-magnesaemia caused by low min. temp., little sun, high rainfall, strong winds.
	Facial Eczema (p. 306) Liver disease	Photo-dynamic substance: phylloerythrin (product of digestion of chlorophyll in ruminants)	Sheep, Cattle, Horses	Outbreaks in autumn after hot dry summer, particularly after rain.
		Poultry Diseases		
Botulism	A few hours after ingestion of toxin birds are extremely weak, difficulty in swallowing, progressive paralysis of legs, wings and neck muscles and finally dying	*Clostridium botulinum* which produces a dangerous toxin (p. 9)	Birds	Poisonous substances related to outside temp.; small quantities produced at 12° C, much toxin produced above 20° C. Disease common during hot summers and in winter near power stations, causing thermal pollution in unfrozen water (often visited by congregations of water fowl).

TABLE 11.4—contd

Name of Disease	Clinical Symptoms	Causative Agents	Animals Involved	Meteorotropic Relationships
Marek's disease	One of the most serious neoplastic chicken diseases of the lymphoid system. Signs of lameness uncoordination etc.	Herpes virus in dander[b] from infected chicken, transmitted by airborne particles	Chicken Pheasants	Conditions favouring airborne transportation.
Kerato-conjuncti-vitis	First signs are tearing; birds tend to keep eyes closed; blindness may develop; animals lose weight	Ammonia fumes over 50 ppm, in poorly ventilated spaces	Chicken	Particularly in winter in inadequately ventilated poultry houses in which chickens are kept on litter or shavings.
Newcastle disease	Virus disease with rather high mortality; begins with coughing and rattling; rapidly spreading through the flock; difficulties in coordination, paralysis, coma; egg production drops to zero in a few days	Virus	Chicken, Turkey and Pheasants. Ducks are resistant	Related to changes in temp. and RH; excessive sunlight reduces transmission.

[b] Dander is composed of epithelial cells of the horny layer of the skin, and it is in this "dandruff" that the Herpes virus is transmitted.

11.4.3. Forecasting of animal diseases

Smith[173] summarized for a number of animal diseases their meteorotropic characteristics and the various methods developed in recent years for forecasting the occurrence of these diseases. Many of these diseases may have an epidemic character. The diseases described by Smith are Foot-and-Mouth Diseases, Pregnancy Toxaemia in ewes, Swayback in ewes, Fascioliasis, Nematodiriasis in lambs etc. It would be beyond the scope of the present book to discuss in detail the various methods used for predicting the occurrence of the above mentioned diseases. The reader is referred to the original W.M.O. publication.[173]

One of the reasons why meteorological forecasting of animal diseases has found less application in the past than the forecasting of plant diseases is that of acquired immunity of animals. Often, after an epidemic of animal diseases, several years are required to build up a new population of susceptible animals.

If one does not take into account the immunological state of the animal a weather forecast may be of little value.

11.5. DESIGN OF ANIMAL HOUSES

The influence of building and floor construction on animal health, on meat and wool quality and on milk production has been discussed in earlier sections of the book in particular in Sections 10.6 and 11.1.3. Architects responsible for the construction of animal accommodation should consider the various physiological and economic consequences of alternative types of construction.

11.6 CROSSBREEDING FOR IMPROVED RESISTANCE AGAINST EXTREME CLIMATIC CONDITIONS

In the previous sections it was shown that various physical environmental factors could affect the general health and reproduction of domestic animals, their meat and wool quality and their milk production. In particular thermal stress and high humidity conditions are often poorly resisted by the European cattle breeds used in tropical countries. By crossbreeding with native cattle the general resistance against climatic stress and parasitic diseases of the offspring can be considerably improved. The problems and benefits related to crossbreeding have been extensively reported.[179-186] Some conclusions are summarized below:

According to Temple[186] experiments tend to show that in the tropics and sub-tropics, cattle with zebu breeding, when properly managed, excel in reproduction and are superior to the exotic breeds. It was also shown that the maternal influence of the zebu is superior to that of the *Bos taurus* in the tropics and sub-tropics because the zebu cow possesses a much greater natural immunity against endemic diseases. Bonsma[180] maintained that calves suckling the zebu mother probably obtained a greater spectrum of immune bodies through the colostrum (first milk secreted before or after the birth of the young) of the highly immune cow.

McCarthy and Hamilton[187] conducted a survey of 28 beef cattle farms in tropical Queensland (Australia), and compared the performance of British and zebu cross cattle. From the productivity and management indices used it was found that the only significant difference between the groups was that zebu cross cattle were marketed at three years of age compared to 3.9 years for British breeds. This is an important aspect of productivity under tropical conditions. British and European breeds of cattle calve at a younger age than zebu type cattle. Tropically adapted cattle are inclined to have a longer lactation anoustrus period.[180,188] It was found that the reconception rate of lactating Africander cows was considerably lower than that of the Sussex cows.

High environmental temperatures affect the fertility of bulls of different breeds in a similar way. Skinner and Louw[189] exposed Friesland and Africander bulls for varying periods to 40 °C. The more heat tolerant Africander bulls were less adversely affected than the Friesland bulls. It is thus evident that the selection of livestock breeds, individuals within breeds and breed combinations with the object to optimize reproduction under tropical conditions requires special basic management and breeding policy considerations.

Not only cattle but also sheep, having a more pronounced seasonal breeding pattern than most domestic animals, are greatly affected by thermal stresses particularly during the breeding season.[190] Special measures such as artificial cooling of rams,[191] shearing of the wool[192] etc. may result in higher semen quality and increased fertility rates. Crossbreeding of various races of sheep may increase the thermal resistance of sheep, both males and females, their fertility, wool quality and quantity. Particularly in Australia much research has been carried out in this field. It was found that the British breeds are least heat tolerant, followed in order among the better-known wool breeds by the Corriedale, Polwarth and Merino; and among the meat breeds by the Dorper, Blackhead Persian and hairy desert breeds (for hot-dry environments) and breeds such as the Barbados for tropical maritime or other warm humid environments.[190,193]

According to Steinbach[194] one-third of the world pig population is found in the tropics, at temperatures constantly above the zone of thermal indifference for this species which, unable to sweat, is highly sensitive to the effects of high ambient temperatures. Nevertheless it has been shown that high fertility rates, two litters per sow per annum, can be obtained under natural conditions in the tropics. Sow fecundity is but slightly depressed by high ambient temperatures. The total number of piglets in a litter does not differ significantly between tropical and temperate sow herds and seasonal changes in the number of piglets born alive have not been observed. According to Steinbach[195] only 2–4% of the variability in litter size can be attributed to climatic changes. However, once they are born sty conditions affect their lives considerably (see p. 273). Although excessive heat stress may adversely affect reproductive behaviour and sexual development in both male and female pigs, a considerable portion of the reduced reproductive efficiency may be due to reduced food intake.[195]

REFERENCES

1. S. W. Tromp, H. D. Johnson and J. J. Bouma (Eds), *Progress in Animal Biometeorology, I*, Swets & Zeitlinger, Amsterdam, The Netherlands, 1976.
2. *idem ibid.*, *Part II*, Swets & Zeitlinger, Amsterdam, The Netherlands, 1976.
3. Findlay *et al.*, *Prog. Physiol. Farm Anim.* **1**, 252 (1954).
4. S. W. Tromp, *Medical Biometeorology*, Elsevier, Amsterdam, The Netherlands, 1963.
5. J. C. D. Hutchinson, in ref. 2, p. 47.
6. V. A. Finch, Thesis, Univ. Nairobi, Kenya, 1973.
7. T. J. Dawson, *Austral. J. Zool.* **20**, 17 (1972).
8. G. D. Brown, in ref. 2, p. 61.
9. K. Cena and J. L. Monteith, in ref. 1, p. 343.

10. C. J. Thwaites, *ibid*. p. 352.
11. R. J. Wurtman, in *Neuroendocrinology*, L. Martini and W. F. Ganong (Eds), Academic Press, New York, USA, 1967, p. 19.
12. J. C. Hoffmann, in ref. 2, p. 80.
13. W. Rowan, Biol. Rev. Camb. Phil. Soc. **13**, 374 (1938).
14. J. Benoit, *Yale J. Biol. Med.* **34**, 97 (1961).
15. F. M. A. Marshall and F. P. Bowden, *J. Expl. Biol.* **13**, 383 (1936).
16. B. E. Piacsek and G. M. Hautzinger, *Biol. Reprod.* **10**, 380 (1974).
17. J. Aschoff, *Cold Spring Harbour Symp. Quant. Biol. (Biol. Lab. Cold Spring Harb. N.Y.)* **25**, 11 (1961).
18. L. Zacharias and R. J. Wurtman, *Science* **144**, 1154 (1964).
19. K. Magee *et al.*, *Life Sci.* **9**, 7 (1970).
20. Dumond, in ref. 2, p. 80.
21. W. Zukowski, A. Kmietowicz-Zukowska and S. Gruszka, *Pregl. Lek. (Warsaw)* **3**, 273 (1965), in Polish.
22. K. Bojlen and M. W. Bentzon, *Human Biol.* **43**, 493 (1971).
23. J. A. Valsik, in *Biometeorology V*, S. W. Tromp and J. J. Bouma (Eds), Swets & Zeitlinger, Amsterdam, The Netherlands, 1972, p. 120.
24. S. W. Tromp and J. J. Bouma (Eds), *Progress in Human Biometeorology, Part IB* Swets & Zeitlinger, Amsterdam, The Netherlands, 1974, p. 512.
25. Cardinali, in ref. 2, p. 85.
26. R. F. Grover, *ibid*., p. 105.
27. P. D. Atland, *ibid*., p. 124.
28. P. R. Morrison, *ibid*., p. 132.
29. P. Bert, *Barometric Pressure* Columbus Book Co., Columbus, USA, 1943.
30. K. G. Johnson, in ref. 1, p. 140.
31. K. L. Knox, *ibid*., p. 189.
32. W. V. Macfarlane, *ibid*., p. 49.
33. R. O. Rawson, *ibid*., p. 81.
34. A. J. F. Webster, *ibid*., p. 218.
35. J. D. Wood, *Amer. J. Physiol.* **219**, 159 (1970).
36. W. Riedel, G. Siaplauras and E. Simon, *Pflügers Arch. Ges. Physiol.* **340**, 59 (1973).
37. K. G. Johnson, *J. Physiol. (London)* **215**, 743 (1971).
38. *idem*, *Int. J. Biometeorol.* **15**, 281 (1971).
39. L. E. Mount, *The Climatic Physiology of the Pig*, Arnold, London, UK, 1968.
40. D. L. Ingram, in ref. 1, pp. 148 and 442.
41. L. E. Mount, *ibid*., p. 227.
42. J. Steinbach, *ibid*., pp. 320 and 393.
43. S. R. Morrison *et al*., *Trans. Amer. Soc. Agric. Eng.* **10**, 691 (1967).
44. S. Brody, *Res. Bull. Mo. Agric. Expl. Stat.* **423**. 3 (1948).
45. H. Heitman and E. H. Hughes, *J. Anim. Sci.* **8**, 171 (1949).
46. D. L. Ingram, *Resp. Physiol.* **8**, 1 (1976).
47. W. Bianca, *Int. J. Biometeorol.* **5**, 3 (1961).
48. E. G. E. Hafez (Ed), *Adaptation of Domestic Animals*, Lee & Febiger, Philadelphia, 1968.
49. W. Bianca, in ref. 48, p. 97.
50. K. F. Johnson, G. M. O. Malaoiy and J. Bligh, *Amer. J. Physiol.* **223**, 604 (1972).
51. D. H. K. Lee, in *Physiological Adaptations, Desert and Mountain,* M. K. Yousef *et al.* (Eds), Academic Press, New York, USA, 1972, p. 107.
52. W. V. Macfarlane, in *Handbook of Physiology, Adaptation to the Environment,* D. B. Dill *et al.* (Eds), Amer. Physiol. Soc., Washington, USA, 1964, p. 509.

53. K. Schmidt-Nielsen, *Desert Animals* Oxford Univ. Press, London, UK, 1964.
54. G. C. Whittow and J. D. Findlay, *Amer. J. Physiol.* **214**, 94 (1968).
55. M. K. Yousef, in ref. 1, p. 108.
56. L. Irving, in ref. 52, p. 361.
57. P. F. Scholander *et al.*, *Biol. Bull.* **99**, 225 (1950).
58. G. C. Whittow, in ref. 1, p. 295.
59. J. Bligh, *Environ. Res.* **1**, 28 (1967).
60. R. W. Bullard *et al.*, *J. Appl. Physiol.* **29**, 159 (1970).
61. K. G. Johnson *et al.*, *Amer. J. Physiol.* **223**, 604 (1972).
62. D. Robertshaw and C. R. Taylor, *J. Physiol. (London)* **203**, 135 (1969).
63. C. R. Taylor, Amer. J. Physiol. **219**, 1136 (1970).
64. M. A. Baker and J. N. Hayward, *J. Physiol. (London)* **198**, 561 (1968).
65. P. M. Daniel *et al.*, *Phil. Trans. Roy. Soc. B* **237**, 173 (1953).
66. C. R. Taylor and C. P. Lyman, *Amer. J. Physiol.* **222**, 114 (1972).
67. G. M. O. Maloiy and D. Hopcraft, *Compt. Biochem. Physiol.* **38A**, 525 (1971).
68. C. R. Taylor, *Nature (London)* **219**, 181 (1968).
69. *idem*, *Sci. Amer.* **220**, 88 (1969).
70. *idem*, *Amer. J. Physiol.* **217**, 630 (1969).
71. P. A. Buxton, *Proc. Roy. Soc. B* **96**, 123 (1924).
72. W. V. MacFarlane *et al.*, *Austral. J. Sci.* **25**, 112 (1962).
73. *idem*, *Nature (London)* **197**, 260 (1963).
74. G. M. O. Maloiy, *Symp. Zool. Soc. (London)* **31**, 243 (1972).
75. D. Robertshaw, in ref. 1, p. 132.
76. B. Schmidt-Nielsen, *Amer. J. Physiol.* **185**, 185 (1956).
77. A. J. F. Webster, in ref. 1, p. 33.
78. B. D. Siebert and W. V. MacFarlane, *Physiol. Zool.* **44**, 225 (1971).
79. G. Kramer, *Naturwissenschaften* **16**, 377 (1950).
80. F. N. Reece and J. W. Deaton, in ref. 1, p. 337.
81. J. D. Rising, *ibid.*, p. 103.
82. H. S. Siegel, *ibid.*, pp. 204 and 239.
83. M. van Kampen, *ibid.*, p. 158.
84. G. C. West, *ibid.*, p. 452.
85. W. O. Wilson, *ibid.*, p. 411.
86. P. C. Harrison, *ibid.*, pp. 301 and 310.
87. R. C. Lasiewski *et al.*, *Compt. Biochem. Physiol.* **19**, 445 (1966).
88. K. Schmidt-Nielsen *et al.*, *Condor* **71**, 341 (1969).
89. P. D. Sturkie, *Avian Physiology*, Cornell Univ. Press, 1965.
90. J. Benoit, *Coll. Int. Ventre Nat. Rech. Sci.* **172**, 121 (1970).
91. I. Assenmacher, *C.R. Soc. Biol.* **245**, 2388 (1957).
92. R. Miline, *C.R. Assoc. Anat.* **87**, 431 (1955).
93. *idem ibid.* **87**, 1035 (1956).
94. F. Hollwich, *Proc. 71st Conf. Dtsch. Ges. Inn. Med.* **26**, 278 (1965).
95. T. H. Bissonette, *Physiol. Zool.* **5**, 92 (1932).
96. *idem*, *Quart. Rev. Biol.* **8**, 201 (1934).
97. H. M. Scott and L. F. Payne, *Poultry Sci.* **16**, 90 (1937).
98. A. R. Ringoen, *Amer. J. Anat.* **71**, 99 (1942).
99. J. Benoit and L. Ott, *Yale J. Biol. Med.* **17**, 27 (1944).
100. A. E. Woodward, J. A. Moore and W. O. Wilson, *Poultry Sci.* **48**, 118 (1969).
101. P. C. Harrison *et al.*, *ibid.* **47**, 878 (1969).
102. *idem*, *J. Reprod. Fert.* **22**, 269 (1969).
103. M. Menaker, *Sci. Amer.* **226**, 22 (1972).
104. M. Van Albada, *Int. J. Bioclimatol. Biometeorol.* **2**, 3A (1958).

105. O. Riddle and W. S. Fisher, *Amer. J. Physiol.* **72**, 464 (1925).
106. J. R. Brett, in ref. 1, p. 328.
107. R. G. Ferguson, *J. Fish Res. Bd. Cda.* **15**, 607 (1958).
108. F. E. J. Fry, in ref. 1, p. 95.
109. C. M. Sullivan, *J. Fish Res. Bd. Cda.* **11**, 153 (1954).
110. D. H. Bennett, *Arch. Hydrobiol.* **68**, 376 (1971).
111. S. A. Spigarelli and M. M. Thommes, *Argonne National Lab. Report No. ANL 7960*, p. 102.
112. F. G. Carey and J. M. Teal, *Proc. Nat. Acad. Sci. (Washington)* **56**, 1464 (1966).
113. F. G. Carey and K. D. Lawson, *Compt. Biochem. Physiol.* **44A**, 375 (1973).
114. E. D. Stevens and F. E. J. Fry, *ibid.* **38A**, 203 (1971).
115. W. O. Haufe, in ref. 4, p. 688.
116. J. H. Fabre, *The Sacred Beetle and Others*, Dodd Mead, New York, USA, 1918.
117. B. P. Uvarov, *Trans. Entomol. Soc. (London)* **79**, 1 (1931).
118. W. G. Wellington, *Ann. Rev. Entomol.* **2**, 143 (1957).
119. L. Schua, *Z. Vergl. Physiol.* **34**, 258 (1952).
120. W. C. Levengood and W. P. Shinkle, *Science* **132**, 34 (1960).
121. *idem ibid.* **133**, 115 (1960).
122. M. G. Maw, *Canad. Entomol.* **93**, 602 (1961).
123. E. Haine and H. König, *Z. Angew. Entomol.* **47**, 459 (1960).
124. R. Ross, in *Tropische Gezondheidsleer,* J. J. van Loghem (Ed.), Kosmos, Amsterdam, The Netherlands, 1899.
125. F. T. Abushama, *Flies, Mosquitoes and Disease in the Sudan*, Khartoum Univ. Press, Sudan, 1974.
126. R. C. Muirhead-Thompson, *Brit. Med. J.* **1**, 114 (1951).
127. T. C. E. Thomas, *ibid.* **2**, 1402 (1951).
128. Gjullin, in ref. 125.
129. J. Brouwer, Thesis, Leiden University, The Netherlands, 1958.
130. R. C. Rainey, *Meteorology and the Desert Locust*, Tech. Note 69, W.M.O., 171, TP 85, 1965.
131. *idem*, *Int. J. Biometeorol.* **14**, 91 (1970).
132. B. Uvarov, in ref. 130.
133. J. L. Cloudsley-Thompson and F. T. E. Abushama, *A Guide to the Physiology of Terrestrial Arthropoda*, Khartoum Univ. Press, Sudan, 1970.
134. R. T. Berg, *Plan. Feeders Day Rept. (Univ. Alberta)* **45**, 23 (1969).
135. R. M. Butterfield and R. T. Berg, *Proc. Austral. Soc. Ann. Prod.* **6**, 298 (1966).
136. D. F. Dowling, *ibid.* **13**, 184 (1970).
137. *idem*, in ref. 1, p. 315.
138. A. J. F. Webster, in *Physiology*, J. D. Robertshaw (Ed), Butterworth, London, UK, 1972, Chap. 2.
139. Ref. 1, p. 131.
140. W. Bianca, *J. Dairy Res.* **32**, 291 (1965).
141. H. D. Johnson, in ref. 1, p. 358.
142. *idem et al.*, *Mo. Agric. Expl. Stn. Res. Bull.* 1962.
143. P. Mahadevan and H. G. Hutchison, *Anim. Prod.* **6**, 331 (1964).
144. Lecky and Mahadevan, in ref. 1, p. 358.
145. J. S. Popoff, *Z. Tierz. Biol.* **10**, 285 (1927).
146. M. A. R. Kelly and I. W. Rupel, *U.S. Dept. Agric. Tech. Bull., 591,* 1937.
147. J. R. Dice, *J. Dairy Sci.* **18**, 447 (1935).
148. *idem ibid.*, **23**, 61 (1940).
149. *idem ibid.*, **25**, 678 (1942).
150. Regan and Richardson, in ref. 1, p. 289.

151. G. L. Hahn, H. D. Johnson, M. D. Shanklin and H. H. Kibler, *Trans. A.S.A.E.* **8**, 332 (1965).
152. G. L. Hahn, *J. Dairy Sci.* **52**, 800 (1969).
153. H. D. Johnson, in *Progress in Human Biometeorology III,* S. W. Tromp and J. J. Bouma (Eds), Swets & Zeitlinger, Amsterdam, The Netherlands, 1972.
154. T. W. Bond, *Proc. 1st Int. Livestock Envir. Symp.* Lincoln, Nebraska, USA, 1974.
155. Montero, in ref. 1, p. 358.
156. G. D. Brown, *Austral. J. Agric. Res.* **22**, 797 (1971).
157. J. M. Doney, *J. Agric. Sci. (Camb.)* **67**, 25 (1966).
158. R. H. Hayman and T. Nay, *Austral. J. Agric. Res.* **12**, 513 (1961).
159. J. C. D. Hutchinson, G. D. Brown and T. E. Allen, in ref. 2, p. 61.
160. J. Slee, in *Biology of the Skin, and Hair Growth,* A. G. Lyne and B. F. Short (Eds), Angus & Robertson, Sydney, Australia, Chap. 33.
161. Williams, in ref. 10.
162. R. B. Symington, *Nature (London)* **184**, 1076 (1959).
163. Walford, in ref. 104, p. 1.
164. K. Tanaka *et al.*, *Poultry Sci.* **44**, 662 (1965).
165. P. N. Winn and E. F. Godfrey, *Int. J. Biometeorol.* **11**, 39 (1967).
166. J. W. Deaton, F. N. Reece and C. W. Bouchillon, *Poultry Sci.* **48**, 1579 (1969).
167. A. R. El-Boushy, P. C. M. Simmons and G. Wiertz. *ibid.*, **47**, 456 (1968).
168. L. Lacassagne and J. D. Jacquet, *Ann. Zootechn.* **14**, 169 (1965).
169. C. G. Payne, in *Environmental Control in Poultry Production,* T. C. Carter (Ed), Oliver & Boyd, London, UK, 1967, p. 40.
170. W. O. Wilson, S. Itoh and T. D. Siopes, *Poultry Sci.* **51**, 1014 (1971).
171. D. Smidt, in ref. 153, p. 87.
172. M. Crawford, in *Biometeorology I,* S. W. Tromp (Ed), Pergamon, London, UK, 1962, p. 162.
173. L. P. Smith, *Weather and Animal Diseases,* Tech. Note 13, W.M.O., Geneva, Switzerland, TP152, 1970.
174. W. Ferguson and D. Branagan, in ref. 153, p. 98.
175. A. E. R. Taylor and R. Muller, *The Effects of Meteorological Factors upon Parasites,* Symp. Brit. Soc. Parasitol. No. 12, Blackwell, London, UK, 1974.
176. W. M. Allcroft, *J. Comp. Path.* **51**, 176 (1938).
177. *idem*, *Brit. Vet. J.* **103**, 75 (1947).
178. t'Hart and Kemp, in ref. 4, pp. 276 and 685.
179. W. G. Bradley and M. K. Yousef, in ref. 1, p. 417.
180. J. C. Bonsama *et al.*, *Emp. J. Exp. Agric.* **21**, 154 (1953).
181. R. E. McDowell, *Improvement of Livestock Production in Warm Climates,* Freemand, San Francisco, USA, 1972.
182. W. V. MacFarlane, in ref. 1, p. 425.
183. *idem et al., Austral. J. Agric. Res.* **9**, 217 (1958).
184. J. E. Vercoe, in ref. 1, p. 434.
185. H. A. W. Venter *et al.*, *Int. J. Biometeorol.* **17**, 147 (1972).
186. R. S. Temple, *World Anim. Prod.* **7**, 39 (1972).
187. W. O. McCarthy and C. P. Hamilton, *Proc. Trop. Agric.* **41**, 293 (1964).
188. G. O. Harwin *et al.*, *Proc. S. Afr. Anim. Prod.* **6**, 171 (1967).
189. J. D. Skinner and G. N. Louw, *J. Appl. Physiol. (London)* **21**, 1784 (1966).
190. C. W. Alliston, in ref. 1, p. 381.
191. G. E. Egli *et al.*, *Proc. Assoc. S. Agric. Workers* **53**, 80 (1956).
192. C. V. Hulet *et al.*, *J. Anim. Sci.* **15**, 607 (1956).

193. N. T. M. Yeates, in ref. 1, p. 387.
194. J. Steinbach, *ibid.*, p. 393.
195. *idem, Int. J. Biometeorol.* **17**, 141 (1973).

CHAPTER 12

PRACTICAL APPLICATIONS OF BIOMETEOROLOGY IN DEVELOPING COUNTRIES

In the previous chapters the many practical applications of biometeorology which would be especially useful in developing countries, or in developed areas with extreme climatic conditions, have been discussed. A considerable part of the arid desert and of the semi-arid areas in the world belong to the developing countries. It is estimated that at present 9 million km^2 of soil are true deserts and more than 48 million km^2 may become deserts in the near future if no serious preventive measures are taken. In other words 43% of the total surface of the earth, and the people living in these areas, are seriously endangered. If one adds to this problem that of the rapidly increasing world population it is understandable that many national and international agencies are trying to find efficient methods both to stop desert formation and to develop the arid and semi-arid zones on earth. This explains the creation of the Arid Zone Committee, around 1952, by Unesco Head Quarters in Paris which was followed by many international conferences on the economic development of developing countries suffering from extreme climatic conditions.

An important aspect of arid-zone research is the study of the effects of weather and climate on plants, animals and humans living in arid and in semi-arid countries. The applications of biometeorology have great impact both on the health of people (and indirectly on their performance capability) and on the economic development of the usually warm, arid, developing countries. Some of these biometeorological aspects will be summarized in the following sections.

12.1. APPLICATIONS OF PLANT BIOMETEOROLOGY

12.1.1. Agricultural meteorology

It has been suggested, particularly by Gloyne,[1] Borghorst[2] and others, that:

(a) In order to study applications in the field of agriculture in most developing countries an up-to-date network of meteorological stations must be established which can record automatically temperature, humidity, windspeed, rain, cloudiness, hours of sunshine etc. The administrations concerned should be adapted as soon as possible to meet the specific needs of agriculture, forestry, hydrology, construction industries, architecture, urban planning, veterinary and medical sciences.
(b) Urgent attempts should be made to achieve the greatest possible degree of standardization of instruments, of auxiliary equipment (e.g. recording and data processing systems), and of techniques of observing, measuring and tabulating meteorological features of the natural environment. The essential characteristics of the instruments to be used are: reliability of operation; stability of response and calibration; reproducibility of response between different instruments of the same kind; ease of maintenance and repair; ready availability of spare parts etc.
(c) In view of the importance of data on solar radiation in all branches of biometeorology, various centres of observation should be established to measure the total solar radiation both visible and ultraviolet.
(d) Because of the phytobiological importance of the deposition of moisture on plant surfaces, other than by rain or snow, arrangements should be made to install a simple type of dew-gauge at a representative selection of reporting stations, and a simple recorder of wetness-duration at agricultural meteorological stations, and at such other stations as is necessary to provide an adequate network of stations.
(e) At a selection of agricultural meteorological stations, the improved recording of soil temperature at depths down to 20 cm should be initiated. As a temporary expedient the use of a combined maximum/minimum thermometer to give the daily extremes may be considered.
(f) In addition to the ordinary anemometer with its measuring head at 2 m, additional instruments of this type should be supplied to agricultural meteorological stations where, for scientific or operational purposes, wind data at other heights, for example 10 m, are required.
(g) Every effort should be made to adhere strictly to the times at which observations should be taken, as are laid down by the W.M.O. for the various types of station. Further advice can be obtained from the agricultural experts in the W.M.O. in Geneva and The Food and Agricultural Organization (F.A.O.) in Rome.

12.1.2. General agriculture

Knowing the specific macro- and microclimatic conditions in a certain developing country it is possible to predict, if one knows the prevailing soil types, the type of plants which should be planted in order to obtain the best crops. Most of the large agricultural institutes in western Europe, USA, Canada and the

F.A.O. are able to give such advice to any developing country which asks for this information. Also advice can be given on the specific weather and climatic conditions triggering certain plant diseases and pests in a developing country and on the methods to fight these plagues.

12.1.3. Forestry

The same applies for forest development. In particular institutes in Europe and the Americas, specialized in tropical tree development, should be consulted. Apart from the micro- and macroclimatic conditions and prevailing soil types the various dangerous indigenous insect species should also be mentioned to the advising organization.

12.1.4. Entomology

In Section 11.1.8 some of the biometeorological aspects of certain groups of insects were discussed. Similar weather relationships of other insects are often known already in the developing countries or need to be studied using the various methods developed in insect research centres in the developed countries.

12.1.5. Plant diseases

In Section 2.6 some plant diseases were discussed, as also were modern methods used to predict the occurrence of the disease at a certain time depending on the prevailing weather conditions. Similar predictions could be made in developing countries if experts in this field were able to study the conditions in the developing country itself. Accurate micro- and macroclimate data must be available and also laboratory studies to reveal which temperature-humidity conditions favour the development of certain plant viruses.

12.1.6. Plant protection

In most developing countries a number of measures could be taken to protect crops against destruction, such as:

(a) A study of existing windbreaks in the country to evaluate their effectiveness in reducing damage by wind and salt and their effect on the distribution of pests and beneficial insects. As a basis for this work attention is drawn to the W.M.O. publication No. 59 *Windbreaks and Shelter-belts*.
(b) Rust and temperature damages are serious on wheat and other crops. The provision of a climatic chamber could accelerate the selection of resistant plants.
(c) Various air pollutants have unfavourable effects on man, animals and plants. By growing small plots of susceptible test plants it should be possible to identify these pollutants and to define their frequency and distribution.
(d) An insect trapping system should be established to observe the seasonal

appearance, abundance, and movement of pests, so as to facilitate the eventual forecasting of pest and virus disease outbreaks and the consequent reduction of preventive spraying. This trapping system should be matched to the agro-meteorological network. Personnel should be recruited and trained, and the system established, as soon as is practicable.

(e) Scientists, both in the advisory and research services as well as in the universities, must be given courses in the principal fungal and bacterial plant diseases which occur in the country; they must learn about their phenology and weather relationships.

(f) The scientific staffing of research centres in developing countries is often insufficient. Apart from a plant pathologist there should be a plant nematologist to study the nematodes, which are serious pests and may act as virus vectors. Also a plant virologist is required to consolidate and expand the work of visiting virologists.

(g) Often important literature is not known to the scientists of the developing countries, due to lack of funds or uniformity in libraries and journal circulation. Publications from world organizations and semi-arid zones are particularly important and need to be readily available.

These and other measures are required to fight the meteorologically induced plant pests in a developing country.

12.2. APPLICATIONS OF ANIMAL BIOMETEOROLOGY

12.2.1. Veterinarian aspects

The principal thermoregulation characteristics of cattle, sheep, pigs and birds were discussed in Section 11.1. Environmental conditions, including animal housing construction, which disturb the thermoregulation of these various animals are harmful to their health, and effect milk, wool and egg production, and meat quality.

Among the factors which could impair animal production in developing countries, nutrition, including the supply of drinking water, seems to be of primary importance in the case of cattle, sheep and pigs. Improvement of nutrition, is desirable in view of the fact that animals in a low state of nutrition, especially when this is combined with poor management and disease control, have an increased susceptibility to climatic stress.

The direct effects of climatic factors upon domestic animals are often of great economic significance. Climatic factors vary from one area to another and it is indeed for the developing arid and semi-arid areas that data are largely unavailable. Integrated studies involving macro- and microclimatic conditions are greatly needed and there is extreme need for analyses of the physiological stresses (including heat, cold, and altitude stresses as determined by temperature, humidity, wind, rain, and reduced oxygen pressure) created by different climates on cattle, sheep, goats, and other domestic animals. Environmental effects on milk production (it is already known that extreme warm humid climates and cold stresses usually reduce milk production), on skin coat and meat quality, and on reproduction should receive intensive study. For animals kept indoors due to shortage of land or to extreme outdoor climatic factors, the microclimatic conditions differ strongly from conditions experienced by animals that have no shelter. For example, the microclimatic conditions in cattle barns may be unfavourable because of a humid warm environment (especially above 20°C) and the presence of such deleterious gases as carbon dioxide and ammonia. Poor floor conditions (such as those associated with uninsulated concrete floors) may have adverse effects, especially on cattle resting on such floors, because the cattle will lose a large amount of heat by conduction to the cold floor.

Some of the very unfavourable effects of adverse climatic factors on sheep are well known, but additional studies are strongly needed. With respect to sheep, a combination of low temperatures with high winds and rain, which penetrates to the skin, destroys the thermal insulation of the fleece and hair coat and may cause serious cold stresses, especially in newborn lambs. Simple shelters in areas with extreme climates may improve conditions that otherwise seriously affect the health of domestic animals. Once climatic factors and the effects of climate become common knowledge, animal breeders will be in a more favourable position to develop breeds adapted to cope with the rigors of extreme climates. Integrated physiological and veterinary research is needed in the study of resistance of different breeds to adverse climatic conditions and in the selection and crossbreeding necessary for the development of more suitable kinds of livestock.

There is sufficient evidence that in developing countries climatic stress is present in certain animals under certain conditions in certain regions. In an effort to mitigate such effects the following suggestions and recommendations were made for the middle East in 1966 by a number of veterinarians.[3]

(a) In the coastal region in the middle East pure-bred European-type cows are kept in summer at environmental temperatures around 30°C (with occasional higher temperatures) combined with relatively high humidities. Since the milk yield in European cows is well known to begin to decline with environmental temperature rising above about 20°C, steps must be taken to improve the microclimate in dairy barns. The feasibility and the economics of measures such as spraying the animals with water, fanning and cooling of the drinking water are recommended. In areas where sheep are grazed in summer consideration should be given to a better and wider distribution of water points with shade where possible.

(b) Although air temperatures in certain developing countries are not extremely low, the combinations of low temperature with high winds and rain which penetrates to the skin thus destroying the thermal insulation of the fleece and the hair coat, may cause cold stress, particularly in the new-born lamb. Simple shelters which protect against wind and rain are called for. In cattle barns efforts should be made to provide adequate floor insulation, since the bare concrete floors normally in use cause excessive conductive heat loss from the lying animal. In this connection the possibility of using industrial by-products or other material for cheap bedding should also be considered.

(c) The practice, prevailing in developing countries in small cattle units, of confining cattle under poor housing conditions during a large part of the year creates an unfavourable microclimate characterized by high humidity and the accumulation of deleterious gases such as ammonia and carbon dioxide. Conditions would be improved by providing more ventilation, moderate sunlight and better sanitation.

(d) Regarding the suitability of various species and breeds to live and produce under the conditions prevailing in the developing countries, it would be desirable to obtain more information on the genetic merits of indigenous breeds and on their response to improved conditions of feeding and management. In this way potentially valuable genetic material could be preserved. Furthermore, if animals from abroad are introduced, this introduction should be made in a systematic way and on a scale large enough to allow general conclusions to be drawn from such experiments. The economics of the distribution of cattle, sheep and goats should be evaluated in terms of forage conservation and animal production and the distribution of these species adjusted accordingly.

(e) Finally studies should be made of the economic importance and control of tick-born diseases, helminthiasis and other diseases whose incidence is influenced by weather and climate.

It is evident that in other areas different recommendations may have to be made although several of the recommendations are also valid for other countries (see also various observations described in Chapter 11).

12.2.2. Avian aspects

Integrated studies required in relation to poultry farming in arid and semi-arid areas should concentrate on problems in at least the following ten aspects in which production is seriously influenced by weather and climate:

(a) Fertility and hatchability, both of which are lowered during hot weather;

(b) Storage of feed, with prevention of deterioration (e.g. prevention of loss of water-soluble vitamin A);

(c) Growth which may be considerably retarded in hot seasons, even to the extent that broiler growing is impossible for several months of the year;

(d) Egg production and quality, both of which can be seriously affected especially under conditions of high temperature and humidity and both of which can be maintained at favourable levels even in areas with large fluctuations in diurnal temperatures and cool nights, if the annual and diurnal thermal fluctuations are considered in the design of poultry houses;

(e) Meat quality, a topic that requires considerably more research;
(f) Selection of birds, with efforts made to select the best-adapted and most-productive birds in testing stations located in different climatic zones;
(g) Housing, with special design for maximum comfort and production in various climates. It has been recommended in the case of laying birds that use be made of 'open type' houses with a concrete floor and an aluminium or corrugated metal roof, the top of the roof being painted white in order to increase heat reflection. The ceiling should be installed at least 30 cm from the roof in order to reduce heat transmission. The space between roof and ceiling should be ventilated. In very hot climates artificial cooling may be required. Each of these measures depends on the type of arid or semi-arid climate, and considerable integrated physiological and avian research is required to solve these economic problems;
(h) Incubation, with special attempts to maintain fairly constant incubation temperature and humidity conditions, especially in areas with considerable fluctuations in temperature and humidity;
(i) Storage of hatching eggs, with the prevention of deterioration in warm climates by storage in well-insulated rooms at a temperature of 12–15°C and a relative humidity of 75–85%;
(j) Storage of poultry and eggs intended for consumption, a topic requiring special research in arid and semi-arid regions.

In view of the previous considerations, in 1966, several recommendations were made for the poultry industry in developing countries, particularly by Van Albada and Maas.[4] In several developing countries this industry could be improved if the following recommendations were acted upon.

(a) It is desirable to make some concentration of breeding farms, multiple farms and hatcheries into relatively large units.
(b) Feed deterioration and/or reduced feed consumption during hot weather can be overcome as far as is necessary: by increasing protein, vitamin and mineral content while keeping constant the energy level; by applying vitamin premixes shortly before use; by pelleting feed mixtures and by adding water-soluble vitamin A to the drinking water at regular intervals.
(c) In regions where outdoor temperatures are extremely high during part of the year, broiler production will be uneconomical and must be stopped temporarily.
(d) In designing poultry houses both diurnal and annual temperature fluctuations should be taken into account.
(e) In controlling light in broiler houses, interactions between light control and climatization have to be considered.
(f) In regions with continuous medium or high temperatures, layer houses should be built with walls allowing ample ventilation and only an aluminium or corrugated iron roof with a weather proof board ceiling sufficiently separated to avoid heat transmission by radiation; layer houses in hot and relatively dry climates must be cooled by means of forced

draught ventilation through a screen of dripping water (e.g. in wood shavings); broiler houses also need controlled climatization.

These and possibly other recommendations apply to most warm developing countries in the world.

12.3. APPLICATIONS OF HUMAN BIOMETEOROLOGY

The various applications of human biometeorology in developing countries can be divided in two major groups: Medical and Architectural–Urban applications.

12.3.1. Medical applications

Amongst which the following seems to be the most important:

(a) *Reduction of the effect of excessive heat stress on the thermoregulatory system.* Such a reduction would result in:

(i) improved general health of the population and a reduction of the incidence of meteorotropic diseases, and

(ii) improved general comfort, increased performance capability, reaction speed etc.

The reduction of thermal stresses can be achieved in three different ways:

(1) By introducing air conditioning the many advantages of which in hot tropical areas, both for general comfort and on various skin diseases etc., were discussed in Section 7.3.

(2) By temporary migration to cooler areas in the country during the hottest season, either to the mountains or to the sea. This has been a common practice in Egypt, Lebanon, India and other middle East countries. However if the migration is not one lasting for several months, the results of which would be beneficial, frequent moves of much shorter duration could create an increase in altitude and thermal stresses.

(3) By using special clothing adapted to great thermal stresses (Section 7.1).

(b) *Local studies must be made in the developing country on the physiological effects of solar radiation (both infrared and ultraviolet), humidity and air movement on a climatically adapted population.* While the physiological effects described in Section 5.2 apply to a certain extent to all countries, the magnitude of the effects, however, depends on the local meteorological conditions of an area.

(c) *Diseases due to extreme heat and cold stress.* As most developing countries are located in hot tropical areas diseases due to extreme heat stress are of particular interest. These were described in Section 6.11 and precautionary measures have been briefly mentioned.

(d) *Influence of local micro- and macroclimates on the growth and spread of bacteria and viruses.* Section 6.7 discussed the close relationship between temperature-humidity and the survival rate of microbes. This was studied in particular by Hemmes[5,6] in The Netherlands and by Schulman and Kilbourne[7] in the USA. Local laboratory studies in a developing country, during which the actual microclimate during different seasons in that country were imitated, would enable the epidemiologist to predict the outbreak of serious infectious diseases in that area and to introduce the necessary preventive measures.

(e) *Apart from infections parasitic diseases are also a great threat for developing countries.* The following is a brief summary of a report by Bruyning, Professor of Parasitology at Leiden University, in The Netherlands[8] which he prepared in 1966 for a symposium on medical problems in developing countries and which is still valid.

In most tropical and sub-tropical countries the health and the socio-economic status of the population are largely influenced by parasitic disease. Malaria, filariasis, bilharziasis, ancylostomiasis, trypanosomiasis and many other parasitic diseases still cause a great economic harm by incapacitating the physical development and working capacity of those infected. In Section 11.1.8 many diseases caused by mosquitoes and flies, and the influence of meteorological factors on them, were discussed. The geographic distribution of parasites as well as the prevalence of the parasitic diseases are governed by several factors.

The endemicity depends on conditions like:

(i) The presence of suitable hosts in quantities which ensure the continuity of their life cycle;
(ii) Environmental conditions which are favourable for the survival of the stages outside the host(s);
(iii) Socio-economic conditions which strongly affect their distribution and prevalence.

Low standards of living, inadequate sanitation (individual as well as communal), promiscuous handling of excreta, ignorance etc. promote the spread of the parasitic diseases and thereby the morbidity of the population. The use of polluted water, insufficiently cooked food, primitive food-handling etc. favour the increase of these diseases. Migrations of populations on a large scale introduce parasitic diseases in new areas.

The importation of African people into the western hemisphere is responsible for the introduction of diseases like bilharziasis and hookworm. Since the environmental conditions are of great importance, the parasitic diseases are in most cases more prevalent in tropical and subtropical countries. In most of these countries the socio-economic conditions of the population are worse than in the developed countries. The combination of favourable environmental conditions for the parasites and a low socio-economic situation of the population causes a high infection-rate and great severity of parasitic diseases.

Parasitic diseases are highly important negative factors in the food production in most areas where hunger is a significant factor. Children and adolescents especially are strongly affected and their physical as well as their mental development is retarded. In this way the vicious circle is maintained: the diseases undermine the health and productivity of the population and deteriorate the socio-economic situation and in doing this also increase their own prevalence.

As mentioned before, the incidence of most parasitic diseases is strongly affected by climatic conditions. This is mainly brought about by the influence of the climate on the vectors and on the free-living stages of the parasites. The endemicity of all diseases which are transmitted by vectors is confined to the areas where their vector(s) occurs. The life-cycles of the vectors, however, are profoundly affected by seasonal and climatic factors. The development of the stages of the parasite outside the human host also depends upon climatic factors. This is relevant for vector transmitted parasites as well as for parasites with free-living stages. Some of the effects of temperature, humidity and rainfall will be briefly mentioned.

(i) *Influence of temperature.* Insects and snails are heterothermal animals which assume nearly the same temperature as their environment. Below certain temperatures the parasitic stages in the vector are not able to develop. In the case of malaria it has been proved that no species of the malaria parasite can develop below a temperature of 15–16°C. At the lower temperatures the extrinsic phase of the development lasts much longer than at tropical temperatures. The longevity of the female anophelines is of greatest importance since the mosquito must live longer than the period required by the extrinsic phase to produce sporozoites in the salivary glands.

In the case of filariasis suitable intermediate hosts of bancroftian filariasis are present in some areas, but the temperature is too low for the development of the extrinsic phase of the parasite. In other areas transmission occurs only during the summer. For ancylostomiasis and other soil-transmitted helminths, the free-living stages of *Ancylostoma duodenale* (a nematode parasite related to hookworms) can stand temperatures approaching 0°C but *Necator americanus* (same as *Ancylostoma americanum*) is more susceptible to low temperatures and is killed by brief periods at temperatures well above freezing. The optimal temperatures for *Necator*, however, are higher than those for *Ancylostoma.* On account of its better tolerance of low temperatures *A. duodenale* has a more northerly distribution than *N. americanus.* Sunlight may kill eggs and larve of nematodes in the soil by direct heat, radiation and desiccation.

(ii) *Influence of humidity.* The relative humidity has a great influence on the lifespan and the activity of arthropods (Section 11.1.9). Low relative humidity may shorten the lifespan of anopheline mosquitoes, which are potent vectors at higher humidity, so much that the extrinsic phase cannot complete its development. During the hot, dry seasons, especially in savannah areas, otherwise dangerous vectors become harmless. Bruce-Chwatt[9] found a striking difference in the sporozoite rate of *Anopheles gambiae* and *A. funestus* during the dry and the rainy season. For both mosquitoes the rate during the dry season was 0.007%, while the rates during the wet season were 3.7% for *A. gambiae* and 2.3% for *A. funestus*.

The relative humidity is not always a good indicator of the drying power of the air; the saturation deficiency is a better index of the damaging action on the arthropods. A small number of the vectors may survive the extreme dry and hot seasons by aestivating in microhabitats with a microclimate with a lower saturation deficiency.

Many species of mosquitoes are natural vectors of filariasis but the climatological factors often determine the distribution and abundance of the potential vectors as well as the development of the extrinsic phase of the parasite. The incidence of bancroftian filariasis is highest in coastal and estuarine areas with high humidity.

(iii) *Influence of rainfall*. Rainfall plays an important role in the transmission of arthropod borne parasitic diseases. It has a direct influence on the temperature as well as on the humidity. Moreover the breeding facilities of many mosquitoes are extended during the wet seasons. When the rains are excessively heavy the transmission may decrease consequent upon the flushing of breeding foci.

Rain also affects the spreading of bilharziasis—in areas with a pronounced dry season the semi-permanent waters dry up for several months. Only a small proportion of the snails survive and these repopulate the habitat. In areas with a low relative humidity dessication can be used as a means of control. Other climatic factors such as temperature and light are also important in the epidemiology of bilharziasis.

The dynamics of transmission of parasitic diseases should be carefully studied in each endemic area. In parasitology generalizations are dangerous and it is of the utmost importance to collect adequate data on the local situation, before any control work is started. A better knowledge of the ecologic factors is needed to understand the measures that must be taken to obtain control of the vectors and of interruption of the parasitic disease. As has been shown the climatic factors are of great importance for the ecology of different vectors. A systematic study of the seasonal cycle, the capacity to withstand low humidity and drying out, the influence of rainfall and extreme temperatures, exposure to direct sunlight etc. must be instituted, since control measures against the vectors or against the extrinsic phases of parasites can only be effective and efficient if they are based on thorough knowledge of the dynamics of transmission. During the pre-eradication surveys climatological information must be collected over long periods.

12.3.2. Architectural and urban applications

In Chapter 10 the great importance of meteorological studies for architecture and urban planning has been pointed out. It is evident that the various considerations mentioned in this section apply particularly to the hot tropical countries. Contrary to what has been said in Section 10.1 for the cold countries, in hot countries measures should be taken to increase air-movements and air turbulence. Windows should be on both sides of a house or building to allow the wind to enter the house freely, windows must be opened in the early morning to allow cool air to enter the house but later in the morning all windows and shutters must be closed to prevent heating up by solar radiation. However, if there is a strong wind the cooling of the body may be increased by opening the windows even though the air temperature is increased.

Spacing between houses and particularly between large buildings should be considerable. The building of high apartments close together in tropical countries is a great mistake.

According to architects who are experts on architectural and urban planning in tropical countries[10-18] at least the following principles must be followed in developing countries:

(a) One should start with site selection and the effect of topography;
(b) Orientation of the house or building in regard to solar effects, wind and temperature;
(c) Solar control, biometeorological evaluation of the effectiveness of shading devices and the effects of trees and other kinds of vegetation must be studied;
(d) Preferable building forms must be evaluated in relation to external thermal stresses;
(e) Wind and air-flow effects must be studied to satisfy biometeorological needs in public spaces and in structures lacking air-conditioning;
(f) Preferably local materials for indoor thermal balance and insulation (both inside and outside) must be selected;
(g) A calculation must be made of the total thermal behaviour of the architectural and urban projects.

This brings us to the end of the present volume. Much remains to be studied, but the major effects of our physical environment on the living world are relatively well known and permit the biometeorologist to make the required predictions once he knows the micro- and macroclimate of a region. If one has lived, like the author, with people of tropical and arid countries, one's affection for his own Western culture undergoes a change in perspective. Instead of admiring its accomplishments only as selfish things to be considered within its borders, they are viewed in the light of the advantages they may offer to those outside, who are geographically not so fortunately situated. Such is in my opinion truly international biometeorology, a philosophy which stresses the solidarity and mutual dependence of all countries in this world.

REFERENCES

1. R. W. Gloyne, *Final Report of the 1st Int. Biometeorol. Conf. of Middle and Far East* Laklouk, Lebanon, 1–13 April 1966, p. 35.
2. A. J. W. Borghorrst, *ibid.*
3. *ibid.*, pp. 24, 25 and 34.
4. M. van Albada and H. J. L. Mass, *ibid.*, p. 26.
5. J. H. Hemmes, Thesis, Utrecht Univ., The Netherlands, 1959.
6. *idem*, K. C. Winkler and S. M. Kool, *Nature (London)* **188**, 430 (1960).
7. J. L. Schulman and E. D. Kilbourne, *ibid*, **195**, 1129 (1962).
8. C. F. A. Bruyning, in ref. 1, p. 31.
9. Bruce-Chwatt, *ibid*.
10. R. Ayoub and V. Olgyay, *ibid.*, p. 27.
11. B. Givoni, *Man, Climate and Architecture*, Elsevier, Amsterdam, The Netherlands, 1969.
12. J. K. Page, *Roy. Inst. Brit. Archit.* **62**, 102 (1955).
13. *idem, Climate and Town Planning*, Overseas Building Notes, 52 Bldg. Res. Stn., Watford, UK, 1958.
14. *idem*, in *Medical Biometeorology*, S. W. Tromp (Ed), Elsevier, Amsterdam, The Netherlands, 1963, p. 655.
15. V. Olgyay, *Design with Climate*, Princeton Univ. Press, New York, USA, 1963.
16. *Teaching the Teachers in Building Climatology*, Proc. Symp. Stockholm, 4–6 September, 1972. Swedish Inst. for Building Res., Doc. D. 20, Sven Byggtjänst, Stockholm, Sweden, 1972.
17. *Planning and Construction in Conformity with the Climate*, Proc. Int. Symp. Vienna, 20–21 October 1976, Forsch. f. Wohren, Banen Planen, Monog. 24, Löwengasse 47, Vienna, Austria, 1976.
18. *Uses of Epidemiology in Housing Programmes and in Planning Human Settlements*, W.H.O. Report on Housing and Health, W.H.O., Geneva, Switzerland, Tech. Rep. 544, 1974.

GLOSSARY

A

Acclimation—according to Burton and Edholm: long-term changes in the body taking place during a life time; according to others a synonym for acclimatization.

Acclimatization—complex of reversible changes in the body resulting from gradual adaptation to certain climatic conditions which increases the efficiency of individual organisms.

Accommodation—adjustment to the 'changes' of environmental conditions.

Adaptation—adjustment of an organism to environmental conditions.

Adrenal gland—gland of internal secretion, near the kidney, producing important hormones. It consists of a 'cortex' with columns of polyhedral cells (with well marked nuclei) and the 'medulla' composed of irregularly arranged cells.

Adrenaline—trade mark for epinephrene. A substance produced by the adrenal gland which may cause a constriction of the blood vessels and capillaries.

Aerosols—a system consisting of very fine (2 nm–5 μm) solid or liquid particles dispersed between the gas molecules of the air. The outer surface of the particles is usually electrically charged.

Air mass—large masses of air in the atmosphere with more or less the same physical and chemical properties extending in a horizontal direction over hundreds or even thousands of kilometres and in vertical sense over more than one kilometre. Their properties depend on the source area of these masses of air.

Albumin—a simple protein soluble in water, formed in the liver, easily diffusible through a cluster of capillaries (glomeruli) of the kidney into the urine.

Allergy—altered reactivity of an individual to a specific substance usually resulting from prior experience with the same or a chemically related substance.

Atmospheric cooling—the combination of fall in temperature and increased air movement. It has a chilling effect on the body either through a direct heat loss or as a result of the evaporation of perspiration of the skin.
Apocrine sweat glands—large sweat glands found mainly in the axillae and round the nipples.

B

Bacteria—smallest unicellular micro-organism of which some species produce diseases. Gram positive bacteria are stained by Gram's method. They die rapidly at high relative humidity; gram negative bacteria are not stained by Gram's method. They die more rapidly at lower humidities.
Bar—unit of air pressure equal to one million dynes/sq cm = 1000 millibars (see barometric pressure).
Barometric pressure—pressure of the atmosphere per sq cm. The normal pressure at sea level (one atmosphere) equals the pressure of a mercury column of 76 cm which is equal to 1033.3 g = 1013 millibars.
Blood clotting (or coagulation) time—coagulation time in seconds after the interaction of thrombin with the soluble fibrinogen of the blood plasma forming the insoluble protein fibrin.
Blood pressure—lateral pressure exerted on the walls of the blood vessels by the circulating blood. Systolic B.P. is the highest arterial pressure during the period of the heart contraction (the systole). It fluctuates considerably during emotional states; diastolic B.P. is the lowest arterial pressure after the systole, less subject to temporary fluctuations.
Blood (or erythrocyte) sedimentation rates (B.S.R.)—speed of sedimentation of the red blood corpuscles in a blood sample (kept fluid by means of an anti-coagulant), placed in a vertical, narrow tube; the blood corpuscles settle to the bottom while the transparent plasma rises to the top. The height of the clear fluid after one hour, expressed in mm, is the B.S.R. (or E.S.R.).

C

Cerebrum—the anterior larger part of the brain.
Climate—the average of weather conditions (characterized by temperature, humidity, windspeed, solar radiation etc.) of any locality observed during thirty years.
Cooling—see atmospheric cooling.
Core temperature—internal body temperature reflected in the rectum temperature.
Cyanosis—a blueish tinge of the skin and mucous membranes caused by lack of oxygen.

D

Diuresis—flow of urine.

E

Eccrine sweat glands—sweat glands distributed over the body surface secreting a dilute solution of sodium chloride, urea and lactate (in exercise).

Electrolytes—certain chemical compounds whose positively and negatively charged atoms are attracted respectively to the negative and positive poles of a battery when an electric current is passed through their aqueous solutions. Examples of such compounds are hydrogen chloride or potassium chloride.

Electrolyte balance—equilibrium between production and excretion of electrolytes.

Electromagnetic system—electric forces exerted between a magnetic field and a conductor, placed in that field, carrying a current.

Electrostatic system—electric force exerted between two charges of electricity without displacement of the charges along the conductor.

Endocrinal gland—gland of internal secretion.

Enzymes—complex organic compounds secreted by living cells which are capable of causing or accelerating certain changes in a substrate for which it is often specific.

Eosinophils—medium sized white blood cells of which the large granules in the interior are stained bright red by eosin fluid; number of eosinophil cells increases during allergic conditions (see allergy).

Epiphysis, Pineal Gland or Photoperiodic Brain—it is located at the posterior lower part of the corpus callosum, a part of the brain which lies at the bottom of the cerebrum; it regulates the timing of physiological events by recording changes in light intensity and wavelength.

Erythrocyte—a blood cell which contains the haemoglobin of the blood.

Exocrine gland—gland of external secretion.

Extra-terrestrial forces—chemical or physical (including gravitational) forces originating outside the earth and its atmosphere.

F

Fibrinogen—a protein formed by the liver. It is important during blood coagulation processes.

Föhn—very warm, dry wind, which often blows in the mountain valleys of Switzerland and Tyrol. It originates if humid air rises against high mountains, causing heavy rains. The dried air reaches the valleys on the other side of the mountains and has many biological effects.

Footcandle—unit of illumination. The amount of light at one foot from a standard candle.

Front—line of intersection of a frontal surface (boundary plane between two air masses) with the surface of the earth.

G

Ganglion—aggregation of neurons occurring outside the central nervous system.

Globulin (gamma-)—a simple protein produced in the liver, insoluble in water, concerned with the phenomena of immunity.

H

Haemoglobin—the oxygen carrying pigment of blood.
Hexosamines—derivatives of mucoproteins particularly found in great quantities in the joints and connective tissues of rheumatic subjects.
Homoiotherms—animals which maintain a relatively constant body temperature (birds and mammals).
Hormones—chemical substances produced by endocrine glands, which, carried to an associated organ by the blood stream, influence its functional activity.
Hypertension—a state of abnormally high blood pressure.
Hyperthermia—a state of abnormally high body temperature.
Hypophysis—see pituitary.
Hypotension—a state of abnormally low blood pressure.
Hypothalamus—the heat regulatory centre located more or less between the thalamus and pituitary in the brain; it seems to have integrated control of the three autonomic nervous systems and the secretion of the anterior and posterior lobes of the pituitary. The control takes place either through a complex network of nerve fibres or by blood vessels carrying a chemical substance from the hypothalamus to the pituitary. It consists of several clusters of nuclei in not clearly defined areas. The posterior nuclei controls the ortho-sympathetic nervous system.
Hypothermia—a state of abnormally low body temperature.
Hypoxia—a state of diminished oxygen or oxygen supply.

I

Infrared—invisible radiation with wavelengths greater than 7500 Å (1Å = 10^{-8} cm = 0.1 nm).
Insensible perspiration—the continuous invisible and intangible loss of water through the outermost layer of the skin which, after evaporation, assists the cooling of the body.
Ions (air-)—electrically charged molecules or clusters of molecules in the air.

L

Leucocytes—white bloodcells. Five types are known: lymphocytes, monocytes, neutrophyls, eosinophils, basophils. Their chief function is to protect the body against micro-organisms causing disease.
Lymphocytes—the smallest of the leucocyte cells.

M

Macroclimate—climatic conditions of the atmosphere above specified areas.

Metabolism—total of all the physical and chemical processes by which compounds are converted into substances required by living cells.
Meteorotropism—sensitivity of a living organism (plants, animals or man) to changing meteorological conditions.
Microclimate—climatic conditions directly surrounding the living organism, both near the surface of the earth or in a limited niche (room, factory or mine).
Millibar—1/1000 of a bar, which is a meteorological unit of atmospheric pressure.
mm Hg (760)—pressure of a mercury column of 760 mm high.
Morbidity—disease incidence in a community.
Mortality—death frequency of a particular disease in a certain population.

N

Neuron—one of the cell units making up the nervous system.
Nervous system—autonomic or involuntary N.S. controls all not-consciously perceived phenomena (heart, intestines etc.). It consists of three nervous systems:

(a) Orthosympathetic N.S. enables the organism to adapt itself quickly to sudden changes in the environment.
(b) Parasympathetic N.S. of which the principal function is the regulation and recovery of the disturbed physiological equilibrium of the organism as a result of environmental disturbing factors.
(c) Peripheral N.S. consists mainly of a peripheral, essential network of small autonomic ganglion cells responsible for the tonus of fluids and their diffusion. They act also as receptors of external physical stimuli which are transmitted to the brain centres along peripheral nerves.

Central or animal N.S. regulates all consciously perceived phenomena controlled by our volition, such as movements, psychic processes, etc.
Sympathetic N.S.—see orthosympathetic N.S.
Vegatative N.S.—see parasympathetic N.S.

P

Parameter (meteorological-)—physical magnitude. It characterizes the meteorological conditions of a specific environment, such as temperature, humidity etc.
Pathology—the scientific study of the alterations produced by disease.
Peripheral—outward part of a surface or body.
Pineal gland—see epiphysis.

pH—symbol of acidity (hydrogen ion concentration); pH7 = neutral; a solution with pH>7 is alkaline; <7 acidic.

Photodynamic substances—usually substances exhibiting the physical phenomenon of fluorescence after irradiation. It causes an increased sensitivity to light stimulation and may produce several skin diseases.

Photo-period—rhythmic phenomena in plants, animals and man created by light irradiation. The period depends on wavelength, light intensity and duration.

Photosynthesis—the formation of carbohydrates in plants from carbon dioxide and water under the influence of light.

Physico-chemical processes—processes controlled both by the laws of physics and chemistry.

Physiology—the science studying the function of various parts and organs of living organisms.

Pituitary or hypophysis—known as the 'leader of the endocrine orchestra'. It is located at the base of the brain. It is attached by a stalk to the hypothalamus. It is composed of two hormone producing lobes, the anterior and posterior, which secrete several important hormones. They regulate the proper functioning of other endocrine glands (thyroid, adrenal gland etc.)

Plasma (blood-)—the liquid portion of the blood.

Poikilotherms—animals which take on the temperature of the air or the radiant environment around them (invertebrates, fish, amphibians, reptiles).

Psycho-somatic—pertaining to the inter-relationship of mind and body.

R

Rhinencephalon or olfactory brain—the portion of the brain located above the hypothalamic-hypophysial system. It is concerned with (a) the sense of smell; (b) the unconscious emotional life (including the moods); (c) the mediation of certain cardiovascular and endocrinal activities and (d) the conscious affective experiences.

S

Serotonin—a chemical substance (also known as 5-hydroxytryptamine) with vasoconstricting properties found in the serum of mammals, in the hypothalamus etc.

Smog—combination of the words 'smoke' and 'fog'.

Stress—the sum of all environmental stimuli of a physical, chemical or psychosocial character which creates a physiological strain (over exertion) of the bodily functions.

T

Temperature (ambient-)—Environmental temperature.

dry bulb—temperature measured with a thermometer recording the air temperature.

wet bulb—temperature of evaporation recorded with a thermometer of which the bulb is covered with muslin wetted with pure water.

Thalamus—centre in the brain below the cortex which receives and transmits sensory impulses, both from and to the cerebral cortex, hypothalamus and other parts of the brain.

Thermogenesis—the production of heat in the body through muscle activity, metabolism, chemical processes in the liver etc.

Thermal receptors—organs in the skin which are able to register thermal stimuli of the environment which create a temperature sensation in the body.

Thermoregulation—regulation of the temperature in the body.

Thrombin—protein, soluble in water, which originates from prothrombin, e.g. in the presence of calcium ions. Thrombin is produced by the liver.

Thyroid—large ductless gland in front of the trachea which serves as a storehouse of iodine and elaborates thyroglobulin and thyroxine which affect growth and metabolism.

Trace elements—chemical elements present or needed in extremely small amounts by plants, animals and man, such as copper, cobalt, zinc etc.

U

Ultraviolet—invisible radiation with wavelength of 1000–3900 Å.

V

Vasoconstriction—constriction of the blood vessels.

Vasodilatation—dilatation of the blood vessels.

Vasomotor centre—centre in the brain controlling the degree of contraction and dilatation of blood arteries, arterioles and probably also blood capillaries. It plays an important role in blood pressure fluctuations. It is located in the floor of the fourth ventricle (cavity of the brain) at the level of the apex of the *Calamus scriptorius* (the pointed lower portion of the floor of the fourth ventricle).

Viruses—infectious, non-living agents smaller than ordinary bacteria and requiring susceptible host cells for multiplication and activity.

W

Weather—the day-to-day changing meteorological conditions.

INDEX

N

O

P